Mathematical Research

Advances in Mathematical Optimization

edited by
J. Guddat et al.

Volume 45

AKADEMIE-VERLAG BERLIN

In this series original contributions of mathematical research in all fields are contained, such as

- research monographs
- collections of papers to a single topic
- reports on congresses of exceptional interest for mathematical research.

This series is aimed at promoting quick information and communication between mathematicians of the various special branches.

In diese Reihe werden Originalbeiträge zu allen Gebieten der mathematischen Forschung aufgenommen wie

- Forschungsmonographien
- Sammlungen von Arbeiten zu einem speziellen Thema
- Berichte von Tagungen, die für die mathematische Forschung besonders aktuell sind.

Die Reihe soll die schnelle Information und gute Kommunikation zwischen den Mathematikern der verschiedenen Fachgebiete fördern

---

Manuscripts
in English and German comprising at least 100 pages and not more than 500 pages can be admitted to this series. With respect to a quick publication the manuscripts are reproduced photomechanically. Authors who are interested in this series please turn directly to the 'Akademie-Verlag'. Here you will get more detailed information about the form of the manuscripts and the modalities of publication.

Manuskripte
in englischer und deutscher Sprache, die mindestens 100 Seiten und nicht mehr als 500 Seiten umfassen, können in diese Reihe aufgenommen werden. Im Interesse einer schnellen Publikation werden die Manuskripte auf fotomechanischem Weg reproduziert. Autoren, die an der Veröffentlichung entsprechender Arbeiten in dieser Reihe interessiert sind, wenden sich bitte direkt an den Akademie-Verlag. Sie erhalten dort genauere Informationen über die Gestaltung der Manuskripte und die Modalitäten der Veröffentlichung.

Advances in Mathematical Optimization

Mathematical Research · Mathematische Forschung

Wissenschaftliche Beiträge
herausgegeben von der
Akademie der Wissenschaften der DDR
Karl-Weierstraß-Institut für Mathematik

Band 45

Advances in Mathematical Optimization

edited by
J. Guddat et al.

Advances in Mathematical Optimization

Invited papers dedicated to Prof. Dr. Dr. h. c. F. Nožička
on occasion of his 70th birthday

---

edited by

Jürgen Guddat
Bernd Bank
Horst Hollatz
Peter Kall
Diethard Klatte
Bernd Kummer
Klaus Lommatzsch
Klaus Tammer
Milan Vlach
Karel Zimmermann

Akademie-Verlag Berlin 1988

Die Titel dieser Reihe werden vom Originalmanuskript der Autoren reproduziert

ISBN 3-05-500543-0

ISSN 0138-3019
Erschienen im Akademie-Verlag, Berlin, DDR-1086 Berlin, Leipziger Str. 3-4

Lizenznummer: 202·100/413/88
Printed in the German Democratic Republic
Gesamtherstellung: VEB Kongreß- und Werbedruck, 9273 Oberlungwitz
Lektor: Dr. Reinhard Höppner
LSV 1085
Bestellnummer: 763 862 6 (2182/45)
03200

PREFACE

The excellent scientist and outstanding university professor

Prof. Dr. Dr. h.c. František Nožička

celebrates on 5-th April, 1988 the 70-th anniversary of his birthday.

Prof. Nožička was a full professor of mathematics at the Charles University in Prague and is a visiting professor at the Section of Mathematics of the Humboldt University in Berlin for more than twenty years. Today he still works as a consulting professor at the Charles University and is fully active at the Humboldt University as a researcher and academic teacher. During the last 30 years František Nožička substantially contributed by remarkable results on several fields of mathematics and, in particular, to the development of theory and methods in mathematical optimization. The numerous papers and books he authored or co-authored testify his creative and restless academic life. At both the universities in Prague and Berlin he was supervisor of hundreds of undergraduate as well as postgraduate students. A large number of national and international conferences, workshops and meetings on mathematical optimization and its applications saw him as an excellent organizer or co-organizer.

During the sixties, when Prof. Nožička was the vice-vector of the Charles University and director of its Computer Center, he essentially influenced the development of the computerized numerical mathematics in Czechoslovakia.

A particular fruitful period of his academic career Prof. Nožička started in the second half of the sixties when he became a visiting professor at the Section of Mathematics of the Humboldt University in Berlin (GDR) beside his full professorship at the Charles University in Prague. In Berlin he earned inestimable merits as the scientific founder and spiritus rector of the optimization group at the Section of Mathematics. By his efforts in theory and applications he decisively influenced the development of mathematical optimization in the whole GDR. In 1978 the Mathematical and Natural Science Faculty of the Humboldt University appreciated his extremely successful academic work by awarding him its doctor honoris causa.

This brief valuation of Prof. Nožička's performances as researcher and university teacher shall not be terminated without mentioning that he never stopped his efforts to bring together theory and applications as close as possible. His great personality as researcher teacher and human being let him become a highly esteemed scientist beloved by students and colleagues in many countries.

On occasion of his 70-th birthday the editors, former students, colleagues and friends of Prof. Nožička, invited researchers in mathematical optimization and closed areas to submitt an article in order to honour Prof. Nožička. The refereed versions of these contributions form the content of this volume. Among the papers summarized here the reader finds full papers and survey articles. The topics treated range from integer programming, nonlinear optimization, stochastic programming, non-differentiable and parametric optimization to solution methods for nonlinear optimization problems, nonlinear operators and certain applications. The book will be of interest for researchers in optimization theory as well as in its applications.

The editors are grateful to the authors for submitting their contributions and to the various referees for their support.
Special thanks belong to the publishing house, Akademie Verlag, and its lecturer in chief R. Höppner.

Berlin, January 1988 The Editors

Contents

FIRST STEPS TO AN ON-LINE CONTROL FOR PUSHER TYPE REHEATING FURNACES

By D.Auzinger[1)], Hj.Wacker[1)]

1. Introduction

1.1. Organisation of the Paper

In steel processing it is very difficult to connect the process of continuous casting directly with the rolling process. Therefore, the slabs have to be reheated for the hot rolling. In our case this is done in pusher type furnaces.
In this paper we first describe an efficient computation of the heat distribution of the slabs. Based on these results we propose a procedure that might be helpful to perform an on-line control for the reheating process. To be precise, given a fixed 'optimal' reheating curve in the furnace, how is the reheating process to be controlled to approximate the optimal heat distribution. Finally, we present some results how to determine the optimal curve.

1.2. Description of the Furnace

Figure 1 gives a length cut through the furnace. It is about 36 m long (x-direction), 7 m high (y-direction) and 13 m broad (z-direction). It consists of 4 zones, each being divided into a upper and a lower part. The slabs, which are blocks of steel of about 1,2 m × 0,2 m × 12 m, are pushed through the furnace with their length-direction in the z-direction of the furnace. The furnace is filled with about 3o slabs, which form a "slab band". Whenever a slab is pushed into the furnace, the slab band moves in x-direction and another slab leaves the furnace at the discharging end, at a temperature of about 1250 °C.
In the first three zones (convective zone, preheating zone, heating zone) the slabs are pushed on water cooled skids, which cause the "skid marks" (compare Fig. 4). To diminish these skid marks, the lower part of zone 4 (soaking zone) is an unheated soaking hearth. The furnace is heated by coke gas and by natural gas. There are huge burners at the upper and the lower part of zones 2 and 3, which are the main heating zones, and there

1) Universität Linz Institut für Mathematik A-4o4o Linz

are a lot of small burners at the upper part of zone 4. The flue gas streams countercurrently to the slabs. In zone 1 there are no burners, but it is heated by the flue gas coming from zone 2. The gas leaves the furnace at the charging point and is used to preheat the air required for the combustion in a recuperator.

Fig. 1
pusher type reheating furnace

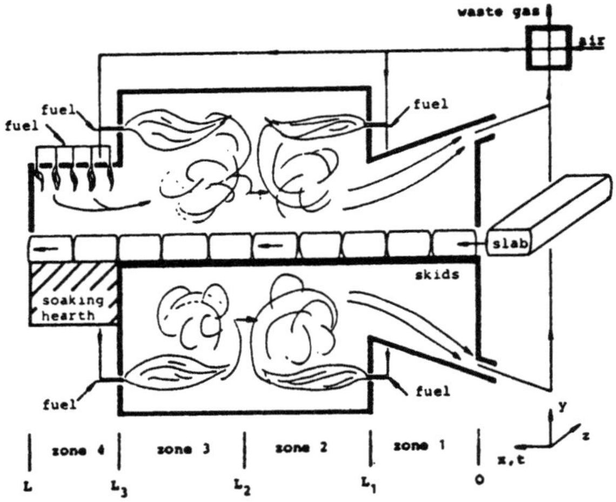

1.3. Basic Assumptions for the Model

Some assumptions are required to idealize and simplify the real reheating process, which is much too complex to be modelled "exactly". The first assumption is the strongest:

(A1) steady operation of the furnace: slabs of the same size and quality are charged at a constant temperature. They are discharged after a constant heating time and at a constant temperature.

(A2) the real movement of the slabs is replaced by a movement with constant velocity v.

(A3) within the slabs no heat conduction in x-direction is considered, as this conductive heat transport is small compared to the other heat transport affects.

(A4) for controlling purposes we assume a 'mixed' temperature for the furnace zones influenced by the flue gas, by the flames of the burners, by the furnace wall and by the slabs. It is further assumed that we can influence this 'mixed' temperature (for instance by choosing the input of the combustion material).

2. Determination of the Heat Distribution

For on-line purposes it is essential to keep the computing time low. Nevertheless, to get more insight into the process, e.g. we want to study the influence of the skids, we decided to use a three dimensional model here.

Computation is done by classical FE-techniques where the structure of the problem is exploited. This is done such that the gradients necessary for the performance of the on-line control are determined just by a simple extension.

2.1. The Mathematical Model

We consider a particular slab being pushed through the furnace, i.e. we use a slab fixed coordinate system. We determine those zone temperatures $W = (W_1^u, W_1^l, W_2^u, W_2^l, W_3^u, W_3^l, W_4^u)$, which give the bulk temperature curve closest to the ideal curve. We work in 3 dimensions: the height of the slabs (y), their breadth (z) and the time t.
At first we reduce the size of the problem by symmetry considerations.

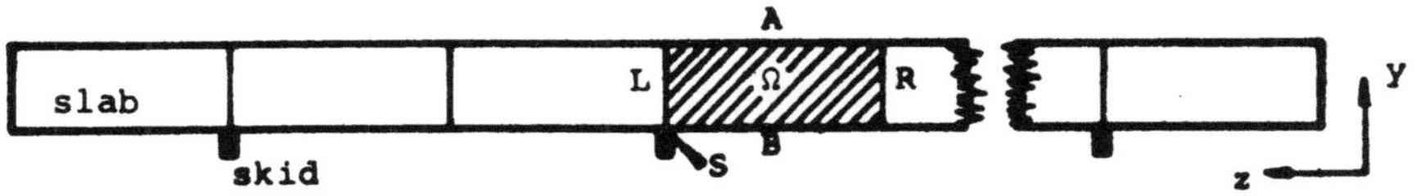

Fig.2 slab

Under the assumption $\left.\frac{\partial T}{\partial z}\right|_{L \cup R} = 0$ it suffices to consider Ω. We get the following model for the heat distribution

$$c_p(T)\rho\, \frac{\partial T}{\partial t} = \mathrm{div}[\lambda(T)\ \mathrm{grad}\ T] \qquad (y,z) \in \Omega,\ t \in (0,t_E) \tag{1}$$

$$T(y,z,t)\Big|_{t=0} = T_A \tag{2}$$

$$\lambda(T)\frac{\partial T}{\partial n}(y,z,t) = R(y,z,t,T) = \begin{cases} 0 & L\cup R,\ \text{zones } 1\text{-}4 \\ h_i^u(W_i^u - T) + r_i^u({W_i^u}^4 - T^4) & A,\ \text{zones } 1\text{-}4 \\ h_i^l(W_i^l - T) + r_i^l({W_i^l}^4 - T^4) & B,\ \text{zones } 1\text{-}3 \\ k^S(T^c - T) & S,\ \text{zones } 1\text{-}3 \\ k^H(T^H - T) & B\cup S,\ \text{zone } 4 \end{cases} \tag{3}$$

h_i^u and h_i^l are the convectivity constants at zone i, r_i^u and r_i^l the radiation constants, k^S resp. k^H the heat exchange coefficients slab-skids-cooling water resp. slab-soaking hearth. For the approximation of the ideal bulk temperature $T_{opt}(t)$ by the bulk temperature $T(\bar{y},\bar{z},t)$ in $L_2\,[0,t_E]$ we obtain the objective

$$f(W) = \int_0^{t_E} [T(\bar{y},\bar{z},t,W) - T_{opt}(t)]^2\, dt = \mathrm{Min!} \tag{4}$$

The control W must satisfy the (technical) box constraints

$$W_{min} \leq W \leq W_{max}. \tag{5}$$

2.2. The Determination of the Heat Distribution

We assume W to be given.

2.2.1. Time Discretization: Trapezoidal Rule

We introduce the time discretization $t_o := 0$, $t_{n+1} := t_n + \Delta t_n$ with Δt_n variable and write $T_n(y,z)$ for the numerical approximation of $T(y,z,t_n)$. To preserve stability we use the trapezoidal rule:

$$\frac{(T_{n+1}-T_n)(y,z)}{\Delta t_n} = \frac{1}{2}\left[\frac{D(T_n)}{r(T_n)} + \frac{D(T_{n+1})}{r(T_{n+1})}\right] \tag{6}$$

with $D(T) := \mathrm{div}(\lambda(T)\,\mathrm{grad}\,T) = \lambda'(T)\nabla T^2 + \lambda(T)\Delta T$ and $r(T) := c_p(T).\rho$.
(6) is a nonlinear elliptic differential equation for T_{n+1}.

2.2.2. Linearization

We solve (6) by a predictor-corrector technique, using a trivial predictor and a Newton corrector (Fig. 6):

predictor P: $\tilde{T}_{n+1} := T_n$

corrector C: $T_{n+1} := \tilde{T}_{n+1} + u$, where $u = u(y,z)$ satisfies

$$\frac{u}{\Delta t_n} = \frac{1}{2}\left[\frac{D(T_n)}{r(T_n)} + \frac{D(T_n)}{r(T_n)} + \frac{D'(T_n)u}{r(T_n)} - \frac{D(T_n)r'(T_n)u}{r^2(T_n)}\right]. \tag{7}$$

(7) is the linearization of (6) at $\tilde{T}_{n+1} = T_n$. We have

$$D'(T)u = \lambda''(T)\nabla T^2 u + 2\lambda'(T)\nabla T \nabla u + \lambda'(T)\Delta T u + \lambda(T)\Delta u.$$

For the linearized boundary condition at boundary A, for instance, we get:

$$\lambda(T)\frac{\partial u}{\partial n} = -\left[h_i^u + 4r_i^u T_n^3 + \lambda'(T_n)\frac{\partial T_n}{\partial n}\right]u, \tag{8}$$

those for B,S,L and R are analogous. This model holds inside each zone. Some care is necessary when passing from one zone to the next one as we have jumps in the right hand side of (3) at these points (see [2], [3]).

2.2.3. Transformation to Standard Form

Multiplying (7) with $2r(T_n)$ and applying the transformation

$$v = v(y,z) := \lambda(T_n(y,z)).u(y,z)$$

to the system (7), (8), we get the standard form

$$-\Delta v + a(y,z)v = b(y,z) \qquad (y,z) \in \Omega \tag{9}$$

$$\frac{\partial v}{\partial n} = \alpha(y,z).v \qquad (y,z) \in \partial\Omega \tag{10}$$

with $a: = \frac{2}{\Delta t_n} \frac{r(T_n)}{\lambda(T_n)} + \frac{r'(T_n)\lambda'(T_n)}{r(T_n)\lambda(T_n)} \nabla T_n^2 + \frac{r'(T_n)}{r(T_n)} \Delta T_n$,

$b: = 2[\lambda'(T_n)\nabla T_n^2 + \lambda(T_n)\Delta T_n]$ and

$\alpha: = \frac{-1}{\lambda(T_n)} [h_i^u + 4r_i^u T_n^3 + \lambda'(T_n) \frac{\partial T_n}{\partial n}]$ for boundary A, e.g.

2.2.4. Galerkin

We use the following ansatz for $v(y,z)$:

$$v(y,z) = \sum_{k=1}^{n} c_k N_k(y,z)$$

with finitly many elements N_k. Then the method of Galerkin (checking equation (9), (10) in the weak sense with respect to each ansatz function N_j) gives the linear system

$$\iint_\Omega -\Delta\left(\sum_{k=1}^{n} c_k N_k\right)N_j \, dydz + \iint_\Omega a(y,z)\left(\sum_{k=1}^{n} c_k N_k\right)N_j \, dydz = \iint_\Omega b(y,z)N_j \, dydz$$

For eliminating the second derivatives, we apply Greens theorem

$$\iint_\Omega \Delta v_1 v_2 \, dydz = -\iint_\Omega \nabla v_1 \nabla v_2 \, dydz + \int_{\partial\Omega} \frac{\partial v_1}{\partial n} v_2 \, ds$$

to all the terms containing the Laplace operator and get:

$$\sum_{k=1}^{n} c_k \{\iint_\Omega \nabla N_k \nabla N_j - \frac{r'}{r} \nabla T(\nabla N_k N_j + N_k \nabla N_j) + \tag{11}$$

$$+ [\frac{2}{\Delta t_n} \frac{r}{\lambda} + (\frac{r'\lambda'}{r\lambda} - (\frac{r'}{r})')\nabla T \nabla T] N_k N_j \, dydz + \int_{\partial\Omega} [\frac{r'}{r} \frac{R}{\lambda} - \alpha] N_k N_j \, ds\} =$$

$$= -2 \iint_\Omega \lambda \nabla T \, N_j \, dydz + 2 \int_{\partial\Omega} R \, N_j \, ds \qquad \text{for } j = 1,\dots,n.$$

(11) is a positive definit linear equation for the n unknowns c_k. For a proper choice of the ansatz functions N_k and for sufficiently large n $\sum_{k=1}^{n} c_k N_k$ is a good approximation of v and T_n+v approximates T_{n+1}. Integration in (11) is performed by Gaussian quadrature formula to gain accuracy.

2.2.5. Finite Element Techniques

We chose the finite element technique for generating the ansatz functions. Ω is divided into 276 triangles and each ansatz function N_k is a piecewise linear function which is zero at all knots but one, where it has the value 1. A knot is a vertex of a triangle; there are 168 knots. This particular form of N_k makes the evaluation of the integrals of (11) very easy. The system matrix of (11) is very sparse, because each integral contains a product of N_k resp. ∇N_k and N_j resp. ∇N_j, which is zero for most of the pairs $(k,j) \in \{1,\ldots,n\} \times \{1,\ldots,n\}$. As the structure of the matrix depends on the numbering of the knots, we used the algorithm of Cuthill-McKee (see [4]) to reduce the band-width. Fig. 3 shows the triangulation - near the skid it is finer to get a good picture of the skid marks - and the numbering of the knots resulting from the Cuthill-McKee algorithm. The band-width of the matrix is m = 12, and thus we need an array of $n(m+1) = 168 \times 13 = 3588$ elements instead of $168^2 = 28224$ elements for its storage. The amount of work for solving the linear equation is about $n(m^2+3m)/2$ operations.

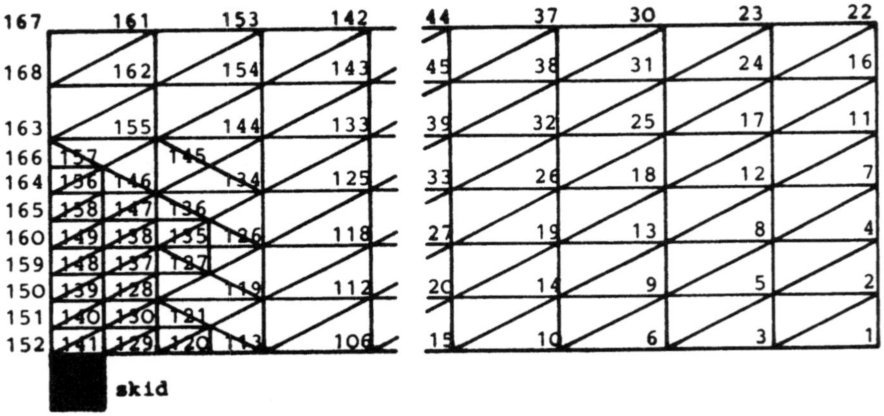

Fig.3 finite element discretization of Ω

2.2.6. Stepsize Control

The accuracy of the results strongly depends on the time stepsize Δt_n. Thus we used the stepsize control proposed in [5], which gave good results. But as the integration has to be performed with both stepsizes Δt_n and $\frac{1}{2}\Delta t_n$, the numerical amount is high. So when we found out that the stepsize sequences are almost the same for all our test examples, we took this typical stepsize sequence and could so reduce the numerical amount drastically with only a small loss of accuracy. On an IBM 3o31 a stepsize controlled solution of our problem needs 17.64 min whereas for a mostly constant stepsize of 2oo [sec] the computation was finished after 4.o5 min. The maximal difference of the solution is less than 1 %.
In addition, for the optimization the fixed step size sequence with partially constant stepsizes gave another substantial reduction of the numerical amount, see chapter 3.

2.2.7. Numerical Results: Temperature Distribution

The following pictures give the temperature distribution of Ω at the end of zones 3 and 4 (before and after the soaking hearth)

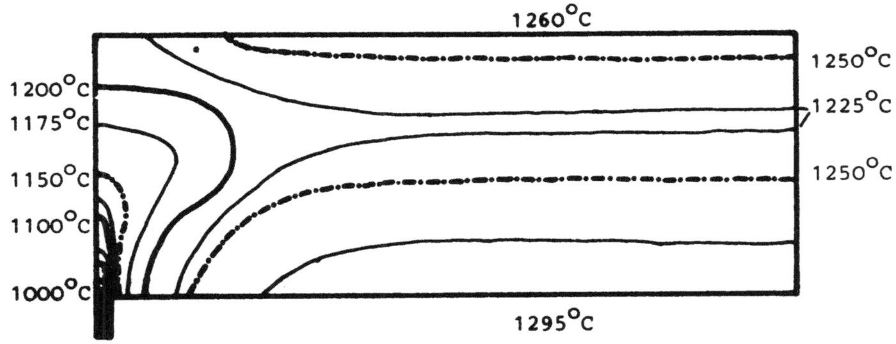

Fig.4 temperature distribution of the slab (Ω) before soaking hearth

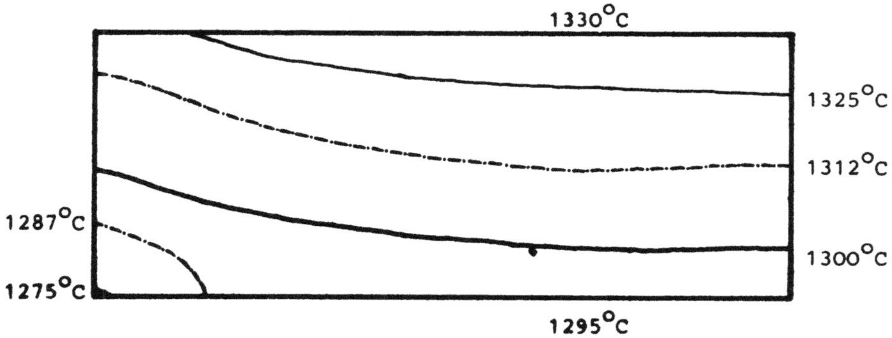

Fig.5 temperature distribution of the slab (Ω) after soaking hearth

3. Furnace Control/Optimization

In section 2.2 we described how to determine T in dependence of W. Now we want to solve

$$f(T(W)) = \text{Min!} \quad W_{min} \leq W \leq W_{max}, \quad W \in \mathbb{R}^7$$

As each function evaluation is very costly because of the computation of T, we decided on a Quasi Newton procedure using the BFGS-formula for the approximation of the Hessian.

3.1. Determination of the Gradients

From (4) we get

$$\frac{\partial f}{\partial W}(T(W)) = \frac{\partial f}{\partial T}\frac{\partial T}{\partial W} = \int_0^{t_E} \underbrace{2[T(\bar{x},\bar{y},t,W)-T_{opt}(t)]}_{=:\,K} \underbrace{\frac{\partial T}{\partial W}}_{=:U} dt \tag{12}$$

To determine U we differentiate the PDE (1), (2), (3) and change the order of differentiation:

$$\frac{\partial U}{\partial t} = -\frac{r'(T)}{r^2(T)} D(T).U + \frac{D'(T)}{r(T)} U \tag{13}$$

$$U(y,z,t)\Big|_{t=0} = 0 \tag{14}$$

$$\lambda'(T)\frac{\partial T}{\partial n} U + \lambda(T)\frac{\partial U}{\partial n} = R'_T(T,W)U + R'_W(T,W) \tag{15}$$

The seven parabolic problems (13), (14), (15) are solved by exactly the same techniques - time discretization, transformation $V_n(y,z)=\lambda(T_n)U_n(y,z)$ - as problem (1), (2), (3). There is no linearization as the problems are already linear. At each time step we have to solve

$$-\Delta V_n + a(y,z)V_n = n(y,z) \tag{16}$$

$$\frac{\partial V_n}{\partial n} = \alpha(y,z)V_n + \beta(y,z) \tag{17}$$

Those parts of the system containing the unknown V_n are exactly the same as those of system (9), (1o). The only difference is that in (9), (1o) the functions a(y,z) and α(y,z), which depend on T, are to evaluate at $\tilde{T}_{n+1} = T_n$, while for system (16), (17) we require the functions at T_{n+1}. But as $T_{n+1} = \tilde{T}_{n+2}$, the system matrix of (16), (17) - after application of Galerkins method and finite element discretization - is exactly the same as that of (11) at the next time step, provided that $\Delta t_n = \Delta t_{n+1}$.

Even the amount of work for the evaluation of the right hand sides can be substantially reduced by a similar consideration. Assume we know (T_n, U_n) and want to determine (T_{n+1}, U_{n+1}) inside a zone (see also Fig. 6)

Step 1: Determine T_{n+1} by solving (11). The system matrix at $\tilde{T}_{n+1} = T_n$ and its Cholesky factorization are already known

Step 2: Determine the Jacobian at T_{n+1}, which is the system matrix for both the systems for V_{n+1} and the system for T_{n+2}.

Step 3: Determine the right hand sides for the systems for V_{n+1} and solve the equations. Back transformation gives U_{n+1}.

Some care is necessary when passing from one zone to the next one, for details see [2]. For changing stepsizes ($\Delta t_{n+1} \neq \Delta t_n$) only little additional amount of work is necessary.

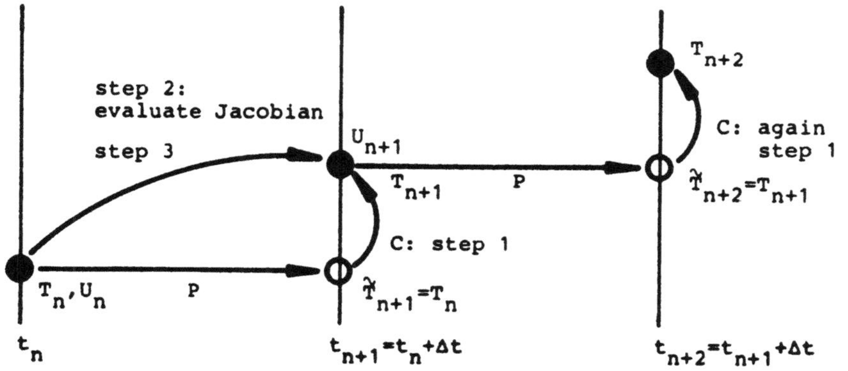

Fig.6 calculation of T_{n+1} and U_{n+1}

3.2. Minimization of f(W)

We determine f(W) and $\frac{\partial f}{\partial W}$ (W) by the trapezoidal rule using just those points where T and U were determined. Then we use the following well known algorithm

Start: $f(W_o)$, $\nabla f(W_o) =: g_o$, $B_o = \beta I$, $k := 0$

S1: $B_k S_k = -g_k \Longrightarrow S_k$

S2: $f(W_k + \alpha S_k) = \text{Min}$, $W_{min} \leq W \leq W_{max} \Longrightarrow \alpha_k$

$W_{k+1} = W_k + \alpha_k S_k$ (Line search is performed by quadratic interpolation)

S3: W_{k+1} optimal $\Longrightarrow$ stop

S4: B_{k+1} by BFGS-Formula, $k = k+1$, GOTO S1

3.3. Numerical Results

Depending on the charging temperature, the quality of steel, the size of the slabs and the pushing velocity v there are different optimal curves $T_{opt}(t) = T_{opt}(\frac{x}{v})$ for the bulk temperature, see chapter 4. We will present the approximation of one typical cold charging optimal curve. In the following table the results are rounded. $\Delta T = \sqrt{\frac{f(W)}{t_E}}$ gives the mean difference of the calculated and the optimal bulk temperature.

Iteration	W_1^u	W_1^l	W_2^u	W_2^l	W_3^u	W_3^l	W_4^u	ΔT
1	600	550	1000	1050	1300	1350	1350	129,96
2	566	521	922	961	1196	1242	1330	58,93
3	563	519	924	965	1209	1260	1336	58,97
line search	656	520	923	963	1202	1251	1335	58,70
4	555	512	915	957	1210	1263	1339	54,81
5	520	484	886	933	1228	1296	1351	41,13
6	487	459	855	908	1236	1318	1360	31,69
7	450	429	816	877	1236	1336	1368	28,97

4. Determination of the Optimal Reheating Curve

We used a model consisting out of three parts:

i) the slabs (two dimensional: (x,y))
ii) the flue gas (one dimensional: y)
iii) the recuperator

Further, we assumed that the flue gas in the zones 2 and 3 is "well-stirred", i.e. one has a uniform temperature of the flue gas in each zone. For zone 1 and 4 we assumed a differential flue gas model.
There is no room here to give a detailled description. The procedure for the solution, however, is similar to that described in Chapter 2 and Chapter 3. First one determines the heating process, i.e. temperatures of the slabs, of the flue gas and of the preheated air, for a given energy input. Then one optimizes the energy input using similar techniques as in Chapter 3 to get the gradients.
We confine ourselves to present two typical pictures for cold and hot charging optimization, for more details see [1].

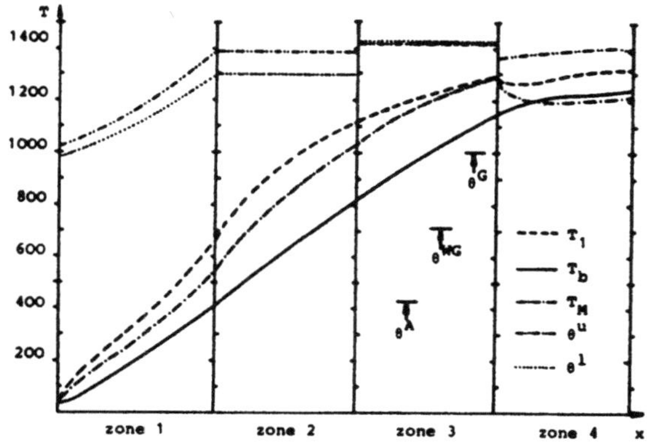

Fig.7 cold charging optimization

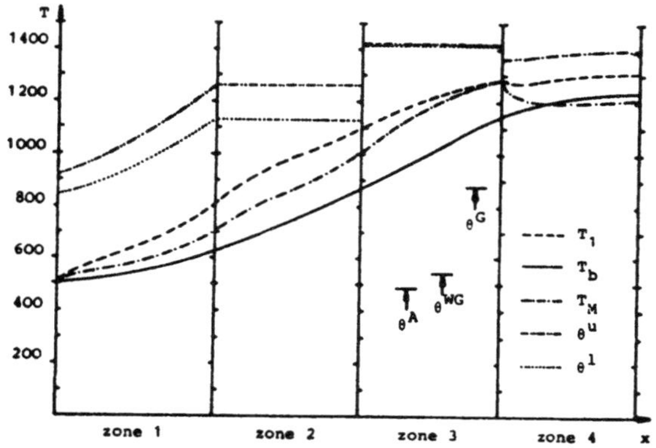

Fig.8 hot charging optimization

Acknowledgements:

The authors are indebted to Dipl.-Ing.B.Lindorfer (VOEST-ALPINE AG) and Dr.Walter Zulehner for their contributions to modelling and to Dr.H.Gfrerer for his advice concerning optimization. Financial support was given by the National Banking Foundations, the Austrian Science Foundation Fonds and the VOEST-ALPINE AG.

References:

[1] D.Auzinger,Hj.Wacker: Optimal Reheating of Slabs in a Pusher Type Furnace, in: H.W.Engl, Hj.Wacker, W.Zulehner (Eds.) Case Studies in Industrial Mathematics

[2] R.Hubmer, "Optimierung eines Stoßofens: Sollkurvenapproximation unter Anwendung der Methode der Finiten Elemente", diploma thesis 1986, Johannes-Kepler-Universität, Linz

[3] M.Lindner, "Hierarchische Formfunktionen angewandt auf ein parabolisches Anfangsrandwertproblem", diploma thesis 1987, Univ.Linz

[4] H.R. Schwarz, "Methode der finiten Elemente", Teubner, Stuttgart 1984

[5] K.Solchenbach, K.Stüben, U.Trottenberg, K.Witsch, "Efficient solution of a nonlinear heat conduction problem by multigrid methods", Sonderforschungsbereich 72, Universität Bonn.

(MIXED-) INTEGER SOLUTIONS OF QUASICONVEX POLYNOMIAL INEQUALITIES[1)]

Bernd Bank and Reinhard Mandel[2)]

Abstract.

Properties of quasiconvex polynomials on R^n will be derived. Based on these properties the existence and stability of (mixed-) integer solutions of quasiconvex polynomial inequalities under weak assumptions is discussed.

1. Introduction.

This paper is intended to complete a stability theory of quasiconvex polynomial (mixed-) integer optimization problems presented in BANK/MANDEL (1987). In particular, it contains the proofs of several results omitted there.

Essentially, the article contains basic properties of quasiconvex polynomials and solution sets of systems of inequalities involving such polynomials. These properties are basic to derive stability (in the sense of lower and upper semi-continuity) of the corresponding multifunction described by such an inequality system if its right-hand sides vary.

It is well-known, that stability of the constraint sets of an optimization problem determines the stability behaviour (i.e. continuity properties of the extremal value and the optimal set) of the problem if data changes are considered. A fairly comprehensive study of this question can be found in BANK/GUDDAT/KLATTE/KUMMER/TAMMER (1982).

First, in Section 2, an analysis of elementary properties of quasiconvex polynomials is given. Here, BELOUSOV's (1977) results are extended to the class of quasiconvex polynomials. Section 3. is devoted to (mixed-) integer solutions of quasiconvex polynomial inequalities. The final section is concerned with the stability behaviour of real and (mixed-) integer solutions of such inequality systems where the right-hand sides changes.

As usual, we understand by a *quasiconvex* function f on a convex set S such one satisfying

---

1) This work was partially supported by the Austrian Science Foundation (Fonds zur Förderung der wissenschaftlichen Forschung), Project S 32/01.

2) Humboldt-Universität zu Berlin, Sektion Mathematik, PSF 1297, BERLIN, 1086 DDR.

$$f(\alpha x+(1-\alpha)y) \leq \max\{f(x),f(y)\},$$

for all $\alpha \in (0,1)$ and $x,y \in S$. The degree of a polynomial $f: R^n \to R$ is denoted by deg f. A multifunction $\Gamma : \Lambda \to 2^{R^n}$ is said to be *upper semicontinuous* (briefly: *u.s.c.*) at $\lambda^0 \in \Lambda$ if

$$\forall \varepsilon > 0 \exists \delta = \delta(\varepsilon) > 0: \Gamma(\lambda) \subset U_\varepsilon \Gamma(\lambda^0) \; \forall \; \lambda \in U_\delta(\lambda^0)$$

and, *lower semicontinuous* (briefly: *l.s.c.*) at $\lambda^0 \in \Lambda$ if

$$\forall \Omega \subset R^n, \Omega \text{ open } \& \; \Omega \cap \Gamma(\lambda^0) \neq \emptyset$$

$$\Longrightarrow \exists \delta = \delta(\Omega): \Omega \cap \Gamma(\lambda) \neq \emptyset \;\; \forall \lambda \in U_\delta(\lambda^0).$$

Further, we say that Γ is *Hausdorff-lower semicontinuous* (briefly: *H-l.s.c.*) at $\lambda^0 \in \Gamma$ if

$$\forall \varepsilon > 0 \; \exists \delta = \delta(\varepsilon): \Gamma(\lambda^0) \subset U_\varepsilon \Gamma(\lambda) \;\; \forall \lambda \in U_\delta(\lambda^0)$$

(U_ε denotes the usual ε-neighbourhood).

Γ is said to be *continuous* at $\lambda^0 \in \Lambda$ if it is both u.s.c. and l.s.c. If Γ is even H-l.s.c. at λ^0, then we say that Γ is *H-continuous* at λ^0. Further, con S denotes the conical hull of a set $S \subset R^n$. By dim S the dimension of a convex set $S \subset R^n$ is abbreviated. If $S \subset R^n$ is a subspace, then $S^\perp$ denotes its orthogonal complement. For the diameter of a bounded set S we shortly write diam S.
As usual, we denote by Z the ring of all integers.
In view of (mixed-) integer points in subsets of R^n we will need rationality assumptions on these sets. Such hypotheses will be posed in the following form: If V is a polyhedral convex cone in R^n and $0 < s \leq n$, then we say '*V satisfies (MIG)*' if there are vectors $v^1,\dots v^r \in Z^s \times R^{n-s}$: $V = \text{con}\{v^1,\dots,v^r\}$ (i.e. V has a basis in $Z^s \times R^{n-s}$).

2. Elementary Properties of Quasiconvex Polynomials.

In this section we present such properties of quasiconvex polynomials which we shall use in the analysis of constraint sets given by a finite system of inequalities whose left-hand sides are quasiconvex polynomials. Thereby we follow a concept similar to that which BELOUSOV (1977,1986) employed for convex polynomials. In particular, Lemma 1. is an extension of his 'main lemma' to quasiconvex polynomials.
Every polynomial $f: R^n \to R$ of degree m has a uniquely defined representation as a sum of forms (i.e. homogeneous polynomials) $p_i(x)$, $i = 0,1,\dots,m$, i.e.

$$f(x) = \sum_{i=0}^{m} p_i(x) \text{ where } p_i(tx) = t^i p_i(x), t \in R. \quad (1)$$

Lemma 1.

Let a quasiconvex polynomial $f: R^n \to R$ of degree m have the representation (1). Then

(i) the form $p_m(x)$ is quasiconvex;

(ii) for each pair x^o, $x^1 \in R^n$ and each $u \in R^n$, $u \neq 0$, the polynomials $f_o: R \to R$ and $f_1: R \to R$ given by

$$f_o(t) = f(x^o+tu) \quad \& \quad f_1(t) = f(x^1+tu)$$

are quasiconvex and of the same degree d.

If $d \geq 1$, then both polynomials have the same leading coefficient.

<u>Proof.</u>

(i) Let $x,y \in R^n$ and $\alpha \in (0,1)$ be arbitrarily fixed.

Then

$$f(t(\alpha x+(1-\alpha)y))=f(\alpha(tx)+(1-\alpha)(ty)) \leq \max \{f(tx),f(ty)\}$$

holds for an arbitrary number $t \in R$. Thus, if $t > 0$, then

$$t^{-m}f(t(\alpha x+(1-\alpha)y)) \leq \max \{t^{-m}f(tx),t^{-m}f(ty)\}.$$

According to (1), it follows that

$$p_m(\alpha x+(1-\alpha)y) \leq \max \{p_m(x),p_m(y)\}$$

if $t \to \infty$.

(ii) Obviously, the proposition is to show for the case that there is no $\bar{t} \in R$ such that $x^o = x^1 + \bar{t}u$. Assume the representations

$$f_o(t) = a_r t^r+\ldots+a_o, \quad a_r \neq 0,$$

$$f_1(t) = b_s t^s+\ldots+b_o, \quad b_s \neq 0.$$

The quasiconvexity of $f_o(t)$ and $f_1(t)$ is clear. Let an $\alpha \in (0,1)$ be arbitrarily fixed. Then there are points $v(\alpha)$ and $w(\alpha)$ in R^n such that

$$x^1 = \alpha x^o + (1-\alpha)v(\alpha) \quad \& \quad x^o = \alpha x^1+(1-\alpha)w(\alpha).$$

Then, the quasiconvexity of f(x) permits the inequalities

$$f(x^1+\alpha tu) \leq \max \{f(x^o+tu),\ f(v(\alpha))\},$$

$$f(x^o+\alpha tu) \leq \max \{f(x^1+tu),\ f(w(\alpha))\}$$

which can also be written in the form

$$b_s \alpha^s t^s+\ldots+b_o \leq \max \{a_r t^r+\ldots+a_o, f(v(\alpha))\}, \qquad (2)$$

$$a_r \alpha^r t^r+\ldots+a_r \leq \max \{b_s t^s+\ldots+b_o, f(w(\alpha))\}. \qquad (3)$$

Since these inequalities must be valid if $|t| \to \infty$, too, it follows that $d = r = s$.

If $d \geq 1$, then it can immediately be concluded from (2) and (3) that sign a_d = sign b_d. The quasiconvexity of f(x) implies that $a_d > 0$ and $b_d > 0$ if d is an even number. We still have to show that $a_d = b_d$. Assume $0 < b_d < a_d$. Then one could choose an $\alpha \in (0,1)$ such that still $a_d \alpha^d > b_d$ holds which would lead, for large t, to a contradiction to (3). The other case can be shown similarly. #

The following simple example shows that the property (ii) in Lemma 1. is a strong consequence of a quasiconvex polynomial.

Let $f(x_1,x_2) = x_1 \cdot x_2$ be and choose $x^o = (1,2)^T$, $x^1 = (0,3)^T$ and $u=(0,1)^T$. Then one gets

$$f_o(t) = 1\cdot(2+t) = 2+t \quad \text{and} \quad f_1(t) = 0\cdot(3+t) = 0,$$

and neither $\deg f_o = \deg f_1$ nor the equality of the leading coefficients holds.

An immediate consequence of Lemma 1. is

Corollary 1.

If a quasiconvex polynomial $f(x)$ is constant (or strictly decreasing or strictly increasing) on a line $S_o = \{x \in R^n \;/\; x = x^o + tu,\; t \in R\}$, x^o, $u \in R^n$ $u \neq 0$, then it shows the same property on every line $S_1 = \{x \in R^n \;/\; x = x^1 + tu,\; t \in R\}$, $x^1 \in R^n$, parallel to S_o. Further, not every leading monomial of a quasiconvex polynomial can be multivariate. The form $p_r(x)$ of lowest degree in the representation of a quasiconvex polynomial by forms is quasiconvex.

Proof.

The first statements are consequences of (ii) in Lemma 1. The latter fact is shown, taking into account that $\lim_{t\to+0} t^{-r} f(tx) = p_r(x)$ holds for the form of lowest degree. #

Now we turn to an analysis of quasiconvex forms.

Lemma 2.

Let $p_m\colon R^n \longrightarrow R$ be a quasiconvex form of degree $m \geq 1$ and, let the sets

$$L \underset{Df}{=} \{x \in R^n \;/\; p_m(x) = 0\} \tag{4}$$

and

$$H \underset{Df}{=} \{x \in R^n \;/\; p_m(x) \leq 0\} \tag{5}$$

be defined. Then

(i) L is a linear subspace in R^n;

(ii) if m is an even number, then L = H;

(iii) if m is an odd number, then L is a hyperplane in R^n and H a half-space.

Proof.

(i) Let U be the subspace of maximal dimension contained in L. Assume that there is an $x^o \in L$ not belonging to U. Obviously, $-x^o \in L$ holds, too. Then, by the quasiconvexity of $p_m(x)$ and $p_m(0) = 0$ we have that $p_m(tx^o) = 0$, for all $t \in R$. According to Corollary 1., the polynomial $p_m(x)$ must be constant on each line S parallel to the line $S_o = \{x / x = tx^o,\ t \in R\}$, too. This means that L contains the linear hull of $U \cup \{x^o\}$, but, this contradicts the choice of U.

(ii) If there would be a point $x \in R^n$ satisfying $p_m(x) < 0$, then $p_m(-x) < 0$ had to hold, too. But, this is impossible because of the quasiconvexity of $p_m(x)$ which yields

$$0 = p_m(0) = p_m(\tfrac{1}{2} x - \tfrac{1}{2} x) \leq \max \{p_m(x),\ p_m(-x)\}.$$

(iii) We show that the orthogonal complement $L^\perp$ of L is of dimension 1. Assume that $L^\perp$ contains two linearly independent vectors x^1 and x^2. Without loss of generality, we suppose that $p_m(x^1) < 0$ and $p_m(x^2) > 0$ (otherwise we might use $-x^1$ instead of x^1). Then there is an $\alpha \in (0,1)$ such that $p_m(\alpha x^1 + (1-\alpha)x^2) = 0$ where $\alpha x^1 + (1-\alpha)x^2 \neq 0$ because of the linear independence of x^1 and x^2. But, this means a contradiction to $L \cap L^\perp = \{0\}$.

Finally, by Corollary 1., we know that if $x^* \in L^\perp$ with $p_m(x^*) < 0$, then $H = \{x \in R^n \ / \ x^{*T}x \geq 0\}$. #

Lemma 3.

Let $p_m\colon R^n \longrightarrow R$ be a quasiconvex form of degree m which has only integer coefficients. Then the linear subspace L from (4) has an integer basis.

Proof.

(By induction on n). Obviously, the cases n = 0 and n = 1 are trivial. Therefore, let $n \geq 2$. Assume there is an $u \in L$, $u \neq 0$. We have $0 = t^m p_m(u) = p_m(tu)$, for all $t \in R$ and, by Corollary 1.,

$$p_m(x+u) = p_m(x) \quad \forall\ x \in R^n. \tag{6}$$

By the same corollary one variable, say x_1, has to appear in the m-th power. Then the form $p_m(x)$ may lexicographically be written as

$$p_m(x) = a_{11}x_1^m + x_1^{m-1}(\sum_{k=2}^{n} a_{1k}x_k) + \\ + x_1^{m-2}(\ldots)+\ldots+a_{nn}x_n^m. \tag{7}$$

Then, $a_{11} \neq 0$ because of Lemma 1. and the assumption that p_m depends on x_1. Substituting x by x + u one obtains from (7) that

$$p_m(x+u) = p_m(x) + x_1^{m-1}(ma_{11}u_1 + \sum_{k=2}^{n} a_{1k}u_k)+\ldots+p_m(u)$$

where, on account of (6),

$$x_1^{m-1}(ma_{11}u_1 + \sum_{k=2}^{n} a_{1k}u_k)+\ldots+p_m(u) = 0.$$

In particular, this implies that

$$ma_{11}u_1 + \sum_{k=2}^{n} a_{1k}u_k = 0, \quad u \in L \tag{8}$$

must hold. The solution set of (8) is a linear subspace U (in R^n) containing L. It has an integer basis, because all coefficients in (8) are integers.

Such a basis of U can be extended to an integer basis of R^n. Now, the original quasiconvex form $p_m(x)$ can be represented by a quasiconvex form $q_m(y_1,\dots,y_{n-1})$ on U with integer coefficients. Proceeding by induction on n proves the proposition. #

Corollary 2.

If the hypotheses of Lemma 3. are fulfilled and m is an odd number, then the hyperplane L from (4) has an integral normal vector.

Proof.

Proceeding as sketched above one constructively finds a linear inequality which has only integer coefficients and describes the half-space H from (5). #

For a quasiconvex polynomial $f: R^n \longrightarrow R$ we define the sets

$$K(f) \underset{Df}{=} \{u \in R^n \ / \ \sup \{f(tu) \mid t \geq 0\} < \infty\} \tag{9}$$

and

$$R(f) \underset{Df}{=} \{u \in R^n \ / \ \sup \{f(tu) \mid t \in R\} < \infty\}. \tag{10}$$

The non-zero vectors u in K(f) and R(f) are the directions of all half-lines in R^n on which the quasiconvex polynomial f(x) decreases or is constant, respectively.

Again underlying for the quasiconvex polynomial f(x) the representation (1) by forms we define an index

$$i_o \underset{Df}{=} \begin{cases} \max\{i \in \{1,\dots,m\} \ / \exists\, u \in R^n: \\ p_m(u)=\dots=p_{i+1}(u) = 0,\ p_i(u) < 0\} & \text{, if it exists} \\ 1 & \text{, otherwise} \end{cases} \tag{11}$$

and the set

$$K^*(f) \underset{Df}{=} \{ u \in R^n \ / \ p_j(u) = 0,\ j = m,\dots,i_o+1\}. \tag{12}$$

Lemma 4.

Let $f: R^n \longrightarrow R$ be a quasiconvex polynomial of degree $m \geq 1$. Further, let K(f), R(f), i_o and $K^*(f)$ be defined by (9), (10), (11) and (12), respectively. Then

$$K(f) = K^*(f) \cap \{u \in R^n \ / \ p_{i_o}(u) \leq 0\}$$

and

$$R(f) = K^*(f) \cap \{u \in R^n \ / \ p_{i_o}(u) = 0\}.$$

Proof.

Let $u^o \in K(f)$, defined by (9). By (1), one has

$$f_o(t) \underset{Df}{=} f(tu^o) = p_d(u^o)t^d + \ldots + p_1(u^o)t + p_o$$

where $d = \deg f_o$. If $d \geq 1$, then $p_d(u^o) < 0$ holds (if f is convex, then even $d = 1$). Therefore,

$$u^o \in K^*(f) \cap \{u \in R^n \;/\; p_{i_o}(u) \leq 0\} .$$

Let now u^o belong to the latter set. Only the case that $p_{i_o}(u^o) = 0$ and $i_o > 1$ has to be considered since $u^o \in K(f)$ is trivial if $p_{i_o}(u^o) < 0$ or $i_o = 1$. The definition (11) of i_o guarantees that $K^*(f) \cap \{u \in R^n \;/\; p_{i_o}(u) < 0\} \neq \emptyset$. Let u^1 be a point belonging to this set. On account of Corollary 1. it suffices to show that $f_o(t) \underset{Df}{=} f(tu^o)$ is not increasing for $t \geq 0$, i.e. that $f_o(t) \leq f(0)$, $t \geq 0$. To see this, suppose that there is a $t_o > 0$ such that $f_o(t_o) > f(0)$. By Lemma 2., one gets

$$u(\alpha) \underset{Df}{=} \alpha u^1 + (1-\alpha)(t_o u^o) \in K^*(f) \cap \{u \in R^n \;/\; p_{i_o}(u) < 0\}$$

for all $\alpha \in (0,1]$. For every fixed α, the polynomial $f_1(t) \underset{Df}{=} f(tu(\alpha))$, $t \in R$ is quasiconvex and it decreases for $t > 0$ because of $p_{i_o}(u(\alpha)) < 0$. This means that $f(u(\alpha)) < f(0)$, for all $\alpha \in (0,1]$. By the continuity of $f(x)$, follows from $\alpha \to 0$ that $f(t_o u^o) \leq f(0)$. But this contradicts the choice of t_o.

The representation of R(f) is an easy consequence since

$$u \in R(f) \Longleftrightarrow u,\ -u \in K(f). \quad \#$$

By Lemma 4., it is clear that K(f) is the recession cone common to all non-empty level sets

$$L_f(\lambda) \underset{Df}{=} \{x \in R^n \;/\; f(x) \leq \lambda\} , \quad \lambda \in R$$

of the quasiconvex polynomial f(x) (we remark, that, in general a quasiconvex function can have bounded as well as unbounded level sets, see e.g. STOER/WITZGALL (1970)).

If the polynomial f(x) is convex, then one obtains from Lemma 4. the same result: $i_o = 1$ as BELOUSOV (1977) obtained for convex polynomials.

Lemma 5.

Let $f: R^n \to R$ be a quasiconvex polynomial.

Then K(f), defined by (9), is a convex polyhedral cone. It has a system of integer generators if all coefficients of f(x) are integers.

Proof.

Underlying the representation (1) and successively applying Lemma 1.(i) and Lemma 2. one obtains the first part of the proposition as follows. The set $\{u \in R^m \;/\; p_m(u) = 0\}$ is a linear subspace in R^n, the set

$\{u \in R^n \ / \ p_m(u) = p_{m-1}(u) = 0\}$ is a subspace of $\{u \in R^n \ / \ p_m(u) = 0\}$ and so on. Finally, the set K(f) is a closed half-space in the subspace $K^*(f)$ or coincides with this set.
The second part of the proposition follows now from Corollary 2. #

By the latter proof it is clear that $K^*(f)$ from (12) is the smallest linear subspace containing the polyhedral recession cone K(f) and, the constancy space R(f) is the largest subspace contained in the recession cone K(f).

Remark.

If the quasiconvex polynomial f(x) has only integer coefficients, then the polyhedral cone K(f) coincides with the solution set of a finite system of linear inequalities having only integer coefficients. In the proofs of Lemma 3. and Corollary 2. an idea is sketched to derive an algorithm which generates such a system.

3. (Mixed-) Integer Solutions of Quasiconvex Polynomial Inequalities

In this section we consider integer points in the solution set of a finite system of quasiconvex polynomial inequalities. Let $f_i: R^n \longrightarrow R$, $i \in I_o = \{1,\dots,m\}$ be quasiconvex polynomials.
Then the set

$$M \underset{Df}{=} \{x \in R^n \ / \ f_i(x) \leq 0, \ i \in I_o\} \qquad (13)$$

is a closed and convex subset of R^n. If we denote by V the recession cone of (13), then it follows from Lemma 5. that

$$V = \bigcap_{i \in I_o} K(f_i) \quad .$$

is a polyhedral convex cone. Also by Lemma 5., the cone V has a system of integer generators if all coefficients of all polynomials f_i, $i \in I_o$ are integers. The single polyhedral cones $K(f_i)$, $i \in I_o$, are characterized by Lemma 4.

For convenience we use the following notation

$$M_J \underset{Df}{=} \{x \in R^n \ / \ f_i(x) \leq 0, \ i \in J\} \qquad (14)$$

if $J \subset I_o$. The corresponding recession cone will be denoted by V_J.
By our considerations above, we know that if there are an index $i_o \in I_o$ and a direction of recession $u^{i_o} \in V$ which is not in $R(f_{i_o})$ (from (10)), then $\inf \{f_{i_o}(x) \ / \ x \in M\} = -\infty$ and the equivalence

$$M = M_{I_o} \neq \emptyset \Longleftrightarrow M_{I_1} \neq \emptyset$$

holds where $I_1 \underset{Df}{=} I_o \setminus \{i_o\}$. Obviously, if there are an index $i_1 \in I_1$ and a vector $u^{i_1} \in V \setminus R(f_i)$, then

$$M_{I_1} \neq \emptyset \Longleftrightarrow M_{I_2} \neq \emptyset$$

where $I_2 \underset{Df}{=} I_1 \setminus \{i_1\}$. After a finite number of such eliminations we arrive at an index set I_r (where $I_r = I_o$ is possible) such that

$$M \neq \emptyset \Longleftrightarrow M_{I_r} \neq \emptyset$$

holds, and the recession cone V_{I_r} is a linear subspace.

Further, since V_{I_r} is a linear subspace, the closed convex set M_{I_r} may be represented as

$$M_{I_r} = (M_{I_r} \cap V_{I_r}^{\perp}) + V_{I_r}$$

where $V_{I_r}^{\perp}$ is the orthogonal complement of the linear subspace V_{I_r}, i.e. the set M_{I_r} is representable as the sum of a compact convex set and the subspace V_{I_r}.

Now, it is easy to verify that

$$I_r = I \underset{Df}{=} \{ i \in I_o \ / \ \inf\{f_i(x) \ / \ x \in M\} > -\infty \} \qquad (15)$$

holds.

Using the index set I from (15) we define, in the sense of (14), the stable set M_I with respect to M.

Let $0 < s \leq n$. Then we define the set M_s of (mixed-) integer points in $M \subset R^n$ by

$$M_s \underset{Df}{=} \{x \in M \ / \ x_1, \ldots, x_s \in Z\}. \qquad (16)$$

The subspace V_I (i.e. the recession cone of the stable set M_I) satisfies (MIG) (see section 1.) if and only if V_I has a (mixed-) integer basis.

Now we turn to a lemma which BELOUSOV (1977) proved for convex polynomials.

Lemma 6.

Let $f_i: R^n \longrightarrow R$, $i \in I_o$, be quasiconvex polynomials. Further, let M, I and M_I be defined by (13), (15) and (14), respectively. If $M \neq \emptyset$, then, for all $\varepsilon > 0$, there exists a point $\bar{x} \in M_I$ such that

$$f_i(x) \leq 0 \quad \forall \, x \in U_\varepsilon(\bar{x}) \quad \forall \, i \notin I.$$

Proof.

We apply the elimination procedure described before. Without loss of generality, we suppose that $I_r = I = \{r+1, \ldots, m\}$ and $\{u^1, \ldots, u^r\}$ are the index set and the vectors, respectively, found by it. Let $\varepsilon > 0$ and $x^o \in M$ be arbitrarily chosen.

As the quasiconvex polynomial $f_r(x)$ decreases on each half-line

$$S_r = \{x \in R^n \ / \ x = y + tu^r, \ t \geq 0\} \ , \ y \in R^n,$$

we find a number $t_r > 0$ sufficiently large such that

$$f_r(x) \leq 0 \quad \forall \ x \in \text{cl } U_\varepsilon \ (x^0 + t_r u^r).$$

Furthermore, the point $x^0 + t_r u^r$ belongs to the set M_I because

$$f_i(x^0 + t_r u^r) = f_i(x^0) \leq 0, \ i \in I.$$

Now, there is a number $t_{r-1} > 0$ sufficiently large such that

$$f_{r-1}(x) \leq 0 \ \forall \ x \in \text{cl } U_\varepsilon \ (x^0 + t_r u^r + t_{r-1} u^{r-1})$$

holds. From

$$f_i(x^0 + t_r u^r + t_{r-1} u^{r-1}) = f_i(x^0 + t_r u^r) \leq 0 \quad \forall \ i \in I$$

and

$$f_r(x) \leq \sup\{f_r(x) \ / \ x \in \text{cl } U_\varepsilon (x^0 + t_r u^r)\} \leq 0,$$

for all $x \in \text{cl } U_\varepsilon (x^0 + t_r u^r + t_{r-1} u^{r-1})$, one concludes that

$$x^0 + t_r u^r + t_{r-1} u^{r-1} \in M_I,$$

$$f_r(x) \leq 0, \ f_{r-1}(x) \leq 0 \quad \forall \ x \in \text{cl } U_\varepsilon (x^0 + t_r u^r + t_{r-1} u^{r-1}).$$

Proceeding in this way, we arrive at a point

$$\bar{x} = x^0 + t_r u^r + \ldots + t_1 u^1 \in M_I$$

satisfying

$$f_i(x) \leq 0 \quad \forall \ i \notin I \ \forall \ x \in \text{cl } U_\varepsilon (\bar{x}). \quad \#$$

Proposition 7.

Let $f_i: R^n \longrightarrow R$, $i \in I_0$, be quasiconvex polynomials. Further, let M, I, M_I and M_s be defined as in (13), (15), (14) and (16), respectively. If the recession cone V_I of M_I satisfies (MIG), then the set M_s is non-empty if and only if the stable set M_I of M contains a (mixed-) integer point.

Proof.

Since the recession cone V_I is a linear subspace, we know that

$$M_I = (M_I \cap V_I^\perp) + V_I.$$

By assumption, the subspace V_I has a basis $\{v^1, \ldots, v^p\} \subset Z^s \times R^{n-s}$. Let $\delta > 0$ be such that

$$\delta > 1/2 \text{ diam } [(M_I \cap V_I^\perp) + \{ x \in R^n \ / \ x = \sum_{i=1}^{p} t_i v^i, \ t_i \in [0,1], \ i=1,\ldots,p\}].$$

Then

$$U_\delta \ (x) \cap \{x \in M_I \ / \ x_1, \ldots, x_s \in Z\} \neq \emptyset$$

holds, for all $x \in M_I$. According to Lemma 6. there is a point $\bar{x} \in M_I$

such that

$$f_i(x) \leq 0 \quad \forall\, x \in U_\delta(\bar{x}) \quad \forall\, i \notin I.$$

Hence, there exists a point $\tilde{x} \in M_I$, $\tilde{x}_1,\ldots,\tilde{x}_s \in Z$ which satisfies $f_i(\tilde{x}) \leq 0$, for all $i \notin I$, too, i.e. $\tilde{x} \in M_s$. #

Example.

Let

$$M = \{x \in R^2 \,/\, \sqrt{2}\, x_1 - x_2 = 0,\ x_1 \geq 1\}.$$

Then one has for the stable set

$$M_I = \{x \in R^2 \,/\, \sqrt{2}\, x_1 - x_2 = 0\} = V_I.$$

The only integer point in M_I is the origin, but the set M_2 is empty. This shows that the assumption on V_I is essential for the validity of Proposition 7.

4. Stability of Solution Sets of Quasiconvex Polynomial Inequalities.

Let $f_i: R^n \to R$, $i \in I_0 \underset{Df}{=} \{1,\ldots,m\}$, be quasiconvex polynomials.

We define the parameter-dependent sets

$$M(\lambda) \underset{Df}{=} \{x \in R^n \,/\, f_i(x) \leq \lambda_i,\ i \in I_0\}, \lambda \in R^m. \tag{17}$$

These closed convex sets determine a multifunction $M: R^m \to 2^{R^n}$ with the domain

$$\Lambda \underset{Df}{=} \{\lambda \in R^m \,/\, M(\lambda) \neq \emptyset\}, \tag{18}$$

i.e. $M: \Lambda \to 2^{R^n}$.

By Lemma 4. and 5., we know that all sets $M(\lambda)$, $\lambda \in \Lambda$, have a common polyhedral recession cone given by

$$V = \bigcap_{i \in I_0} K(f_i).$$

First we prove, that the multifunction M is lower semicontinuous on its domain Λ. This result is contained in BANK/MANDEL (1987) without a proof.

Proposition 8.

Let $f_i: R^n \to R$, $i \in I_0$, be quasiconvex polynomials. Further, let $M(\lambda)$ and Λ be defined as in (17) and (18), respectively. Then the multifunction $M: \Lambda \to 2^{R^n}$ given by $M(\lambda)$ is l.s.c. on Λ.

Proof.

The proof makes use of the Theorems 3.1.3. and 3.1.6. from the monography BANK/GUDDAT/KLATTE/KUMMER/TAMMER (1982). Let $\lambda^0 \in \Lambda$ be an arbitrary point. We define the set of indices

$$J \underset{Df}{=} \{ i \in I_0 \,/\, f_i(x) = \lambda_i^0 \quad \forall\, x \in M(\lambda^0) \}$$

and the parameter-dependent sets

$$\Gamma(\lambda) \underset{Df}{=} \{ x \in R^n \,/\, f_i(x) \leq \lambda_i,\ i \in J \},$$

$$M^0(\lambda) \underset{Df}{=} \{ x \in R^n \,/\, f_i(x) < \lambda_i,\ i \notin J \},$$

for $\lambda \in \Lambda$. By Theorem 3.1.6., it suffices to show that the multifunction $\Gamma : \Lambda \to 2^{R^n}$ defined by $\Gamma(\lambda)$ is l.s.c. at λ^0. By the choice of J and Corollary 1., the set $\Gamma(\lambda^0)$ is an affine subspace. Hence, the recession cone $\bar{V}$ common to all sets $\Gamma(\lambda)$, $\lambda \in I$, is a linear subspace, Therefore,

$$\Gamma(\lambda) = (\Gamma(\lambda) \cap \bar{V}^{\perp}) + \bar{V}$$

and, obviously, it suffices to show that the compact-valued multifunction $\bar{\Gamma} : \Lambda \to 2^R$ defined by $\bar{\Gamma}(\lambda) \underset{Df}{=} \Gamma(\lambda) \cap \bar{V}^{\perp}$ is l.s.c. at λ^0. By Theorem 3.1.3., one obtains that $\bar{\Gamma}$ is u.s.c. at λ^0. Now, taking into account that $\bar{\Gamma}(\lambda^0)$ is a singleton, it is clear that $\bar{\Gamma}$ is ls.c. at λ^0, too. #

An immediate consequence of the proof of Proposition 8. is

Corollary 3.

Under the hypotheses of Proposition 8., the multifunction M is continuous on Λ if the recession cone V is a linear subspace. Moreover, the domain Λ of M is a closed set.

In general, a multifunction M defined by (17) with quasiconvex polynomials $f_i(x)$, $i \in I_0$, is not u.s.c., even not if all polynomials $f_i(x)$, are convex. This shows a counterexample due to SHIRONIN (1980), see also SHIRONIN's part in BANK/BELOUSOV et al. (1986).

If one imposes the additional assumption

$$\text{(CPC)} \quad \begin{cases} \text{There are a } \lambda^0 \in \Lambda \text{ and a compact set} \\ C_0 \subset R^n \text{ such that } M(\lambda^0) = C_0 + V. \end{cases}$$

on the multifunction M, then M is continuous on Λ. More precisely, in BANK/MANDEL (1987) it is shown that (CPC) is a necessary and sufficient condition for the existence of a continuous multifunction $C: \Lambda \to 2^{R^n}$ with compact convex images which permits the representation

$$M(\lambda) = C(\lambda) + V \qquad \forall\, \lambda \in \Lambda. \tag{19}$$

From the latter fact follows, that the multifunction M is even H-continuous on Λ if (CPC) is fulfilled. In this case the closedness of Λ is obvious. We will later see that the domain Λ of M from (17) is closed, too, even if (CPC) is not satisfied (Proposition 10.).

In view of the stability of (mixed-) integer points in $M(\lambda)$ we associate (as in the previous section for M) to $M(\lambda)$ its stable set. For this purpose we introduce the set of indices

$$I(\lambda) \underset{Df}{=} \{ i \in I_0 \ / \ \inf \{ f_i(x) \ / \ x \in M(\lambda) \} > -\infty \} , \tag{20}$$

for a fixed $\lambda \in \Lambda$. Actually, the set $I(\lambda)$ does not depend on the choice of $\lambda \in \Lambda$, which is a consequence of the fact that neither the recession dirgctions $u \in K(f_i)$ nor the constancy directions $u \in R(f_i)$, $i \in I_0$, depend on $\lambda \in \Lambda$. Therefore, from Section 3. one concludes the following two corollaries.

Corollary 4.

If the assumptions of Proposition 8. are fulfilled and $I(\lambda)$ is the set of indices defined in (20), then

$$I = I(\lambda), \ \lambda \in \Lambda \ . \tag{21}$$

Applying the elimination procedure from the previous section to $M(\lambda')$, for any $\lambda' \in \Lambda$, one obtains I.

Now, using the set of indices I from (21), we can associate to a set $M(\lambda), \lambda \in \Lambda$ its stable set $M_I(\lambda)$ by

$$M_I(\lambda) \underset{Df}{=} \{ x \in R^n \ / \ f_i(x) \leq \lambda_i , \ i \in I \} \tag{22}$$

From the previous section, it is clear that the recession cone $V_I = \bigcap_{i \in I} K(f_i)$ of $M_I(\lambda)$, $\lambda \in \Lambda$, is a linear subspace.

On account of Corollary 4., one derives from Lemma 6.

Corollary 5.

Let the assumptions of Proposition 8. be fulfilled, and let I and $M_I(\lambda)$ be defined as in (21) and (22), respectively. If $\lambda \in \Lambda$, then, for every $\varepsilon > 0$, there exists a point $x^* \in M_I(\lambda)$ such that

$$f_i(x) \leq \lambda_i \quad \forall \ x \in U_\varepsilon(x^*) \quad \forall \ i \notin I .$$

In particular, Corollary 5, implies that

$$\Lambda = \{ \lambda \in R^m \ / \ M_I(\lambda) \neq \emptyset \}$$

holds. Therefore, we can assign to the multifunction $M: \Lambda \rightarrow 2^{R^n}$ its stable mapping $M_I: \Lambda \longrightarrow 2^{R^n}$ defined by the sets $M_I(\lambda)$ from (22).

Proposition 9.

Under the hypotheses of Proposition 8. the stable mapping M_I defined by the closed convex sets $M_I(\lambda)$ from (22) is H-continuous. The common domain Λ of M and M_I is a closed set.

Proof.

Since the subspace V_I is constant, for all $\lambda \in \Lambda$, the multifunction M_I satisfies (CPC). Hence, M_I is H-continuous and Λ is closed. #

Let $0 < s \leq n$. As in (16) we define the sets $(M(\lambda))_s$ and $(M_I(\lambda))_s$ of (mixed-) integer points in the sets $M(\lambda)$ and $M_I(\lambda)$ from (17) and (22), respectively. Further, let

$$B = \{\lambda \in \Lambda / \ (M(\lambda))_s \neq \emptyset\}.$$

In BANK/MANDEL (1987) the following was shown. If the hypotheses of Proposition 8. and (CPC) are fulfilled and, furthermore, if V satisfies (MIG), then there is a compact-valued, u.s.c. multifunction $C_s : B \longrightarrow 2^{R^n}$ such that

$$(M(\lambda))_s = (C(\lambda))_s + V_s \quad \forall \ \lambda \in B$$

holds where V_s denotes the set of (mixed-) integer points in V.

Proposition 10.

Let the hypotheses of Proposition 8. be fulfilled. Further, let $M_I(\lambda)$, V_I, $(M(\lambda))_s$, $(M_I(\lambda))_s$ and B be defined as before. If V_I satisfies (MIG), then

(i) $(M(\lambda))_s \neq \emptyset \Longleftrightarrow (M_I(\lambda))_s \neq \emptyset$,
(ii) B is a closed set.

Proof.

The first part can be shown following the same way as for Proposition 7. (remark, δ depends on λ). The second part is a consequence of the fact that V_I is a constant subspace with a (mixed-) integer basis. Namely, by the remarks preceding this proposition, B is closed if Λ is closed, but we know this from Proposition 9. #

5. Conclusion.

We presented in this paper an analysis of such properties of quasiconvex polynomials which are essential for the existence of optimal points and the stability of (mixed-) integer optimization problems involving quasiconvex polynomials as well as constraints and objectiv function. We showed that the multifunction defined by quasiconvex polynomial inequalities with variable right-hand sides is lower semicontinuous on its domain. In particular, we developed a criterion under which the considered sets contain (mixed-) integer points even if the right-hand sides of the constraints vary. Moreover, we verified that the set of all right-hand sides leading to consistent (mixed-) integer quasiconvex polynomial optimization problems is a closed set. By our considerations, we demonstrated that the assumptions on the recession cone of the feasible sets (requireing the existence of (mixed-) integer generators) cannot be weakend.

Acknowledgements.

We would like to thank Rainer E. Burkard for valuable comments and careful reading of the manuscript.

References.

Bank, B., E.G. Belousov, R. Mandel, J.N. Cheremnych and V.M. Shironin (1986). Matematyceskije programirowanje: voprozy rasreshimosti i ustoičivosti. Iszd. MGU, Moskwa (russian).

Bank, B., J. Guddat, D. Klatte, B. Kummer, K. Tammer (1982). Non-linear Parametric Optimization. Akademie-Verlag, Berlin. ((1983) Birkhäuser Verlag, Basel·Boston·Stuttgart).

Bank, B., R. Mandel (1987). Nonlinear Parametric Integer Programming. In Guddat et al. (Eds.) Proceedings of the International Conference on 'Parametric Optimization and Related Topics'. Plaue/DDR, Okt. 1985, Akademie-Verlag, Berlin.

Bank, B., R. Mandel (to appear). Parametric Integer Optimization. Akademie-Verlag, Berlin.

Belousov, E.G. (1977). Vvedenie v vypuklyj analiz i zeločislennoje programmirovanie, Iszd. MGU, Moskwa (russian).

Mandel, R. (1985). Beiträge zur Theorie ganzzahliger Optimierungsaufgaben. Dissertation (B), Humboldt-Universität zu Berlin.

Shironin, V.M. (1980). O raspoloženii zeločislennych toček v neograničennych vypuklych mnoshestvach euklidivo prostranstwa. Diss. (kand. fiz. mat. nauk). Ekon. Fak. MGU. Moskwa (russian).

Stoer, J., C. Witzgall (1970). Convexity and Optimization in finite dimensions I. Springer-Verlag, Berlin-Heidelberg-New York.

AN EXACT PENALTY FUNCTION METHOD FOR STRICTLY CONVEX QUADRATIC PROBLEMS +

Heinz Bernau ++

Abstract: A solution method for positive definite quadratic programming is described in which an exact penalty function is to be minimized. An advantage of this algorithm is that, in case of inconsistent constraints, an l_1-solution of the quadratic problem will be obtained for a given value of the penalty parameter. This property is very important if the method is used in successive quadratic programming methods, since if there are non-feasible subproblems, the l_1-solutions may also be used as search directions. For the development of the algorithm this fact was the principal motivation.

1. Introduction

In successive quadratic programming methods (SQP-methods) to solve the nonlinear programming problem

$$\begin{aligned} &\text{minimize} && f(x) , \\ &\text{subject to} && g_i(x) = 0 , \quad i \in E, \\ & && g_j(x) \leq 0 , \quad j \in I, \end{aligned} \tag{1.1}$$

where the index sets E and I are finite, the subproblem solved to determine a search direction at the current estimate x often takes the form [1], [8], [11], [14]:

$$\begin{aligned} &\text{minimize} && \tfrac{1}{2}d^T Bd + \nabla f(x)^T d , \\ &\text{subject to} && g_i(x) + \nabla g_i(x)^T d = 0 , \quad i \in E, \\ & && g_j(x) + \nabla g_j(x)^T d \leq 0 , \quad j \in I, \end{aligned} \tag{1.2}$$

where B is an nXn symmetric positive definite matrix, $\nabla f(x)$ and $\nabla g_i(x)$, $i \in E \cup I$, denote the gradients of f(x) and $g_i(x)$ respectively, and superscript T denotes the transpose. A disadvantage of the application of the subproblems (1.2) is that the linearizations of the constraints of (1.1) may introduce inconsistencies, even though the problem (1.1) is consistent [10], [15].

To avoid this difficulty Fletcher [4], [5], [6] proposes a trust-region algorithm to minimize the exact penalty function

$$\Phi(x,v) = vf(x) + \sum_{i \in E} \left| g_i(x) \right| + \sum_{j \in I} \max \left\{ 0 , g_j(x) \right\} , \tag{1.3}$$

of the problem (1.1), where $v > 0$ is a penalty parameter. In this algorithm to determine search directions the following subproblem has to be solved

---

+This research was supported by Hung. Res. Found. OTKA/1044

++ Computer and Automation Institute Hungarian Academy of Sciencies
H-1502 Budapest XI , Kende u. 17-23

$$\text{minimize} \quad v(\tfrac{1}{2}d^TBd + \nabla f(x)^Td) + \sum_{i\in E}\left|g_i(x) + \nabla g_i(x)^Td\right|$$
$$+ \sum_{j\in I} \max\left\{0,\ g_j(x) + \nabla g_j(x)^Td\right\}, \qquad (1.4)$$
$$\text{subject to} \quad \|d\|_\infty \le p,$$

where $p>0$ is the radius of the trust-region [1], [4], [5]. In this way we get feasible subproblems, but the correct choice of the parameter v is uncertain. With the choice of the parameter v we have to guarantee that a local minimizer of the function $\Phi(x,v)$ will also be a local solution of problem (1.1) [2], [4], [9].

In [1] an SQP-method is given which solves subproblems of type (1.4) in such a way that the penalty parameter v is adjusted to be as large as possible subject to the property that the solution of problem (1.4) is feasible for the problem (1.2). If the problem (1.2) has no feasible solution the solution of (1.4) corrsponding to $v=v_{min}$ is used as search direction, where v_{min} is a fixed positvve small value. In the following section we describe the method to determine search directions of this type.

2. An exact penalty function method:

Independently from the notations used earlier we consider the following quadratic programming problem

$$\begin{aligned} &\text{minimize} && \tfrac{1}{2}x^TBx + c^Td, \\ &\text{subject to} && a_i^Tx - b_i = 0, \quad i\in E, \\ & && a_j^Tx - b_j \le 0, \quad j\in I, \\ & && p\le x\le q, \end{aligned} \qquad (2.1)$$

where B is an nXn symmetric positive definite matrix, c,p,q and a_i, $i\in E\cup I$ are n-vectors and the index sets E and I are finite.

Let's introduce the following notations

$$f(x) = \tfrac{1}{2}x^TBx + c^Tx,$$
$$Q(x) = \sum_{i\in E}\left|a_i^Tx - b_i\right| + \sum_{j\in I}\max\left\{0,\ a_j^Tx - b_j\right\} \qquad (2.2)$$
$$\Phi(x,v) = vf(x) + Q(x), \quad v>0.$$

Disregarding the bound constraints in (2.1), the function $\Phi(x,v)$ is an exact penalty function for this problem [1], [4] and an exact penalty function algorithm can be based on the following problem

$$\begin{aligned} &\text{minimize} && \Phi(x,v), \\ &\text{subject to} && p\le x\le q. \end{aligned} \qquad (2.3)$$

The following lemma investigates relations between the problems (2.1) and (2.3) (see also [4], [9]).

<u>Lemma 2.1:</u> For $v=\bar{v}$ let $\bar{x}$ be a solution of the problem (2.3). If $\bar{x}$ is feasible for problem (2.1) then

i.) $\bar{x}$ is the optimal solution of (2.1) ,

ii.) for every v, $0 < v \leq \bar{v}$, $\bar{x}$ is an optimal solution of the corresponding problem (2.3).

Proof: i.) immediately follows from the fact that $Q(x)=0$ for every feasible x of problem (2.1).

If ii.) would not be true, then there exist $\hat{v}$, $0 < \hat{v} < \bar{v}$, and $\hat{x}$, $p \leq \hat{x} \leq q$, with $\hat{v}f(\hat{x}) + Q(\hat{x}) < \hat{v}f(\bar{x})$. As $Q(x) \geq 0$ for every x, from this it follows

$$0 \leq Q(\hat{x}) < \hat{v}(f(\bar{x}) - f(\hat{x})) \leq \bar{v}(f(\bar{x}) - f(\hat{x})) ,$$

where the last inequality comes from $0 < \hat{v} < \bar{v}$. Reordering these ineqalities we get $\bar{v}f(\hat{x}) + Q(\hat{x}) < \bar{v}f(\bar{x})$, which contradicts the optimality of $\bar{x}$ for $v=\bar{v}$, and the lemma is proved. By this lemma it can be seen that independently from the value of v every solution of the problem (2.3), which is feasible for the problem (2.1), is at the same time optimal for the problem (2.1). Furthermore it turns out that the set of values v, for which the solution of (2.3) is also a solution of (2.1), is an intervall, and for all these values of v the problem (2.3) has the same solution. (Note, that the problems (2.1) and (2.3) are strictly convex and therefore their solutions, if they exist, are unique.)

The solution method for problem (2.3) is based on the following optimality conditions for this problem [1], [4]:

Theorem 2.1: A point x^*, $p \leq x^* \leq q$, is the optimal solution of the problem (2.3) if and only if there exist multipliers λ_i^*, $i \in E \cup I$, $\mu^* \in R^n$ such that

$$\begin{aligned}
& v(Bx^* + c) + \sum_{i \in E \cup I} \lambda_i^* a_i + \sum_{j=1}^{n} \mu_j^* e_j = 0 , \\
& |\lambda_i^*| \leq 1, \quad \lambda_i^* = \operatorname{sign}(a_i^T x^* - b_i) \quad \text{if} \quad a_i^T x^* - b_i \neq 0 , \quad i \in E , \\
& \left.\begin{aligned} 0 \leq \lambda_j^* \leq 1 , \quad & \lambda_j^* = 1 \quad \text{if} \quad a_j^T x^* - b_j > 0 , \\ & \lambda_j^* = 0 \quad \text{if} \quad a_j^T x^* - b_j < 0 , \end{aligned}\right\} \quad j \in I , \qquad\qquad (2.4) \\
& \mu_i^* \leq 0 \quad \text{if} \quad x_i^* = p_i , \qquad \mu_i^* \geq 0 \quad \text{if} \quad x_i^* = q_i \\
& \mu_i^* = 0 \quad \text{if} \quad p_i < x_i^* < q_i , \qquad\qquad i=1,2,\dots,n,
\end{aligned}$$

where e_j denotes the j-th unit vector.

Here it is remarkable that the values of the multipliers, belonging to nonzero linear functions are uniquely determined by the signs of these function values. This property will be used to define a sequence of equality constrained quadratic problems, where the last of these problems has the same solution as the problem (2.3) for an eventually modified value of v. (At the definition of the subsequent problems the value of v will be decreased, if the current problem indicates that for the actual value of v the solution of the problem (2.3) will not be feasible for problem (2.1)).

The next lemma proves the existence of such an equality constrained quadratic problem, for which the optimal solution is identical with the solution of problem (2.3). Let's introduce

$$J(x) = \{ i \mid (p_i - x_i)(q_i - x_i) = 0 \}$$

the index set of the variables, which in x attain one of their bounds.

Lemma 2.2: Let x^* and λ^*, μ^* be the solution of problem (2.3) and the corresponding multipliers respectively. Then x^* is a solution of the following quadratic problem

$$\begin{aligned} &\text{minimize} \quad v(\tfrac{1}{2}x^TBx + c^Tx) + \sum_{j\in\bar{Z}} \alpha_j(a_j^Tx - b_j) , \\ &\text{subject to} \quad a_i^Tx - b_i = 0 , \quad i \in Z , \\ &\qquad\qquad\qquad x_r = x_r^* , \quad r \in J(x^*) , \end{aligned} \tag{2.5}$$

where

$$\begin{aligned} &Z = \{ i \in E \cup I \quad a_i^Tx^* - b_i = 0 \} , \quad \bar{Z} = E \cup I \setminus Z, \\ &\alpha_j = \lambda_j^* \quad , \quad j \in \bar{Z} . \end{aligned} \tag{2.6}$$

Proof: The Kuhn-Tucker conditions of the problem (2.5) take the form [4]

$$\begin{aligned} &v(Bx + c) + \sum_{j\in\bar{Z}} \alpha_j a_j + \sum_{i\in Z} \lambda_i a_i + \sum_{r\ J(x\ )} \mu_r e_r = 0 , \\ &a_i^Tx - b_i = 0 \quad , \quad i \in Z , \\ &x_r = x_r^* , \quad r \in J(x^*), \end{aligned} \tag{2.7}$$

where λ_i, $i \in Z$, μ_r, $r \in J(x^*)$ are the Lagrange multipliers of the constraints of problem (2.3). From the definitions of Z in (2.6) and of $J(x^*)$ it follows that x^* satisfies the last two conditions in (2.7). As $\alpha_j = \lambda_j^*$, $j \in \bar{Z}$, it is easy to verify that for $x = x^*$, $\lambda_i = \lambda_i^*$, $i \in Z$, and $\mu_r = \mu_r^*$, $r \in J(x^*)$, the first condition of (2.7) is satisfied. (Notice, that by (2.4) $\mu_t^* = 0$, $t \in \{1,2,...,n\} \setminus J(x^*)$). This shows that x^* is part of a Kuhn-Tucker point of the problem (2.5), and the lemma is proved.

In the algorithm a special "active constraint" strategy is applied to find out the index sets Z and $J(x^*)$. A similiar approach is given in [4], but there the parameter v is fixed. Furthermore the here used handling of linearly dependent active constraints seems to be new.

To get a unique decision, whether a solution of a problem of type (2.5) is also a solution of the problem (2.3), in the following we investigate, under which conditions a Kuhn-Tucker point of problem (2.5) yields a solution of the optimality conditions (2.4). The next lemma contains a first result for this question, however a strong regularity condition with respect to problem (2.5) is used.

Lemma 2.3: Let x^* be a solution of the quadratic problem

$$\begin{aligned} &\text{minimize} \quad v(\tfrac{1}{2}x^TBx + c^Tx) + \sum_{j\in\bar{Z}} \alpha_j(a_j^Tx - b_j) , \\ &\qquad a_i^Tx - b_i = 0 \quad , \quad i \in Z, \\ &\qquad x_r = s_r , \quad r \in J, \end{aligned} \tag{2.9}$$

where $Z \subseteq E \cup I$, $\bar{Z} = E \cup I \setminus Z$, $J \subseteq \{1,2,...,n\}$, s_r, $r \in J$, is one of the bounds p_r or q_r, and the constants α_j have for $j \in \bar{Z} \cap E$ the values -1 or 1, and for $j \in \bar{Z} \cap I$ the values 0 or 1, respectively. Furthermore, assumed that

i.) $p \leq x^* \leq q$,

ii.) the vectors a_i, $i \in Z$, and e_r, $r \in J$ are linearly independent , iii.) $a_j^T x^* - b_j \neq 0$, $j \in \bar{Z}$,

iv.) $\alpha_j = \text{sign}(a_j^T x^* - b_j)$, $j \in \bar{Z} \cap E$,

$\alpha_j = 0$ if $a_j^T x^* - b_j < 0$,

$\alpha_j = 1$ if $a_j^T x^* - b_j > 0$, $\quad j \in \bar{Z} \cap I$,

then x^* is an optimal solution of problem (2.3) if and only if the Lagrange multipliers λ_i, $i \in Z$, μ_r, $r \in J$, of the problem (2.9) satisfy the conditions

a.) $-1 \leq \lambda_i \leq 1$, $i \in Z \cap E$, $\quad 0 \leq \lambda_i \leq 1$, $i \in Z \cap I$,

b.) $\mu_r \leq 0$ if $s_r = p_r$, $\quad \mu_r \geq 0$ if $s_r = q_r$.

Proof: ($\rightarrow$) if x^* is a solution of the problem (2.3), then it follows from lemma 2.2 that x^*, λ^*, μ^* from (2.4) satisfy the Kuhn-Tucker conditions of the problem (2.9). (Remark, that by the assumptions iii.) and iv.) the problems (2.5) with (2.6) and (2.9) agree.) After assumption ii.) the the problem (2.9) has unique Lagrange multipliers, therefore $\lambda_i = \lambda_i^*$, $i \in Z$, and $\mu_r = \mu_r^*$, $r \in J$, holds, where λ_i^* and μ_r^* are the multipliers from (2.4), and the relations a.) and b.) are obviously satisfied.
($\leftarrow$) if the Lagrange multipliers of problem (2.9) satisfy the conditions a.) and b.), then from the assumption iii.) it immediately follows that x^*, $\lambda_j^* = \alpha_j$, $j \in \bar{Z}$, $\lambda_i^* = \lambda_i$, $i \in Z$, $\mu_r^* = \mu_r$, $r \in J$, $\mu_t^* = 0$, $t \notin J$, is a solution of the optimality conditions (2.4), i.e. x^* is the optimal solution of problem (2.3), and the lemma is proved.

Remark 2.1: The assumptions ii.) and iii.) guarantee that for problem (2.3) in its optimal solution the "active" constraints, i.e. the vanishing linear functions and the bound constraints satisfied as equality, are linearly independent. In this case the multipliers for problem (2.3) and the Lagrange multipliers of the corresponding problem (2.9) are unique and therefore the conditions (2.4) and the Kuhn-Tucker conditions combined with conditions a.) and b.) from lemma 2.3 completely agree. Without these regularity assumptions in general this equivalence may be not true.

Remark 2.2: With respect to the determination of the greatest value of v with the property of lemma 2.1, for the case that the solution x of problem (2.9) is also feasible for the problem (2.1), we seperately investigate some conclusions of the lemma 2.3. In this case from assumption iii.) it follows that $\bar{Z} \cap E = 0$ and after iv.) holds $\alpha_j = 0$, $j \in \bar{Z}$, i.e. there will not be nonzero fixed multiplier in the optimality conditions (2.4). By condition a.) of lemma 2.3 holds $|\lambda_i| \leq 1$, $i \in Z$. let

$$K = \max \{ |\lambda_i| \mid i \in Z \} \tag{2.10}$$

be the absolutely maximal value of these multipliers. Then it is easy to verify that for

$$\hat{v} = v/K, \quad \hat{\lambda}_i = \lambda_i/K, \; i \in Z, \quad \hat{\mu}_r = \mu_r/K, \; r \in J, \quad \hat{\mu}_t = 0, \; t \notin J, \quad \hat{x} = x^* \tag{2.11}$$

the optimality conditions (2.4) are also satisfied, i.e. x^* is also optimal for the problem (2.3) corresponding to $v = \hat{v}$. So it follows that, if the value K in (2.10) is less than 1, the value of the parameter v can be increased until $\hat{v}$, and the property of lemma 2.1 for the solution of problem (2.3) will be still valid. Since under the assumptions of lemma 2.3 the multipliers of (2.4) are unique and the maximal absolute value of the $\hat{\lambda}_i$ from (2.11) is 1, the value $\hat{v}$ from (2.11) must be the largest value of v, such that for the solution of problem (2.3) lemma 2.1 applies.

In the following lemma, without using the assumption iii.) of lemma 2.3, a similiar result will be proved. (Assumption ii.) of lemma 2.3 will automatically be satisfied in the algorithm).

<u>Lemma 2.4:</u> Let x^* and λ, μ be the solution of problem (2.9) and the Lagrange multipliers, respectively. Assumed, that

i.) x^* is feasible for the problem (2.1),

ii.) $\alpha_j = 0$, $j \in \bar{Z}$,

iii.) $\lambda_i \geq 0$, $i \in Z \cap I$

$$\mu_r \geq 0 \quad \text{if } s_r = q_r \;, \qquad \mu_r \leq 0 \quad \text{if } s_r = p_r \;, \quad r \in J, \tag{2.12}$$

then x^* is for $\hat{v} = v/K$ the solution of problem (2.3), where K is the in (2.10) defined value.

<u>Proof:</u> After (2.12) the Lagrange multipliers λ_i, $i \in Z$, μ_r, $r \in J$, have the "right" sign to fulfil the conditions (2.4). The transformations (2.10) and (2.11) guarantee that the absolute values of $\hat{\lambda}_i$, $i \in Z$ are not greater than 1. Replacing in (2.7) the index set $J(x^*)$ by J, and taking into account that $\alpha_j = 0$, $j \in \bar{Z}$, we get the relation

$$v/K(Bx^* + c) + \sum_{i \in Z} (\lambda_i/K) a_i + \sum_{r \in J} (\mu_r/K) e_r = 0,$$

i.e. $\hat{v}, \hat{\lambda}, \hat{\mu}, x^*$ from (2.11) satisfy the optimality conditions (2.4), and the lemma is proved.

<u>Remark 2.3:</u> This result is very useful for the algorithm. It shows that the algorithm can be stopped, if for the Kuhn-Tucker solution of a problem of type (2.9) the above given conditions hold, because in this case x^* is a solution of problem (2.3) for $\hat{v} = v/K$, and as x^* is feasible for problem (2.1), x^* is also optimal for (2.1). Note, if $K > 1$ holds, then for the current value of v the point x^* is not a solution of the problem (2.3), i.e. without solving completely the correspondig problem (2.3) the algorithm can be terminated. Furthermore, it can be easily verified that, if x^* is feasible for problem (2.1), the assumptions ii.) and iii.) of lemma 2.3 are equivalent to the assumption that the active constraints of problem (2.1) are linearly independent. By remark 2.2 so it follows that for such problems the value $\hat{v}$, determined by (2.10) and (2.11) is the largest value with the property of lemma 2.1.

Using these connections between the problem (2.3) and the problems of type (2.9) the following active-constraint algorithm can be formulated.

Let x^k, $p \leq x^k \leq q$, be the current estimate, which is feasible for the quadratic problem

$$\text{minimize } v^k(\tfrac{1}{2}x^T Bx + c^T x) + \sum_{j \; \bar{Z}^k} \alpha_j^k (a_j^T x - b_j) \quad ,$$

subject to $\quad a_i^T x - b_i = 0\,, \qquad i \in Z^k\,,$ (2.13)

$\qquad\qquad\qquad x_r = s_r\,, \qquad r \in J^k\,,$

where $Z^k \subseteq E \cup I$, $\bar{Z}^k = E \cup I \setminus Z^k$, $J^k \subseteq \{1,2,\ldots,n\}$, s_r, $r \in J^k$, is one of the bounds p_r or q_r, and for the constants α_j^k, $j \in \bar{Z}^k$ holds

$$\alpha_j^k = 0 \quad \text{if } a_j^T x^k - b_j \neq 0\,, \quad j \in \bar{Z}^k\,,$$

$$\left.\begin{array}{ll} \alpha_j^k = -1 & \text{if } a_j^T x^k - b_j < 0, \\ \alpha_j^k = 1 & \text{if } a_j^T x^k - b_j > 0, \end{array}\right\} \quad j \in \bar{Z}^k \cap E\,,$$

$$\left.\begin{array}{ll} \alpha_j^k = 0 & \text{if } a_j^T x^k - b_j < 0, \\ \alpha_j^k = 1 & \text{if } a_j^T x^k - b_j > 0, \end{array}\right\} \quad j \in \bar{Z}^k \cap I\,.$$

Furthermore, we assume that the constraints of this problem are linearly independent. Let N denotes the matrix, which columns contain the vectors a_i, $i \in Z^k$, and e_r, $r \in J^k$. By the linear independence of the constraints this matrix has full rang, and the optimal solution $\bar{x}^k$ of the problem (2.13) and the corresponding Lagrange multipliers can be given in the form

$$\bar{x}^k = x^k + d^k = x^k - H\nabla\varphi^k(x^k)\,,$$
$$(\lambda^T, \mu^T) = -N^+\nabla\varphi^k(x^k)\,, \tag{2.14}$$

where $\nabla\varphi^k(x^k)$ denotes the gradient of the objective function of the problem (2.13), and the matricies H and N^+ are defined as follows [4], [7], [12]:

$$N^+ = (N^T B^{-1} N)^{-1} N^T B^{-1}\,,$$
$$H = B^{-1} - B^{-1} N (N^T B^{-1} N)^{-1} N^T B^{-1}\,. \tag{2.15}$$

The next point x^{k+1} is chosen on the half line

$$x(\tau) = x^k + \tau d^k, \qquad \tau \in [0, \infty]\,,$$

in the following way. Let $\bar{\tau}$ be the least nonnegative value of τ, for which in $x(\tau)$ one of the linear functions $a_j^T x - b_j$, $j \in \bar{Z}^k$, changes its sign, or one of the variables x_t, $t \in \{1,2,\ldots,n\} \setminus J^k$, attains one of its bounds. As special case, if $a_{jo}^T x^k - b_{jo} = 0$ and $a_{jo}^T d^k \neq 0$, for some $jo \in \bar{Z}^k$, then $\bar{\tau} = 0$ has to be chosen. If on this half line there is no such point, we set $\bar{\tau} = \infty$.

Depending on the value of $\bar{\tau}$ the point x^{k+1} is given by

$$x^{k+1} = x^k + \tau_k d^k\,, \quad \text{with } \tau_k = \min\{1.0\,, \bar{\tau}\}\,.$$

The definition of the quadratic problem, belonging to x^{k+1}, depends on the value of $\bar{\tau}$ and some properties of x^{k+1}. Therefore, we distinguish the following 3 cases:

a.) $\bar{\tau} \leq 1.0$, i.e. in x^{k+1} vanishes one of the linear functions, belonging to $\bar{Z}^k$, or a variable, not fixed in problem (2.13), attains one of its bounds. Corresponding to the general active constraint strategy [4] the index set Z^k or J^k has to be extended by the constraint, determining the value of $\bar{\gamma}$. At this extension it must be checked, whether the new constraint is linearly independent from the constraint of problem (2.13). If this is the case, the new quadratic problem is obtained from problem (2.13) by this extension, i.e. to one of the index sets Z^k or J^k the corresponding index is added. Otherwise, i.e. at linear independence of the new constraint, we consider the next 2 cases seperately

i.) $\bar{\gamma}$ is determined by the linear function $a_{jo}^T x - b_{jo}$, in this case the constraints will not be changed. In the new problem only α_{jo} is set to zero. (Remark, that the vector d^k satisfies the equation $N^T d = 0$. As a_{jo} is linearly dependent from the columns of N, $a_{jo}^T d^k = 0$ holds, i.e. in exact arittmetic this function will not take part at the determination of $\bar{\gamma}$. The choice of α_{jo} is based on lemma 2.4).

ii.) $\bar{\gamma}$ is determined by a variable x_{to}. In this case e_{to} is linearly dependent from the columns of N. Then there must be at least on column, in which the to-th component is nonzero. Between this columns we choose that with absolutely largest to-th component. This column obviously corresponds to linear function. Replacing this constraint with the constraint $x_{to} = s_{to}$, we have the same situation as in case i.), and the algorithm continues as for case i.)

b.) $\bar{\gamma} > 1.0$ and (2.12) is not fulfilled. In this case $x^{k+1} = \bar{x}^k$ is the optimal solution of problem (2.13), but not of problem (2.3). It can be proved [4] that deleting a constraint, for which (2.12) is not satisfied, the objective function of (2.13) can be decreased. Therefore to get the new quadratic problem, we delete one of the constraints, not satisfying (2.12). If this constraint is an inequality constraint of (2.1), the corresponding α_j constant is set to zero. (Note, that in the optimal solution of the next quadratic problem this inequality will be satisfied [3]).

c.) $\bar{\gamma} > 1.0$ and (2.12) is fulfilled. As in case b.) x^{k+1} is optimal solution of (2.13), and we distinguish the following 2 cases:

i.) x^{k+1} is feasible for problem (2.1). Then by lemma 2.4 x^{k+1} is also optimal for this problem, and (2.10),(2.11) give an estimate for the largest value of v, which is exact, if the active constraints of problem (2.1) are linearly independent.

ii.) x^{k+1} is not feasible for problem (2.1). If $v^k > v_{min}$ the value of v will be reduced, i.e. we set $v^{k+1} = v/L$, $L > 1.0$, and the so modified problem (2.13) is again solved. If $v^k \leq v_{min}$, the point x^{k+1} is accepted as l_1-solution of problem (2.1).

The algorithm starts with $v^1 = 1.0$, $Z^1 = \emptyset$, $J^1 = \emptyset$, and arbitrary chosen x^1, $p \leq x^1 \leq q$, i.e. $\bar{x}^1$ is the unconstraint minimum of the corresponding objective function in (2.13).

In the following we show the the finiteness of this algorithm. From the definition of d^k in (2.14) it follows that the method is not increasing w.r.t. the objective functions of the problems (2.13). Furthermore note, that the number of nonzero α_j in these objectives is monotonically de-

creasing, i.e. we cannot return to an earlier used quadratic problem. Therefore after a finite number of occurencies of a.) or b.) we must get case c.), where the algorithm terminates or the value of v is reduced. As only finite different values of v are possible, the algorithm ends in a finite number of steps.
A first implementation of the algorithm, using the same Cholesky and QR factorizations as in the program ZQPCVX [12], [13], was tested with encourage results in an l_1SQP method [1].

Literature:

[1] Bernau, H.: Sequentielle quadratische Programmierungsverfahren, Working paper, MTA SZTAKI MO/72, (1987).

[2] Charalambous, C.: On conditions for optimality of the nonlinear l_1-problem, Math. Progr. 17, (1979), 123-135.

[3] Fletcher, R.: A general quadratic programming algorithm, Journ. of the Inst. of Math. and its Appl., Vol. 7, (1971), 76-91.

[4] Fletcher, R.: Practical Methods of Optimization - Constraint Optimization, John Wiley and Sons, Chichester-New York-Toronto, (1981).

[5] Fletcher, R.: Numerical experiments with an exact l_1-penalty function method, in Mangasarian, O.L., Meyer, R.R., Robinson, S.M. (eds.), Nonlinear Programming 4, Academic Press, New York, (1981), 99-129.

[6] Fletcher, R.: An l_1-penalty method for nonlinear constraints, in Boggs, P.T., Byrd, R.H., Schnabel, R.B. (eds.), Numerical Optimization 1984, SIAM, Philadelphia, (1985), 26-40.

[7] Goldfarb, D., Idnani, A.: A numerically stable dual method for solving strictly convex quadratic programs, Math. Progr. 27, (1983), 1-33.

[8] Han, S.P.: Superlinearly convergent method for nonlinear programming, Math. Progr. 11, (1976), 263-282.

[9] Han, S.P., Mangasarian, O.L., Exact penalty functions in nonlinear programming, Math. Progr. 17, (1979), 251-279.

[10] Powell, M.J.D.: A fast algorithm for nonlinearly constraint optimization, in Watson, G.A. (ed.), Numerical Analysis Dundee 1977, Lecture Notes in Math. 630, Springer-Verlag, Berlin, (1978), 144-157.

[11] Powell, M.J.D.: Variable metric methods for constraint optimization, in Bachem, A., Grötschel, M., Korte, B. (eds.), Mathematical Programming, Springer-Verlag, Berlin, (1983), 288-311.

[12] Powell, M.J.D.: ZQPCVX a Fortran subroutine for convex quadratic programming, Report DAMTP83/NA17, University of Cambridge, (1983)

[13] Powell, M.J.D.: Corrections and Extensions to the Fortran listing of ZQPCVX, Report, University of Cambridge, (1984).

[14] Schittkowski, K.: The nonlinear programming method of Wilson, Han and Powell with an augmented Lagrangian type line search function, Num. Math. 38, (1981), 83-117.

[15] Tone, K.: Revision of constraint approximation in successive QP method for nonlinear programming problems, Math. Progr. 26, (1983), 144-152.

DOUBLY STOCHASTIC MATRICES AND OPTIMIZATION

Miroslav Fiedler [1)]

Let us recall first that doubly stochastic matrices are square non-negative matrices all of whose row - as well as column sums are equal to one.

In terms of optimization problems, the set of all n-by-n doubly stochastic matrices is identical with the set of all feasible solutions of the transportation problem (without specifying the objective function)

$$x_{ik} \geqq 0 , \quad i, k = 1, \dots, n,$$
$$\sum_{k=1}^{n} x_{ik} = a_i, \quad i = 1, \dots, n,$$
$$\sum_{i=1}^{n} x_{ik} = b_k, \quad k = 1, \dots, n$$

satisfying $a_i = b_k = 1$, $i, k = 1, \dots, n$.

In fact, the famous Birkhoff's theorem [2] stating that this set is equal to the convex hull of all permutation matrices is a consequence of similar more general statements about the m-by-n transportation problem (as stated, e. g., in [1]) saying that the extremal points of the corresponding convex polyhedron correspond to those feasible solutions the matrix (x_{ik}) of which does not contain any circuit, i. e. a non-void set of an even number of indices $i_1, j_1, i_2, j_2, \dots, i_s, j_s$ $(1 \leqq i_k \leqq m, 1 \leqq j_k \leqq n)$ such that all i_k's are mutually distinct, all j_k's are mutually distinct and

$$x_{i_k, j_k} \neq 0, \quad k = 1, \dots, s,$$

$$x_{i_{k+1}, j_k} \neq 0, \quad k = 1, \dots, s - 1, \quad x_{i_1, j_s} \neq 0 .$$

If we assign to the m-by-n matrix (x_{ik}) a bipartite graph $G(V_1, V_2, E)$ with two sets of vertices $V_1 = \{1, \dots, m\}$, $V_2 = \{1', \dots, n'\}$ and the set of edges $E \subset V_1 \times V_2$, $E = \{(i, k'), i \in V_1, k' \in V_2; x_{ik} \neq 0\}$ then the bigraph corresponds to a feasible solution without circuits if and only if its every component is a tree.

It is also well known that a doubly stochastic matrix enjoys several important properties. We shall list them in Theorem 1. Multiplication of a real vector by a doubly stochastic matrix averages, in a sense, the coordinates of this vector. It is related to the so called

1) Mathematics Institute Acad., Žitná 25, 115 67 Praha, Czechoslovakia

"majorization" in the set R_n of real n-dimensional column vectors. If $y \in R_n$, $z \in R_n$, $y = (y_1, y_2, \dots, y_n)^T$ and $y_{i_1} \geqq y_{i_2} \geqq \dots \geqq y_{i_n}$ for a permutation $(i_1, \dots, i_n)$ of $(1, \dots, n)$, $z = (z_1, z_2, \dots, z_n)^T$ and similarly $z_{k_1} \geqq z_{k_2} \geqq \dots \geqq z_{k_n}$, then "y majorizes z", denoted by $y \succ z$, if

$$y_{i_1} \geqq z_{k_1},$$
$$y_{i_1} + y_{i_2} \geqq z_{k_1} + z_{k_2},$$
$$\dots$$
$$y_{i_1} + y_{i_2} + \dots + y_{i_{n-1}} \geqq z_{k_1} + z_{k_2} + \dots + z_{k_{n-1}},$$
$$y_{i_1} + y_{i_2} + \dots + y_{i_n} = z_{k_1} + z_{k_2} + \dots + z_{k_n}.$$

The relation $\succ$ is reflexive and transitive. In addition, the vector $e = (\frac{1}{n}, \frac{1}{n}, \dots, \frac{1}{n})^T$ is majorized by every vector in R_n.

We shall also need the notion of an orthostochastic matrix. If $U = (u_{ik})$ is a real orthogonal n-by-n matrix, then the matrix $S = (u_{ik}^2)$ is clearly doubly stochastic. Matrices obtained in such a manner are called orthostochastic. Not all doubly stochastic n-by-n matrices are orthostochastic if $n \geqq 3$. For instance, the doubly stochastic matrix

$$\begin{pmatrix} 0 & \frac{1}{2} & \frac{1}{2} \\ \frac{1}{2} & 0 & \frac{1}{2} \\ \frac{1}{2} & \frac{1}{2} & 0 \end{pmatrix}$$

is not orthostochastic: in the previous notation, the corresponding vectors (u_{11}, u_{12}, u_{13}) and (u_{21}, u_{22}, u_{23}) would not be orthogonal.

Theorem 1 ([6]). Let $y \in R_n$, $z \in R_n$. Then the following are equivalent:

1^0 $y \succ z$;

2^0 there exists a doubly stochastic matrix D such that

$$z = D y ;$$

3^0 there exists an orthostochastic matrix S such that

$$z = S y ;$$

4^0 there exists a real symmetric matrix with the eigenvalues $y_1, \dots, y_n$ and the diagonal entries $z_1, \dots, z_n$.

The proof was given in [6]. It was substantially simplified in [4].

We intend to add here a maybe new geometric equivalent characterization.

Theorem 2. With the same notation as in Theorem 1, $y \succ z$ if and only if either $y = e$ $(= (\frac{1}{n}, \ldots, \frac{1}{n})^T)$, and then $z = e$ as well, or $y \neq e$ and to the points

$$\hat{Y}_1, \ldots, \hat{Y}_n, \quad \hat{Z}_1, \ldots, \hat{Z}_n$$

in R_1 with the respective (single) coordinate $y_1, \ldots, y_n, z_1, \ldots, z_n$ there exist in some R_{n-1} containing R_1 points $Y_1, \ldots, Y_n, Z_1, \ldots, Z_n$ with the following properties:

1° $\hat{Y}_i, \hat{Z}_k$ are orthogonal projections of Y_i, Z_k on R_1, $i, k = 1, \ldots, n$;

2° the points $Y_1, \ldots, Y_n$ are linearly independent, thus forming vertices of an $(n-1)$-simplex Σ_y, the convex hull of $Y_1, \ldots, Y_n$;

3° all points Z_k are contained in Σ_y;

4° the arithmetic mean of the points Z_k (or, the centre of gravity of these points) coincides with the centre of gravity of Σ_y.

Proof. The first statement about $y = e$ being trivial, let us show that the given geometric condition in the case $y \neq e$ is equivalent to condition 2° in Theorem 1.

Let first such points Y_i, Z_k exist. Choose an orthonormal coordinate system in R_{n-1} in such a way that R_1 is its first axis and y_i, z_k will be the first coordinates of Y_i, Z_k. By 3°, Z_k is a convex combination of $Y_1, \ldots, Y_n$:

$$Z_k = \sum_{m=1}^{n} s_{km} Y_m, \quad k = 1, \ldots, n$$

where $s_{km} \geq 0$, $\sum_{m=1}^{n} s_{km} = 1$.

By 4°, we have for the arithmetic mean of the Z_k's

$$\left(\frac{1}{n}\sum_{k=1}^{n} Z_k =\right) \frac{1}{n}\sum_{k,m=1}^{n} s_{km}Y_m = \frac{1}{n}\sum_{m=1}^{n} Y_m .$$

Therefore, the linear independence of the Y_i's yields

$$\sum_{k=1}^{n} s_{km} = 1, \quad m = 1, \ldots, n.$$

Thus the matrix $S = (s_{ij})$ is doubly stochastic and - using the first coordinates only - satisfies $z = Sy$.

Conversely, if $z = Sy$ for a doubly stochastic matrix $S = (s_{ij})$ and $y \neq e$, one can choose in an $(n-1)$-dimensional space R_{n-1} containing R_1 n linearly independent points $Y_1, \ldots, Y_n$ whose projections on R_1 are $\hat{Y}_1, \ldots, \hat{Y}_n$.

It is then easily seen that the points Z_k defined by

$$Z_k = \sum_{m=1}^{n} s_{km} Y_m$$

satisfy the conditions 1° - 4°.

Let us recall that doubly stochastic matrices have been generalized to the rectangular case. These m-by-n doubly stochastic matrices are nonnegative matrices all of whose row sums are equal to one and all column sums are equal to n/m. We shall denote by D_{mn} the set of all such matrices.

In [3], Čihák studied a special matrix E_{mn} in D_{mn} which has, in a sense, analogous properties as the identity matrix in the class D_{nn} of usual doubly stochastic matrices. This "unit" matrix E_{mn} can be obtained by the following construction: one starts by completing the first column of E_{mn} from top to the bottom, then the second column from top to the bottom etc, putting on every place the largest possible number (having in mind, of course, the row-sums and column-sums conditions). In other words,

$$E_{mn} = (e_{ik}),\ i = 1, \ldots, m;\ k = 1, \ldots, n$$

where

$$e_{ik} = \min \Big(1 - \sum_{j<i} e_{jk},\ \frac{n}{m} - \sum_{s<k} e_{is}\Big),$$

$$i = 1, \ldots, m;\ k = 1, \ldots, n .$$

Examples.

$$E_{73} = \frac{1}{3}\begin{pmatrix} 3 & 0 & 0 \\ 3 & 0 & 0 \\ 1 & 2 & 0 \\ 0 & 3 & 0 \\ 0 & 2 & 1 \\ 0 & 0 & 3 \\ 0 & 0 & 3 \end{pmatrix},$$

$$E_{n+1,n} = \frac{1}{n}\begin{pmatrix} n & 0 & 0 & \ldots & 0 \\ 1 & n-1 & 0 & \ldots & 0 \\ 0 & 2 & n-2 & \ldots & 0 \\ 0 & 0 & 3 & \ldots & 0 \\ \ldots & \ldots & \ldots & \ldots & \ldots \\ 0 & 0 & 0 & \ldots & 1 \\ 0 & 0 & 0 & \ldots & n \end{pmatrix};$$

$E_{37} = \frac{3}{7} E_{73}^T$ where T means transposition.

To present some of Čihák's results from [3], let us denote

$$V_n = \{x \in R_n,\ x = (x_1, \ldots, x_n)^T;\ x_1 \geq x_2 \geq \ldots \geq x_n\};$$

we denote by V_n^d the dual of V_n, i. e.

$$V_n^d = \{x \in R_n;\ x \cdot y \geq 0 \quad \forall y \in V_n\}.$$

Here, $x \cdot y$ means the inner product of x and y. It is easily seen that V_n is a convex wedge, the line generated by e is $V_n \cap (-V_n)$; V_n^d is a convex $(n-1)$-dimensional cone generated by the vectors $c_1 = (1, -1, 0, \ldots, 0)$, $c_2 = (0, 1, -1, 0, \ldots, 0)$, ..., $c_{n-1} = (0, 0, \ldots, 1, -1)$.

Further, we introduce in $\mathbf{D}_{mn}$ a partial ordering $\gg$ by:

If $Q_1 \in \mathbf{D}_{mn}$, $Q_2 \in \mathbf{D}_{mn}$, then $Q_1 \gg Q_2$ if

$$Q_1 y - Q_2 y \in V_m^d \qquad \forall y \in V_n,$$

i. e. if

$$(Q_1 y) \cdot x \geq (Q_2 y) \cdot x \qquad \forall y \in V_n,\ x \in V_m.$$

Theorem 3 ([3]). Let m, n be positive integers. Then:

a) The unit matrix E_{mn} is the unique maximal element in $\mathbf{D}_{mn}$ in the ordering $\gg$.

b) E_{mn} is centrally symmetric, i. e. for $E_{mn} = (e_{ik})$, $e_{ik} = e_{m+1-i,n+1-k}$, $i = 1, \ldots, m$; $k = 1, \ldots, n$.

c) E_{nn} is the identity matrix of order n.

In the following theorem which generalizes the equivalence of 1° and 2° in Theorem 1, we assume that the vectors y, z belong to V_n, V_m respectively instead of to R_n, R_m. The general result is easily obtained using the permutations which transform the vectors from R to V.

Theorem 4 ([3]). Let $y \in V_n$, $z \in V_m$. Then the following are equivalent:

1° there exists a matrix $D \in \mathbf{D}_{mn}$ such that

$$z = D\,y\,;$$

2° $E_{mn} y - z \in V_m^d$.

One can ask whether the matrix E_{mn} with all its row- and column-permutations generate $\mathbf{D}_{mn}$ as convex hull.

The answer - in general - is negative. For instance, the following matrix

$$F_{73} = \begin{pmatrix} 1 & 0 & 0 \\ 1 & 0 & 0 \\ \frac{1}{3} & \frac{1}{3} & \frac{1}{3} \\ 0 & 1 & 0 \\ 0 & 1 & 0 \\ 0 & 0 & 1 \\ 0 & 0 & 1 \end{pmatrix}$$

is an extreme point of the polyhedron $\mathbf{D}_{73}$, different from any $PE_{73}Q$ where E_{73} is the matrix from Example and P, Q are permutation matrices.

(This can be also seen by considering the bipartite graphs of F_{73} and E_{73}. Both are trees. The first has four vertices of degree 3 while the second only three.)

In some cases, i. e. if $|m - n| \leq 1$, the analogue of Birkhoff's theorem is valid.

We shall turn now to another type of matrices in $\mathbf{D}_{mn}$, in this case always $m \geq n$, which surprisingly are related to the problem of parametric approximations of polygonal lines.

For $m \geq n$, we define a matrix G_{mn} as identity of order n if $m = n$, and as the product

$$G_{mn} = E_{m,m-1} E_{m-1,m-2} \cdots E_{n+1,n}$$

if $m > n$.

It follows immediately that for $m \geq k \geq n$,

$$G_{mn} = G_{mk} G_{kn} .$$

In addition, G_{mn} belongs to $\mathbf{D}_{mn}$ since $E_{n+1,n}e_n = e_{n+1}$ where $e_n = (\frac{1}{n}, \ldots, \frac{1}{n})^T$ (with n coordinates) as well as $e_m^T E_{m,m-1} = e_{m-1}^T$.

Let us consider the following geometric construction in the class of polygonal lines in some space R_s.

Let $\mathbf{P} = P_1 P_2 \ldots P_n$ ($P_1, \ldots, P_n$ points in R_s) be an n-polygonal line in R_s. If A is a real m-by-n matrix, $A = (a_{ij})$, all of whose row sums are equal to one, then the transform of $\mathbf{P}$ by A is the m-polygonal line $\mathbf{Q} = Q_1 Q_2 \ldots Q_m$ in R_s the vertices Q_k of which are given by

$$Q_k = \sum_{j=1}^{n} a_{kj} P_j \ , \quad k = 1, \ldots, m \ . \tag{*}$$

This transform is affine, e. g. in the sense that if each of the points P_i is moving with constant speed - which may be zero - in a constant direction then the same will be true for each of the points Q_k.

Assume, for a moment, that $P_1, \ldots, P_n$ are linearly independent and that $Q_1, \ldots, Q_{n+1}$ are obtained by (*) for $A = E_{n+1,n}$. It follows from Example that $Q_1 = P_1$, $Q_{n+1} = P_n$ and Q_i belongs to the segment $P_{i-1}P_i$ for $i = 2, \ldots, n$. In addition, the arithmetic mean of the Q_i's coincides with the arithmetic mean of the P_i's. These conditions, in fact, determine the Q_i's uniquely. Repeating this process (even if the vertices of the new polygonal line may not be linearly independent) we obtain the m-polygonal line obtained from $P_1 \ldots P_n$ by applying the matrix G_{mn}.

This method to obtain to a given n polygonal line $P_1 P_2 \ldots P_n$ the new m-line $Q_1 Q_2 \ldots Q_m$ with $m > n$ is known (e. g. [5]). However, the following explicit formula for each entry of the matrix G_{mn} seems to be new.

Theorem 5. Let $m \geq n \geq 2$. Then the (i, k)-th entry $(1 \leq i \leq m, 1 \leq k \leq n)$ of G_{mn} is given by

$$g_{ik} = \binom{i-1}{k-1}\binom{m-i}{n-k}\Big/\binom{m-1}{n-1} .$$

The proof is easily obtained by induction. We shall mention, however, that the number m can tend to infinity (n is fixed). In this case we obtain as a limit the curve

$$P(t) = \sum_{k=1}^{n}\binom{n-1}{k-1} t^{k-1} (1-t)^{n-k} P_k , \; 0 \leqq t \leqq 1.$$

This curve which is an algebraic curve of degree $n - 1$ is known as Bézier curve corresponding to the n-polygonal line $P_1 P_2 \dots P_n$. For $n = 3$, it is a parabola touching the lines P_1P_2 at P_1 and P_3P_2 at P_3.

The final topic concerns normal matrices, i. e. square complex matrices A satisfying $AA^* = A^*A$ where A^* means complex conjugate matrix to A^T. An equivalent characterization is that A has the form $A = U L U^*$ where L is a complex diagonal matrix (its diagonal entries are eigenvalues of A) and U a unitary matrix, i. e. $UU^* = I$.

If $U = (u_{ik})$ is unitary, then $S = (|u_{ik}|^2)$ is again doubly stochastic. We shall call such a matrix unistochastic.

Theorem 6. Let y, z be complex column n-vectors, $y = (y_1, \dots, y_n)^T$, $z = (z_1, \dots, z_n)^T$. Then the following are equivalent:

1^0 there exists a unistochastic matrix S such that

$$z = Sy ;$$

2^0 there exists a normal n-by-n matrix with the eigenvalues $y_1, \dots, y_n$ and the diagonal entries $z_1, \dots, z_n$.

Proof. Follows immediately, both ways, from the fact that the diagonal entries a_{ii} of the matrix $A = U L U^*$ are equal to

$$a_{ii} = \sum_{k=1}^{n} |u_{ik}|^2 l_k , \; i = 1, \dots, n.$$

The problem to characterize orthostochastic and unistochastic matrices is open. One can ask to find a similar geometric characterization to that of Theorem 2. The above counterexample of the 3-by-3 doubly stochastic matrix which is not orthostochastic (and also not unistochastic) shows easily that the conditions $1^0 - 4^0$ in Theorem 2 applied to complex vectors (hence to R_2 instead of R_1) do not suffice.

References

[1] Bílý, J., Fiedler, M., Nožička, F.: Die Graphentheorie in Anwendung auf das Transportproblem. Czech. Math. J. 8 (1958), 94-121.

[2] Birkhoff, G.: Tres observaciones sobre el algebra lineal. Univ. Nac. Tucumán Rev. Ser. A 5 (1946), 147-151.

[3] Čihák, P.: On an exposed element of a set of doubly stochastic rectangular matrices. Comm. Math. Univ. Car. 11 (1970) 1, 99-113.

[4] Fiedler, M.: On a theorem by A. Horn. In: Mathematical structures, computational mathematics, mathematical modelling, Sofia 1975, 251-255.

[5] Goldman, R. N.: An urnful of blending functions. IEEE Comp. Graphics and Appl. 3 (1983) 7, 49-54.

[6] Horn, A.: Doubly stochastic matrices and the diagonal of a rotation matrix. Amer. J. Math. 76 (1954), 620-630.

[7] Marshall, A. W., Olkin, I.: Inequalities: Theory of majorization and its applications. Academic Press, N. York 1979.

ON A GENERALIZATION OF F.JOHN THEOREM FOR CONSTRAINED EXTREMUM PROBLEMS

F. Giannessi[1)]

Abstract. A necessary optimality condition is proved for constrained extremum problems having a finite-dimensional image. Both the classic F.John and Euler conditions are derived as corollaries. The result is obtained by exploiting the concept of image of a constrained extremum problem [3,4,5, 9,13,14].

1. Introduction

Assume we are given the positive integer m, a non-empty subset X of a Banach space Y whose norm is denoted by $\|\cdot\|$, and the real-valued functions $\varphi: X \to \mathbb{R}$ and $g: X \to \mathbb{R}^m$. Consider the problem:

$$\min \varphi(x) \quad , \quad x \in R \triangleq \{x \in X: g(x) \geq 0\} \ . \tag{1}$$

A crucial aspect of our analysis is based on the concept of image of (1), which will now be briefly recalled. Obviously $\bar{x} \in R$ is a local minimum point of (1) iff there exists a neighbourhood (open sphere of centre $\bar{x}$ and radius $\rho > 0$) $N \triangleq \{x \in Y: \|x-\bar{x}\| < \rho\}$ of $\bar{x}$, such that system (2),

$$f(x) \triangleq \varphi(\bar{x})-\varphi(x) > 0 \ ; \ g(x) \geq 0 \ ; \ x \in X_N \triangleq X \cap N \tag{2}$$

is impossible. Hence, in order to state local optimality for (1) we are led to prove disjunction of sets $\mathcal{H}$ and $\mathcal{K}_N$, where:

$$\mathcal{H} \triangleq \{(u,v) \in \mathbb{R} \times \mathbb{R}^m : u>0;\ v \geq 0\},\ F(x) \triangleq (f(x),g(x)),\ \mathcal{K}_N \triangleq F(X_N).$$

The set $\mathcal{K} \triangleq F(X)$, which will be called the image of X under F, represents $\mathcal{K}_N$ when[2)] $N \supseteq X$. To prove directly whether or not $\mathcal{H} \cap \mathcal{K}_N = \emptyset$ is generally impracticable; then we try to show such a disjunction by proving that the two sets lie in two disjoint halfspaces or, more generally, in two disjoint level sets, respectively. This separation approach is exactly equivalent to looking for a system which is in the alternative with (2). In this order of ideas the set $\mathcal{K}_N$ plays a key role. We are referred to Refs. [4,5,6,13] for details about sets and concepts which will be used here.

In the sequel, besides the finite-dimensional case where $X \subseteq \mathbb{R}^n$, we will

---

1) Professor, Department of Mathematics, Faculty of Sciences, University of Pisa, Via Buonarroti 2, PISA, Italy.

2) Both here and in the sequel, $\supseteq$ and $\supset$ will denote "contains" and "strictly contains", respectively; analogously for $\subseteq$ and $\subset$.

consider another particular case, namely, the one where[1] $X = C^1[a,b]$, with $a,b \in \mathbb{R}$, X is equipped with the norm $\|z\|_\infty = \max_{t\in[a,b]} |z(t)|$, and where:

$$\varphi(x) = \int_a^b \psi_0(t,x(t),x'(t))dt \ ; \ g_i(x) = \int_a^b \psi_i(t,x(t),x'(t))dt, i \in I, \qquad (3)$$

the functions $\psi_i: \mathbb{R}^3 \to \mathbb{R}$, $i \in \{0\} \cup I$, with $I \triangleq \{1,\dots,m\}$, being given. Fixed endpoint conditions can be included in the definition of X.

The problems which can be reduced to scheme (1) share the characteristic of having a finite-dimensional image. Hence certain problems, for instance of geodesic-type, escape from (1). In [5] it is shown how the present approach can be modified by means of multifunctions theory, in order to handle problems having an infinite-dimensional image; these will be analyzed in a subsequent paper.

The generalization of the concept of stationary point, which keeps up with that of necessary conditions, has received much attention [12]. The crucial part is the kind of convergence which is required. It seems impossible to handle every problem with a unique kind of convergence. Our present aim is to enlarge as much as possible the class of problems for which a Lagrangian-type necessary condition can be established by means of the simplest generalization of the concept of limit, namely lower (or upper) limit. This does not avoid the introduction of a more complicated notion of convergence, but reduces its use to a class as small as possible. In the remaining part of this section, f denotes a generic function and not that previously introduced.

Definition 1. $\bar{x} \in X$ will be called a lower semistationary point of a problem of type: $\min_{x\in X} f(x)$, with $f: X \to \mathbb{R}$, iff

$$\liminf_{x \to \bar{x}} \frac{f(x)-f(\bar{x})}{\|x-\bar{x}\|} \geq 0 \ . \qquad (4)$$

An upper semistationary point is defined by replacing, above, liminf and $\geq$ with limsup and $\leq$, respectively. A point which is both upper and lower semistationary is a stationary point; it obviously fulfils (4) as equality.

The above definition is motivated by the following property [6] :

Proposition 1. (i) If $\bar{x}$ is a local minimum point of f on X, then (4) holds. (ii) If X and f are convex, then a lower semistationary point of f on X is

---

[1] $C^r[a,b]$ denotes the set of real-valued functions of which the first r derivatives are continuous.

a global minimum point.

The concept of stationarity expressed by (4) is equivalent to that mentioned in [12, page 58]. In fact, the latter requires the existence of a neighbourhood N of $\bar{x}$ and of a function $\varepsilon : X \times X \to \mathbb{R}$, with $\lim_{x \to \bar{x}} \varepsilon(\bar{x}; x-\bar{x})/\|x-\bar{x}\| = 0$, such that:

$$f(x) \geq f(\bar{x}) + \varepsilon(\bar{x}; x-\bar{x}) \ , \quad \forall x \in X_N \ .$$

When $x \neq \bar{x}$, this inequality is equivalent to

$$[f(x)-f(\bar{x})]/\|x-\bar{x}\| \geq \varepsilon(\bar{x}; x-\bar{x})/\|x-\bar{x}\| \ , \quad \forall x \in X_N \ ,$$

and hence to (4).

Starting with (4) instead of equivalent formats, it is possible to achieve general theoretical results and also some "calculus" for a wide class of functions. If f is differentiable, it is easy to show that the left-hand side of (4) collapses to the minimum directional derivative of f. The concept of stationarity expressed by Definition 1 is not, of course, the only one conceivable for the general case. It is probably the simplest, and it enables one to achieve a necessary condition for a wide class of constrained extremum problems, including convex, differentiable, and some (even if not all) discontinuous ones. The concept of stationarity can be strengthened to embrace other classes of problems.

2. Some Properties of a Finte-Dimensional Image

Both here and in the sequel $T(\bar{h}, \mathscr{K})$ will denote the tangent cone[1)] to $\mathscr{K}$ at $\bar{h} = F(x)$ and $G(\bar{h}; \mathscr{K}) \triangleq \{h \in \mathbb{R}^{1+m} : h = \bar{h} + \alpha(\hat{h} - \bar{h}),\ \hat{h} \in \mathscr{K},\ \alpha \in [0, +\infty[\}$, the cone generated by $\mathscr{K}$ at $\bar{h}$; moreover $\langle \cdot, \cdot \rangle$ will denote scalar product and, given a cone C, $C^* \triangleq \{z : \langle y, z \rangle \geq 0,\ \forall y \in C\}$ will be its positive polar. To problem (1) we associate the generalized difference quotient:

$$Q(x; \theta, \lambda) \triangleq [w(f(x), g(x); \theta, \lambda) - w(f(\bar{x}), g(\bar{x}); \theta, \lambda)]/\|x-\bar{x}\|,$$

where

$$w(u, v; \theta, \lambda) \triangleq \theta u + \langle \lambda, v \rangle \ , \ \theta \in \mathbb{R} \ , \ \lambda \in \mathbb{R}^m \ . \qquad (5)$$

---

1) The tangent cone to set $S \subset \mathbb{R}^\nu$ at $\bar{s} \in \mathrm{cl}\, S$ is defined as the set of $\bar{s}+s$ for which there exists a sequence $\{s^r\} \subseteq S$, such that $\lim_{r \to +\infty} s^r = \bar{s}$, and a positive sequence $\{\alpha_r\} \in \mathbb{R}$, such that $\lim_{r \to +\infty} \alpha_r(s^r - \bar{s}) = s$. We stipulate that the tangent cone is empty if S is empty; $S \neq \emptyset$ implies that the tangent cone is non-empty and closed. Recall that $F(x) = (f(x), g(x))$, where f is the function defined in Section 1. The symbols not defined in this section are those defined in Section 1.

Under a suitable assumption (5) is a class of weak separation functions, as defined in [4,5]. A necessary condition for (1) will be achieved by considering the upper semistationary points of $w(f(x),g(x);\theta,\lambda)$ on X. More general results can be obtained by adopting a nonlinear class of separation functions in place of (5). Recall that N is a neighbourhood (open sphere) of $\bar{x}$ and set[1] $\mathscr{L}(x;\theta,\lambda) \stackrel{\Delta}{=} \theta\varphi(x)-\langle\lambda,g(x)\rangle$, with $\theta\in\mathbb{R}$, $\lambda\in\mathbb{R}^m$.

<u>Proposition 2</u>. <u>There exist multipliers</u> $\bar{\theta}\in\mathbb{R}$ <u>and</u> $\bar{\lambda}\in\mathbb{R}^m$, <u>with</u> $(\bar{\theta},\bar{\lambda})\neq 0$, <u>such that</u>:

$$-(\bar{\theta},\bar{\lambda}) \in [G(\bar{h};\mathscr{K}_N)-\bar{h}]^*, \tag{6}$$

<u>iff we have</u>:

$$Q(x;\bar{\theta},\bar{\lambda}) \leq 0 \ , \quad \forall x\in X_N\setminus\{\bar{x}\} \ . \tag{7}$$

<u>Proof</u>. (6) is equivalent to $\langle(\bar{\theta},\bar{\lambda}),h-\bar{h}\rangle \leq 0$, $\forall h\in G(\bar{h},\mathscr{K}_N)$. This inequality, considering that $h\in G(\bar{h};\mathscr{K}_N)$ iff either $h=\bar{h}$ or $\exists\alpha>0$ such that $k\stackrel{\Delta}{=}\bar{h}+\alpha(h-\bar{h})\in$ $\in\mathscr{K}_N$, holds iff $\langle(\bar{\theta},\bar{\lambda}),k-\bar{h}\rangle \leq 0$, $\forall k\in\mathscr{K}_N$, or:

$$\langle(\bar{\theta},\bar{\lambda}),(f(x)-f(\bar{x}),g(x)-g(\bar{x}))\rangle \leq 0 \ , \quad \forall x\in X_N,$$

and hence (7) follows. This completes the proof. □

Condition (6) does not imply that $\bar{x}$ be a minimum point of (1) (take for instance $x=\mathbb{R}$, $\varphi(x)=x^2-1$, $g(x)=x$, $\bar{x}=1$). Nevertheless (6) is a strong requirement which, however, cannot easily be weakened in the general case.

The set $\mathscr{K}_N$ can be extended without losing anything from the viewpoint of optimality conditions; consequently the polar cone which appears in (6) is not enalarged, but the lost elements will turn out to be useless multipliers. Denote by $\mathscr{E}(S)\stackrel{\Delta}{=}S-C$ the <u>conic extension</u> of set $S\subseteq\mathbb{R}^{1+m}$ with respect to $\mathbb{R}^{1+m}_+$ (see [5]). It is easy to show (see the proof of Theorem 3.1 of [4]) that, from the optimality viewpoint, $\mathscr{K}$ can be equivalently replaced with $\mathscr{E}(\mathscr{K})$. The same is not true if, before (or after) applying the conic extension, $\mathscr{K}$ is replaced, even locally, with the cone generated by it, unless a suitable assumption is made (for instance, the continuity of F).

To achieve a necessary condition for a wide class of problems (including some discontinuous one, as will be shown by an Example) we have to

---

[1] The vector difference is denoted by −, while the difference between sets is denoted by \. Note that in (6) we have $\bar{h}\in\mathscr{K}_N$ and this is crucial for achieving the thesis.

restrict the class of problems (1); this is done by the following condition; int and conv denote interior and convex hull, respectively.

<u>Condition C</u>. <u>Let</u> $\bar{x} \in R$ <u>and set</u> $\bar{h}=F(\bar{x})$. <u>The following statement holds</u>: <u>if system</u> (2) <u>is impossible</u>, <u>then</u>

$$(\text{int}\,\mathcal{H}) \cap \text{conv}\,\mathcal{E}[G(\bar{h};\, \mathcal{K}_N)] = \emptyset. \tag{8}$$

Because of its obvious convexity, the set conv $\mathcal{E}$ of (8) (and hence $\mathcal{E}$) admits a supporting halfspace at $\bar{h}$, so that (8) implies the following condition:

$$\{\mathcal{E}[G(\bar{h};\, \mathcal{K}_N)-\bar{h}]\}^* \neq \{0\}\ . \tag{10}$$

The Theorem of next section will show that (8) is fulfilled in the convex case; it may also be satisfied, however, in some discontinuous cases, as will be shown by the Example of next section. Condition C requires the check of the optimality of $\bar{x}$. This is obviously a drawback; however this is only apparent. In fact, no difficulty arises in establishing a necessary condition (see the Lemma of next section), where the impossibility of (2) is an assumption. As it concernes the check of (8), the Example of next section will shows how is possible, in practice, to overcome the above drawback. However, Condition C (and the corresponding necessary condition, i.e. the Lemma of next section) is conceived as a source for deriving necessary conditions for certain classes of functions, like the Theorem of next section.

Condition (8) is obviously equivalent to claim the existence of a closed halfspace containing $\mathcal{K}_N$, the set int $\mathcal{H}$ being contained in its complement. In comparison with this statement, (8) has the advantage that the existence of a separating hyperplane is expressed in terms of cone generated and conic extension, which simplify the proof of the Theorem and let us put Condition C in terms of φ, g and X. As it concerns this latest aspect, let us note that Condition C is conceived an expressed in the image space, and it will be used in this form, otherwise the proofs would be unnecessarily complicated. This does not exclude the possibility that the final version should be in terms of the given problem. In other words, we point out the importance of passing to the image space, where the analysis and the proofs are carried on more easily than in the space where the given problem runs, and eventually to interpret in this space the results found.

Now it will be shown that, within a suitable class of functions, we can reduce ourselves to a convex extended image. This will mean that we can handle, as regarding necessary conditions, some nonconvex problems in the same way as convex ones.

Definition 2. Let L and $\mathscr{C}$ denote respectively the sets of linear and sublinear functions. A function[1)] f will be said $\mathscr{C}$-differentiable at $x = \bar{x}$ iff there exist the following two functions: $g \in \mathscr{C}$, which satisfies the inequalities

$$f^+(\bar{x};z) \triangleq \limsup_{t \downarrow 0} \frac{f(\bar{x}+tz)-f(\bar{x})}{t} \leq g(\bar{x};z) \quad , \quad \forall z \in Z \triangleq X-\{\bar{x}\} \tag{10a}$$

and function $\varepsilon: X^2 \to \mathbb{R}$ which fulfils the condition:

$$\lim_{z \to 0} \frac{\varepsilon(\bar{x};z)}{\|z\|} = 0 \ , \tag{10b}$$

such that we have

$$f(x) = f(\bar{x}) + g(\bar{x};z) + \varepsilon(\bar{x};z) \quad , \quad \forall z \in Z \ , \tag{10c}$$

and for every couple of functions $(\tilde{g},\tilde{\varepsilon})$ which fulfil condition (10a-c) and with $\tilde{g} \in \mathscr{C}$ we have

$$\text{epi } \tilde{g} \subseteq \text{epi } g \tag{10d}$$

Function $g(\bar{x};z/\|z\|)$ will be called the upper $\mathscr{C}$-derivative of f at $\bar{x}$ in the direction z. A L-differentiable function is obviously a differentiable one. Note that, when f is locally Lipschitzian, then f^+ is the upper Dini derivative.

Definition 3. A function $f: X \to \mathbb{R}$ will be said to be of class $\mathscr{S}^-$ at $\bar{x}$ iff there exist a neighbourhood N of $\bar{x}$ and a $\mathscr{C}$-differentiable function $\sigma_f: X \to \mathbb{R}$, whose directional $\mathscr{C}$-derivative is closed and, when $Y \neq \mathbb{R}^n$, bounded from above by a finite constant, such that $f(x) \geq \sigma_f(x)$, $\forall x \in X_N$, and $f(\bar{x}) = \sigma_f(\bar{x})$; σ_f will be called a lower support of f at $\bar{x}$. f will be called of class $\mathscr{S}^+$ at $\bar{x}$ iff $-f$ is of class $\mathscr{S}^-$; $\Sigma_f \triangleq \sigma_{-f}$ will be called an upper support of f at $\bar{x}$.

3. A Necessary Condition

First of all we will establish a necessary condition, which holds under a suitable hypothesis and which is expressed in terms of the image of (1) (even if it might be put in terms of (1); see [6]) to stress the importance of the image to carry on the analysis. Such a condition represents a source for deriving practically meaningful theorems. Denote by $\mathscr{L}(x;\theta,\lambda) \triangleq \theta\varphi(x)-\langle\lambda,g(x)\rangle$ the generalized Lagrangian function associated with (1).

---

1) In Definitions 2,3 and only here f denotes a generic function and not that previously introduced.

Lemma. Let $\bar{x} \in X$ and $\bar{h}$ be the image of $\bar{x}$ through function $F(x)=(\varphi(\bar{x})-\varphi(x), g(x))$, and assume that there exists a neighbourhood N of $\bar{x}$ such that Condition C is fulfilled. If $\bar{x}$ is a minimum point of problrm (1), then there exist multipliers $\bar{\theta} \in \mathbb{R}$ and $\bar{\lambda} \in \mathbb{R}^m$, such that:

$$\liminf_{x \to \bar{x}} \frac{\mathcal{L}(x;\bar{\theta},\bar{\lambda})-\mathcal{L}(\bar{x};\bar{\theta},\bar{\lambda})}{\| x-\bar{x} \|} \geq 0 \quad ; \tag{11a}$$

$$g(\bar{x}) \geq 0 \quad ; \quad \bar{\theta} \geq 0 \quad ; \quad \bar{\lambda} \geq 0 \; ; \quad (\bar{\theta},\bar{\lambda}) \neq 0 \; ; \tag{11b}$$

$$\langle \bar{\lambda}, g(\bar{x}) \rangle = 0 \quad . \tag{11c}$$

The multiplier-vector $(\bar{\theta},\bar{\lambda})$ belongs to the opposite of the polar cone in (9).

Proof. Since $G(\bar{h};\mathcal{X}_N) \subseteq \mathcal{E}[G(\bar{h};\mathcal{X}_N)]$, we have:

$$\{\mathcal{E}[G(\bar{h};\mathcal{X}_N)-\bar{h}]\}^* \subseteq [G(\bar{h};\mathcal{X}_N)-\bar{h}]^* \; .$$

From this inclusion and Proposition 2 - recalling that $w=\theta\varphi(\bar{x})-\mathcal{L}$ and noting that (7) obviously implies (4) where f is replaced by $\mathcal{L}(\bar{x};\bar{\theta};\bar{\lambda})$ - we deduce that for every $-(\bar{\theta},\bar{\lambda})$ which belongs to the left-hand side of (9) (which holds by assumption), the inequalities:

$$[\mathcal{L}(x;\bar{\theta},\bar{\lambda})-\mathcal{L}(\bar{x};\bar{\theta},\bar{\lambda})]/\|x-\bar{x}\| \geq 0, \quad \forall x \in X_N \setminus \{\bar{x}\},$$

and hence (11a), hold. Then, since $g(\bar{x}) \geq 0$ is obvious, (11a,b) are proved, if we show that:

$$\{\mathcal{E}[G(\bar{h};\mathcal{X}_N)-\bar{h}\,]\}^* \cap [\mathbb{R}_+^{1+m} \setminus \{0\}] \neq \emptyset \; . \tag{12}$$

Really we prove that the polar set of (12) is contained into $\mathbb{R}_+^{1+m} \setminus \{0\}$. Because of the definition of (positive) polar, the set $C \stackrel{\Delta}{=} \mathcal{E}[G(\bar{h};\mathcal{X}_N)-\bar{h}]$ is contained in the halfspace defined by $\langle \pi,h \rangle \geq 0$ whatever $\pi=(\pi_0,\ldots,\pi_m) \neq 0$ may be in the left-hand side of (9). It follows that $\mathbb{R}_-^{1+m}$ (being contained in C) is contained in the halfspace $\langle \pi,h \rangle \geq 0$, and this inclusion implies $\pi \leq 0$, so that (12) holds. Now, ab absurdo, let us assume that (11c) does not hold. Then, observing that the opposite of a couple $(\bar{\theta},\bar{\lambda})$ fulfilling (11a,b) is precisely the above π, and recalling that $\bar{h}=(\bar{h}_0,\bar{h}_1,\ldots,\bar{h}_m) = f(x)=(f(\bar{x})=0,g(\bar{x}))$ is the image of $\bar{x}$, so that $g_i(\bar{x})=\bar{h}_i$, we see that $\forall(-\bar{\theta},-\bar{\lambda}) \in C^*$ there exists an index $i=1,\ldots,m$ (depending on $\bar{\theta},\bar{\lambda}$) such that $\lambda_i g_i(\bar{x}) > 0$ or $\pi_i \bar{h}_i < 0$. Then, considering that $\pi \leq 0$ and $\bar{h} \geq 0$, we find $\langle \pi,\bar{h} \rangle < 0$, $\forall \pi \in C^*$, and hence (since $0 \in C$):

$$-h \in \text{int}(C^*)^* = \text{int cl conv } C = \text{int conv } C.$$

Denoting by $S(\bar{h})$ an open sphere with center at $\bar{h}$ and with suitably small radius, from the last relations we deduce that $S(-\bar{h}) \subset \text{conv } C$ and then we achieve the inclusion $S(0) \subset \text{conv } \mathcal{E}[G(\bar{h};\mathcal{X}_N)]$, which implies $(\text{int } \mathcal{H}) \subset$

$\subset \text{conv}\, \mathscr{B}[G(\bar{h}; \mathscr{K}_N)] \neq \emptyset$; this relation contradicts the assumption that Condition C is fulfilled, since $\bar{x}$ is a minimum point, so that there exists a neighbourhood N of $\bar{x}$, such that $\mathscr{K} \cap \mathscr{K}_N = \emptyset$. The complementarity condition (11c) and hence (11) follow. □

From the above Lemma it is possible to derive (see [6] for the proof) a necessary condition, which embodies several known ones. To this end assume that $\mathscr{L}$ is $\mathscr{G}$-differentiable and let

$$\mathscr{L}(x;\theta,\lambda) = \mathscr{L}(\bar{x};\theta,\lambda) + \mathscr{L}^c(\bar{x};z;\theta,\lambda) + \varepsilon^c(\bar{x};z;\theta,\lambda)$$

be the corresponding expansion (10c). When φ, g_i, $i \in I$ are differentiable, then $\mathscr{L}^c$ collapses to the variation of $\mathscr{L}$, which we denote by $V_{\mathscr{L}}$.

Theorem. Let $\bar{x} \in X$ and assume that φ be of class $\mathscr{S}^-$ and that g_i, $i \in I$ are of class $\mathscr{S}^+$ at $\bar{x}$. If $\bar{x}$ is a minimum point of (1), then (11) hold and (11a) becomes:

$$\inf_{z \in S} \mathscr{L}^c(\bar{x};z;\bar{\theta},\bar{\lambda}) \geq 0 \tag{13a}$$

where $\mathscr{L}^c$ is the directional $\mathscr{G}$-derivative of $\mathscr{L}$ and $S=\{z : \|z\|=1\}$. (13a) implies[1]:

$$0 \in \bar{\theta}\, \partial_G \varphi(\bar{x}) - \sum_{i \in I} \bar{\lambda}_i \partial_G g_i(\bar{x}) \ , \tag{13b}$$

which becomes:

$$0 \in \bar{\theta}\, \partial \varphi(\bar{x}) - \sum_{i \in I} \bar{\lambda}_i \partial g_i(\bar{x}) \ , \tag{14}$$

if in particular φ, $-g_i$, $i \in I$ are convex (and X is convex). When φ, g_i, $i \in I$ are differentiable, then (13a) collapses to $V_{\mathscr{L}} = 0$ along $x=\bar{x}$, which in case (3) becomes:

$$\Psi'_x(t,\bar{x},\bar{x}';\bar{\theta},\bar{\lambda}) - \frac{d}{dt}\Psi'_{x'}(t,\bar{x},\bar{x}';\bar{\theta},\bar{\lambda}) = 0, \tag{15}$$

where Ψ is the integrand of the Lagrangian function:

$$\mathscr{L}(x;\theta,\lambda) \triangleq \int_a^b \Psi(t,x,x';\theta,\lambda)dt, \quad \Psi \triangleq \theta\psi_0(t,x,x') - \sum_{i \in I} \lambda_i \psi_i(t,x,x');$$

if in addition $X=\mathbb{R}^n$, then (13a) collapses to:

$$\mathscr{L}'_x(\bar{x};\bar{\theta},\bar{\lambda}) = 0 \ , \tag{16}$$

where $\mathscr{L}'_x$ is the gradient of the Lagragian function $\mathscr{L}(x;\bar{\theta},\bar{\lambda}) \triangleq \bar{\theta}\varphi(x) - \langle \bar{\lambda}, g(x) \rangle$ at $x=\bar{x}$.

Condition (16)-(11b,c) is the well-known F.John necessary condition and collapses to that of Karush and Kuhn-Tucker when $\bar{\theta}=1$ [10]. Condition (15)-(11b,c) is the equally well-known necessary condition for isoperimetric

---

[1] When (10c) holds, we set $\partial_G f(\bar{x}) = \partial g(\bar{x};0)$, ∂ being the ordinary subdifferential.

problems, (15) being the classic Euler equation [2,7] . Condition (14)-(11b,c) is the well-known necessary condition for convex problems [10,12].

The presence of equality constraints in (1) may be trivially handled by replacing each of them with two inequalities and then applying the Theorem to find (11) with obvious modifications.

The following example shows that the above Lemma is not only a source for deriving theorems, but also a useful condition in some cases.

Example. In (1) let us set $X=\mathbb{R}$; $\varphi(x)=3+x^2$ if $x<1$ and $x\neq 0$, $\varphi(0)=1, \varphi(1)=0$, and $\varphi(x)=x^2-2x$ if $x>1$; $m=1$; $g(x)=2-x$ if $x<1$, $g(1)=0$, and $g(x)=-1-x$ if $x>1$; $N=X$; $\|\cdot\| = |\cdot|$. Consider (11a) and note that $\mathscr{L}(x;\bar{\theta},\bar{\lambda})=\bar{\theta}x^2+\bar{\lambda}x+3\bar{\theta}-2\bar{\lambda}$ if $x<1$ and $x\neq 0$, $\mathscr{L}=\bar{\theta}-2\bar{\lambda}$ if $x=0$, $\mathscr{L}=0$ if $x=1$, and $\mathscr{L}=\bar{\theta}x^2+(-2\bar{\theta}+\bar{\lambda})x+\bar{\lambda}$ if $x>1$; also note that (11b) requires $\bar{x}\leq 1$. Simple calculations lead us to state that (11a) is fulfilled with $(\bar{\theta},\bar{\lambda})\geq 0$ iff $\bar{\lambda}=-\bar{\theta}\bar{x}$ if $\bar{x}<1$ and $\bar{x}\neq 0$, whatever $\bar{\theta}$ and $\bar{\lambda}$ may be if $\bar{x}=0$, with $(\bar{\theta}/2)\leq\bar{\lambda}\leq 4\bar{\theta}$ if $\bar{x}=1$. When $\bar{x}<1$, since $g(\bar{x})>0$, (11c) implies $\bar{\lambda}=0$; then, if $\bar{x}\neq 0$, (11a) requires $\bar{\theta}=0$ so that the last one of (11b) is violated, while if $\bar{x}=0$, then (11a) requires merely $\bar{\theta}\geq 0$ and thus (11) are fulfilled with $\bar{\theta}>0$ and $\bar{\lambda}=0$. When $\bar{x}=1$, since $g(\bar{x})=0$, (11c) is satisfied by any $\bar{\lambda}$; it follows that (11) are fulfilled if $(\bar{\theta}/2)\leq\lambda\leq 4\bar{\theta}$ and $(\bar{\theta},\bar{\lambda})\neq 0$. In conclusion, the only points which satisfy condition (11) are $\bar{x}=0$ and $\bar{x}=1$ (which are easily recognizable, by direct inspection, as the only local minima). Now let us consider $\bar{x}=0$, so that $\bar{h}=(0,2)$, and study Condition C. We find:

$\mathscr{K}=\{(u,v)\in\mathbb{R}^2: u=-v^2-4v-2; v<-2\}\cup\{(1,0),(0,2)\}\cup\{(u,v): u=-v^2+4v-6; v>1; v\neq 2\}$;

$G(\bar{h};\mathscr{K})=\{(u,v): u\leq -(1/2)v+1; v<2\}\cup\{(u,v): u\leq -2\sqrt{2}v+4\sqrt{2}; v>2\}$;

$\mathscr{E}[G(\bar{h};\mathscr{K})-\bar{h}]=\{(u,v): u\leq \min[-(1/2)v, -2\sqrt{2}v]\}$;

$\{\mathscr{E}[G(\bar{h};\mathscr{K})-h]\}^*=\{(u,v): v\leq 0;\ 2v\leq u\leq(\sqrt{2}/4)v\}\neq\{0\}$.

Since $m=1$, Condition C collapses to the following one: if $g(\bar{x})\geq 0$ and $g_p(\bar{x})>0$ for some p, and if $\exists x'\in X_N$ such that $h'\overset{\Delta}{=}F(x')\notin\mathscr{H}$ and the segment $[\bar{h},h']$ intersects $\operatorname{int}\mathscr{H}$, then $\mathscr{H}\cap\mathscr{K}_N\neq\emptyset$. This condition can be equivalently formulated this way: if $g(\bar{x})\geq 0$ and $g_p(\bar{x})>0$ for some p, and if $\exists x'\in X_N$ and $\exists\alpha\in]0,1[$ such that $\varphi(x')<\varphi(\bar{x}), g(x')<0$ and $(1-\alpha)g(\bar{x})+\alpha g(x')>0$, then (2) is possible. We can now observe that at $\bar{x}=0$ Condition C is satisfied in a small enough neighbourhood of $\bar{x}$, since $[\bar{h},h']\cap\mathscr{H}=\emptyset$, or since $\varphi(x')<\varphi(\bar{x})$ becomes $2+(x')^2<0$ and is impossible with $X_N\subset]-\infty,1[$.

Note that the elements of the left-hand side of (9) are those of the rays generated at 0 by the convex combinations of $(-1,-1/2)$ and $(-1,-2\sqrt{2})$. (The multipliers $(\bar{\theta},\bar{\lambda})$ which satisfy (11) are the opposite of the above elements). Hence the assumptions of the Lemma are fulfilled, so that the necessary condition is verified at $\bar{x}=0$. Now let us consider $\bar{x}=1$, so that $\bar{h}=(0,0)$, and study Condition C. We find:

$$\mathcal{K}=\{(u,v):u=-v^2-4v-3;v<-2\}\cup\{(0,0),(-1,2)\}\cup\{(u,v):u=-v^2+4v-7;v>1;v\neq 2\};$$

$$G(\bar{h};\mathcal{K})=\{(u,v):u<-(1/2)v;v<0\}\cup\{(u,v):-4v<u<(4-2\sqrt{7})v;v>0\}\cup\{(u,v):u=-(1/2)v;v\geq 0\};$$

$$\mathcal{H}[G(\bar{h};\mathcal{K})-\bar{h}]=\{(u,v):u<-(1/2)v\}\cup\{(u,v):u=-(1/2)v;v\geq 0\};$$

$$\{\mathcal{H}[G(\bar{h};\mathcal{K})-\bar{h}]\}^*=\{(u,v):u=2v;v\leq 0\}\neq\{0\}.$$

As before Condition C at $\bar{x}=1$ is satisfied in a small enough neighbourhood of $\bar{x}$, since hère $[\bar{h},h']\cap\mathcal{K}=\emptyset$, or since $g(\bar{x})=0$.
Note that the elements of the left-hand side of (9) are those of the ray generated at 0 by (-1,-1/2). (The multipliers $(\bar{\theta},\bar{\lambda})$ which satisfy (11) are the opposite of the above elements). Hence the assumptions of the Lemma are again satisfied, so that the necessary condition is verified at $\bar{x}=1$ too. Note that, at the global minimum point $\bar{x}=1$, φ is not even semicontinuous. The study of (11a) has been performed on the whole feasible region to clarify the condition expressed by the Lemma; except for points $\bar{x}=0$ and $\bar{x}=1$, where the functions are discontinuous, we could have employed a tool simpler that (11a). Note that the above problem, if $X=\mathbb{R}$ is replaced by $X=\{x\in\mathbb{R}:0\leq x<1\}$, is an instance of the so-called fixed-charge problem.

References

[1] Ben-Tal, A., and J. Zowe: Necessary and Sufficient Optimality Conditions for a Class of Nonsmooth Minimization Problems. Mathematical Programming, 24 (1982), 70-91.

[2] Bliss, G.A.: Lectures on the Calculus of Variations. The University of Chicago Press, Ill, (1945).

[3] Clarke, F.H.: Optimization and Nonsmooth Analysis. John Wiley, New York (1984).

[4] Giannessi,F.: Theorems of the Alternative and Optimality Conditions. Journal of Optimization Theory and Applications, 42 (1984), 331-365.

[5] Giannessi,F.: Theorems of the Alternative for Multifunctions with Applications to Optimization. General Results. Journal of Optimization Theory and Applications, 55 (1987) 1, 233-256.

[6] Giannessi,F.: Semidifferentiable functions and necessary optimality conditions. Journal of Optimization Theory and Applications. To appear.

[7] Hestenes,M.R.: Calculus of Variations and Optimal Control Theory. John Wiley, New York, New York, (1966).

[8] Hirriart-Urruty,J.B.: Tangent Cones, Generalized Gradient, and Mathematical Programming in Banach Spaces. Mathematics of Operations Research, 4 (1979) 1, 79-97.

[9] Ioffe,A.D.: Necessary Conditions in Nonsmooth Optimization. Mathematics of Operations Research, 9 (1984), 159-189.

[10] Mangasarian,O.L.: Nonlinear Programming.McGraw-Hill, New York, New York, (1969).

[11] Rockafellar,R.T.: Convex Analysis. Princeton University Press,(1970).

[12] Rockafellar,R.T.: The Theory of Subgradients and its Applications to Problems of Optimization. Convex and Nonconvex Functions. Heldermann Verlag, Berlin, (1981).

[13] Tardella,F.: On the Image of a Constrained Extremum Problem and Some Applications to the Existence of the Minimum. Journal of Optimization Theory and Applications. To appear.

[14] Warga,J.: Controllability and Necessary Conditions in Unilateral Problems without Differentiability Assumptions. SIAM Journal on Control and Optimization, <u>14</u> (1976), 546-573.

ON GLOBAL OPTIMIZATION BASED ON PARAMETRIC OPTIMIZATION

Jürgen Guddat[1], Hubertus Th. Jongen[2]

Abstract.

It will be considered a special parametric optimization problem obtained by applying the concept of embedding to the problem of global optimization. The included analysis allows a deeper insight into the essential difficulties of this problem on the one hand. On the other hand it is shown that the application of pathfollowing methods to walk on a curve of generalized critical points can be helpful to solve this problem.

1. Introduction

We consider the problem of global optimization:
Find a global minimizer of a given function $F \in C^3(\mathbb{R}^n, \mathbb{R})$ subject to $\mathbb{R}^n$ and K respectively, where K is a convex and compact subset of $\mathbb{R}^n$, e. g. $K := \{x \in \mathbb{R}^n \mid \|x\|^2 \leq q\}$ and q is a given positive number. These problems are symbolized briefly by

$$\text{glob min } \{F(x) \mid x \in \mathbb{R}^n\}, \tag{1.1}$$

$$\text{glob min } \{F(x) \mid x \in K\}. \tag{1.2}$$

The goal of this paper is to discuss this problem from the view of solution algorithms for nonlinear parametric optimization problems of the following form:

$$P(t): \quad \min \{f(x,t) \mid x \in M(t)\}, \quad t \in [0,1] \text{ resp. } t \in \mathbb{R} \tag{1.3}$$

where

$$M(t) = \{x \in \mathbb{R}^n \mid h_i(x,t) = 0,\ i \in I,\ g_j(x,t) \leq 0,\ j \in J\} \tag{1.4}$$

[1] Humboldt-Universität, Sektion Mathematik, PSF 1297
DDR - 1086 Berlin

[2] RWTH Aachen, Lehrstuhl C für Mathematik, Templergraben 55,
D - 5100 Aachen

with

$$I = \{1,\dots,m\}, \quad m<n, \quad J = \{1,\dots,s\}$$

we assume that $f, h_i, g_j \in C^3(\mathbb{R}^n \times \mathbb{R}, \mathbb{R})$, $i \in I$, $j \in J$.
Before we clarify the notion of a solution algorithm for $P(t)$, $t \in [0,1]$, we introduce the Karush-Kuhn-Tucker system (briefly KKT-system):

$$D_x f(x,t) + \sum_{i\in I} \lambda_i D_x h_i(x,t) + \sum_{j\in J} \mu_j D_x g_j(x,t) = 0,$$

$$h_i(x,t) = 0, \quad i \in I, \quad g_j(x,t) \le 0, \quad \mu_j \ge 0,$$

$$\mu_j \cdot g_j(x,t) = 0, \quad j \in J.$$

Here, $D_x f$ represents the row vector of first partial derivatives. $(x(t), \lambda(t), \mu(t))$ is called Karush-Kuhn-Tucker point (briefly KKT-point) and $x(t)$ is called stationary point (Note that the notations are not unique in the literature).
By a solution algorithm for $P(t)$, $t \in [0,1]$ we understand the solution of the following problem: Find a local minimizer (stationary point) $x(t)$ for $P(t)$, $t \in [0,1]$ (assume that such a point exists for all $t \in [0,1]$); or, more precisely, find a discretization

$$0 = t_0 < \dots < t_{i-1} < t_i < \dots < t_N = 1$$

of the interval $[0,1]$ and for each t_i a local minimizer (stationary point) $x(t_i)$ of $P(t_i)$.
To this end we assume that a local minimizer (stationary point) $x(0)$ is known.
Chapter 2 includes a brief survey on solution algorithm for $P(t)$, $t \in [0,1]$. Here, we propose the continuation technique, also called pathfollowing or homotopy technique, (cf. e. g. [2], [3], [4]) to follow numerically a KKT-curve. The existence of KKT-curves is secured under the assumptions proposed by M. Kojima and R. Hirabayashi [12], H. Th. Jongen et al. [8], [9], but such curve does not necessarily lead to a KKT-point corresponding to $t = 1$ (cf. Fig. 1.1).

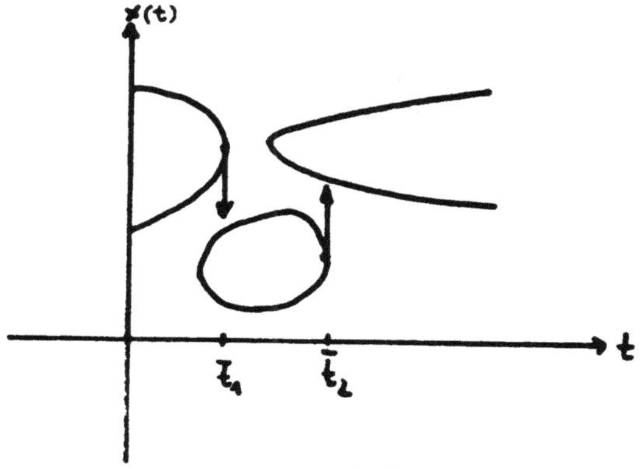

Fig. 1.1

This is the reason why we proposed to jump from one KKT-curve to another one by using additional information on such turning points (cf. J. Guddat et al. [6]). Such information is given by characterizing indices at the singularities of the problem P(t) (cf. H. Th. Jongen et al. [8], [9]).
Now we ask the question: How can we use solution algorithms for parametric optimization problems to solve Problem (1.2)? We propose the following simple strategy (see Fig. 1.2):

Step 1: Compute a stationary point $\hat{x}$ for $\min \{F(x) \mid x \in K\}$.

Step 2: Find a point belonging to

$$E(\varepsilon) := \{x \in K \mid F(x) \leq F(\hat{x}) - \varepsilon\} \qquad (1.5)$$

with $\varepsilon > 0$ sufficiently small.

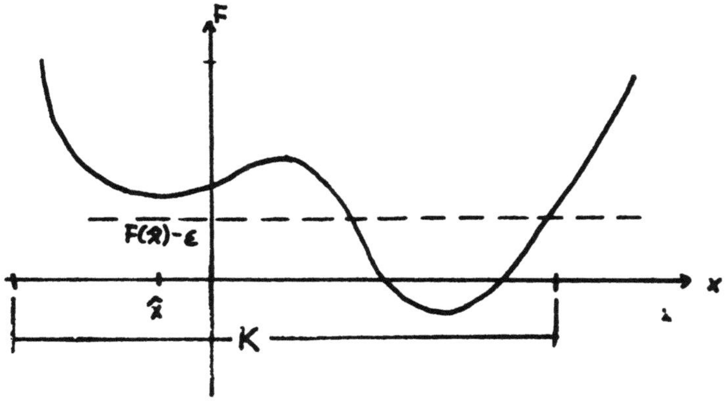

Fig. 1.2

Of course, Step 2 is the difficult one and the subject of our discussion in this paper.

We consider an arbitrary point $x^o \in K$ and an arbitrarily chosen function $f \in C^3(\mathbb{R}^n, \mathbb{R})$ with the following property

(A1) x^o is a global minimizer for $\min\{f(x) \mid x \in \mathbb{R}^n\}$ and x^o is the only stationary point for $\min\{f(x) \mid x \in \mathbb{R}^n\}$ (e. g. $f(x) = \|x - x^o\|^2$).

Of course, Step 2 can be realized if we can compute a stationary point of the following problem

$$(P): \quad \min\{f(x) \mid x \in M\} \tag{1.6}$$

where

$$M := \{x \in K \mid g(x) \leq 0\},$$

$$g(x) := F(x) - F(\hat{x}) + \varepsilon$$

and

(A2) $M \neq \emptyset$

is assumed.

Thus, we consider the problem of finding a stationary point of the non-convex problem (P) using a starting point without any additional information on $x^o \in \mathbb{R}^n$ like

- $x^o \in M$,
- x^o lies in a sufficiently small neighbourhood of a stationary point of (P),
- a Lagrange vector corresponding to x^o lies in a sufficiently small neighbourhood of an "optimal" Lagrange vector.

We call such a method a globally convergent algorithm. In case we were able to find such a algorithm, we would have a deterministic approach for solving the problem of global optimization. Up to now there has not been a really powerful deterministic approach to solve this problem. We propose the well-known concept of embedding: Choose a one-parametric optimization problem $P(t)$, $t \in [0,1]$ of the form (1.3) with the following properties:

(V1) A local minimizer x^o of $P(0)$ is known,

(V2) $P(1) \equiv (P)$.

Firstly, we propose the following embedding

$$\hat{P}(t): \quad \min\{f(x,t) \mid x \in \hat{M}(t)\}, \quad t \in [0,1] \tag{1.7}$$

where

$$\hat{M}(t) := \{x \in K \mid g(x,t) \leq 0\}, \tag{1.8}$$

$$f(x,t) := f(x), \quad (1.9)$$

$$g(x,t) := g(x) + (t-1)g(x^0), \quad (1.10)$$

and

(A3) $g(x^0) > 0$

is satisfied.

Of course, unless (A3) is satisfied, we have $F(x^0) \leq F(\hat{x}) - \varepsilon$ and Step 2 is realized.
With respect to (A2) we note that $\hat{x}$ solves the Problem (1.2) if and only if $E(\varepsilon)$ defined by (1.5) is empty for $\varepsilon > 0$ sufficiently small. Here, we do not discuss the question: How can be checked whether $E(\varepsilon)$ is empty or not?
We observe that $\hat{P}(t)$ has the following properties:

(C1) x^0 is a global minimizer for $\hat{P}(0)$ (cf. (A1)),

(C2) $\hat{P}(1) \equiv (P)$,

(C3) $\hat{M}(t)$ is non-empty and $M(t) \subseteq K$ for all $t \in (-\infty, 1]$ (cf. (A2) and (A3)).

(C4) $\hat{\psi}(t)$ is non-empty for all $t \in (-\infty, 1]$ (cf. (C3) and the compactness of K), where $\hat{\psi}(t)$ is the set of all global minimizers for $P(t)$.

Therefore, the necessary conditions for the development of a solution algorithm for $\hat{P}(t)$, $t \in [0,1]$ are satisfied. Of course, such method will be a globally convergent algorithm for the problem (P).
Now we can reify the main question of our investigation: How useful are continuation methods with jumps (as a solution algorithm for $\hat{P}(t)$, $t \in [0,1]$) to solve Problem (1.2). The answer is included in Chapter 3 for several embeddings. As noted above, Chapter 2 includes a brief abstract on pathfollowing methods with jumps (cf. J. Guddat et al. [6]) and on the singularity theory (cf. H. Th. Jongen et al. [8], [9]) as the theoretical foundation for the possible jumps.
In Chapter 3 it will be shown in particular that it can appear a singularity for the embedding $\hat{P}(t)$ defined by (1.7) where we do not have any appropriate proposal for a possible jump yet. This analysis gives a deeper insight into the essential difficulty of solving the problem of global optimization by a deterministic approach.
In case such a singularity appears we propose to leave the "forward" strategy (increasing t, continuation methods with jumps) and follow the curve with decreasing t, too.
We recall that, I. Diener [1] discussed the problem of global optimization also from the view of homotopy by connecting all critical points

of a smooth function.

2. Singularities in parametric optimization, pathfollowing with jumps - an abstract

We consider the general problem $P(t)$ defined by (1.3). A point $z = (x,t)$ is called a generalized critical point (briefly g.c. point) if $x \in M(t)$ and if the set $\{D_x f, D_x h_i, i \in I, D_x g_j, j \in J_0(z)\}|_z$ is linearly dependent. Here, $J_0(z)$ denotes the index set of active inequality constraints, i. e. $J_0(z) = \{j \in J \mid g_j(z) = 0\}$. In particular, $z = (x,t)$ is a g.c. point whenever x is a local minimizer for $P(t)$. Let Σ denote the set of g.c. points for $P(t)$, $t \in \mathbb{R}$. In [9] the local structure of Σ is completely described for $(f, h_1, \dots, h_m, g_1, \dots, g_s)$ belonging to a C^3_s-open and dense subset $\mathcal{F}$ from $C^3(\mathbb{R}^n \times \mathbb{R}, \mathbb{R})^{1+m+s}$; the topology C^3_s refers to the strong (or Whitney-) C^3-topology (cf. [7], [11]). For omitted details on the set $\mathcal{F}$ we refer to [9].
From now on we assume

(B1) $(f, h_1, \dots, h_m, g_1, \dots, g_s) \in \mathcal{F}$.

The points of Σ can be divided into five types.
A point $\bar{z} = (\bar{x}, \bar{t})$ is of Type 1 if the following conditions are satisfied (then, $\bar{x}$ is also called a non-degenerate critical point for $P(\bar{t})$, cf. [10], [14]):

$$D_x f + \sum_{i \in I} \bar{\lambda}_i D_x h_i + \sum_{j \in J_0(\bar{z})} \bar{\mu}_j D_x g_j \Big|_{\bar{z}} = 0 \tag{2.1}$$

$\{D_x h_i, i \in I, D_x g_j, j \in J_0(\bar{z})\}|_{\bar{z}}$ is linearly independent (2.2.a)
(linear independence constraint qualification, briefly LICQ)

The numbers $\bar{\mu}_j, j \in J_0(\bar{z})$ do not equal to zero (2.2.b)
(strict complementarity condition)

$D^2_x L(\bar{z})|_{T(\bar{z})}$ is nonsingular. (2.2.c)

Condition (2.2.c) needs some explanation:
$D^2_x L$ is the Hessian for the Lagrangian $L = f + \sum_{i \in I} \bar{\lambda}_i h_i + \sum_{j \in J_0(\bar{z})} \bar{\mu}_j g_j$, the numbers $\bar{\lambda}_i$, $\bar{\mu}_j$ being taken from (2.1). Furthermore, $T(\bar{z})$ denotes the tangent space of $M(\bar{t})$ at $\bar{x}$, i. e.

$T(\bar{z}) = \{\xi \in \mathbb{R}^n \mid D_x h_i(\bar{z})\xi = 0,\ i \in I,\ D_x g_j(\bar{z})\xi = 0,\ j \in J_0(\bar{z})\}.$
Now, $D_x^2 L(\bar{z})\big|_{T(\bar{z})}$ represents $V^T D_x^2 L(\bar{z}) V$, where V is a matrix whose columns (n-vectors) form a basis for $T(\bar{z})$.
If $\bar{z}$ is a point of Type 1, then the local behaviour of Σ around $\bar{z}$ is completely determined by means of four characteristic numbers: the number of negative/positive scalars $\bar{\mu}_j$, $j \in J_0(\bar{z})$, and the number of negative/positive eigenvalues of $D_x^2 L(\bar{z})\big|_{T(\bar{z})}$.
The set Σ is composed of one-dimensional C^2-manifolds and it is the closure of the set of all points of Type 1; moreover, the points of Type 2-5 constitute a discrete subset of Σ. The points of Type 2-5 represent three basic degeneracies (compare also the interesting paper [14]). In fact, Type 2, Type 3 and Type 4, 5 refer to the violation of (2.2.b), (2.2.c) and (2.2.a), respectively. All possible changes in the four characterizing numbers, when passing points of Type 2-5 along Σ, are given in [9].
Now, we turn to the problem of finding a discretization

$$0 = t_0 < \dots < t_{i-1} < t_i < \dots < t_N = 1$$

of the interval $[0,1]$ and for each t_i a local minimizer $x(t_i)$ for $P(t_i)$. The information on $x(t_{i-1})$ will be used for $x(t_i)$.
Let Σ^i be the set of all g.c. points of Type i ($i \in \{1,\dots,5\}$) and let Σ_{loc} be the set of all $z = (x,t)$, where x is a local minimizer for $P(t)$. We further assume:

(B2) $(x(0),0) \in \Sigma_{loc} \cap \Sigma^1$

(B3) $M(t)$ is non-empty for all $t \in [0,1]$ and there exists a compact set $C \subset \mathbb{R}^n$ which contains $M(t)$ for all $t \in [0,1]$.

Let $\bar{\Sigma}_{loc}$ denote the closure of Σ_{loc}. Based on the investigations in [8] we have the following possibilities for the local structure of $\bar{\Sigma}_{loc}$, as depicted in Fig. 2.1

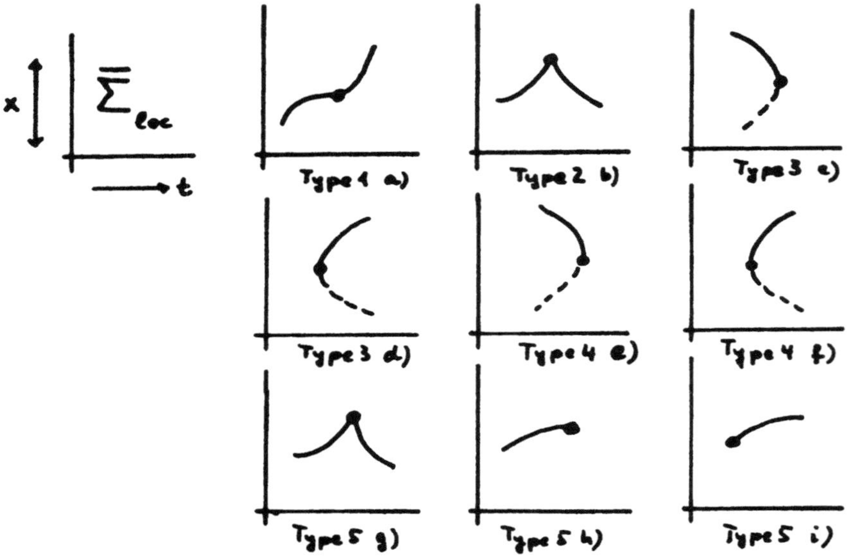

Fig. 2.1

The point $\bar{z} = (\bar{x},\bar{t})$ is identified by an exposed point, whereas the dotted line represents the curve of g.c. points which are not local minimizers. The dotted line in Fig. 2.1.c),d) represents even a curve of stationary points. We also have a curve of stationary points in Fig. 2.1.e),f) if $J_o(\bar{t}) = \emptyset$. At the breakpoint in Fig. 2.1.b), for increasing t, either a new inequality constraint becomes active, or an active inequality constraints becomes inactive. The different behaviour at a point of Type 5 in Fig. 2.1.g) and h) and i) respectively depends on the satisfying of the Mangasarian-Fromovitz constraint qualification (briefly MFCQ) at $\bar{x} \in M(\bar{t})$. We recall that MFCQ is fulfilled at $\bar{x} \in M(\bar{t})$ if both the set $\{D_x h_i,\ i \in I\}\big|_{\bar{z}}$ is linearly independent and if there exists a vector $\xi \in \mathbb{R}^n$ solving the system

$$D_x h_i(\bar{z})\xi = 0,\ i \in I, \quad D_x g_j(\bar{z})\xi < 0,\ j \in J_o(\bar{z}).$$

In the case that MFCQ is satisfied, we have the situation described in Fig. 2.1.g). The local behaviour near to the breakpoints in Fig. 2.1.b) and g) looks similarly. At the breakpoint in Fig. 2.1.g) the number of active constraints becomes exactly $n+1$, but apart from the breakpoint the number of active constraints equals n. At $\bar{z} \in \Sigma^2$, the LICQ is satisfied, whereas only the weaker MFCQ is fulfilled at $\bar{z} \in \Sigma^5$. Then the following behaviour of the Lagrange parameter μ_j, $j \in J_o(\bar{z})$ is typical:

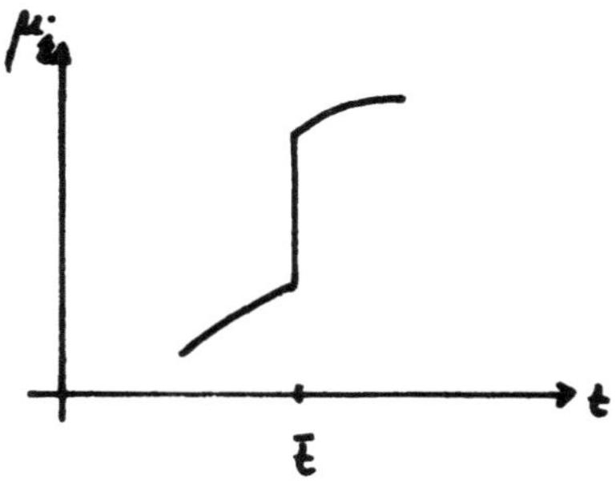

Type 4 (MFCQ is fulfilled)

Fig. 2.2

Also the local behaviour near to the points of Type 3 and 4 looks similarly. Type 3 is a classical quadratic turning point, but the Lagrange vectors uniquely corresponding to points near to the point of Type 4 have the following typical behaviour:

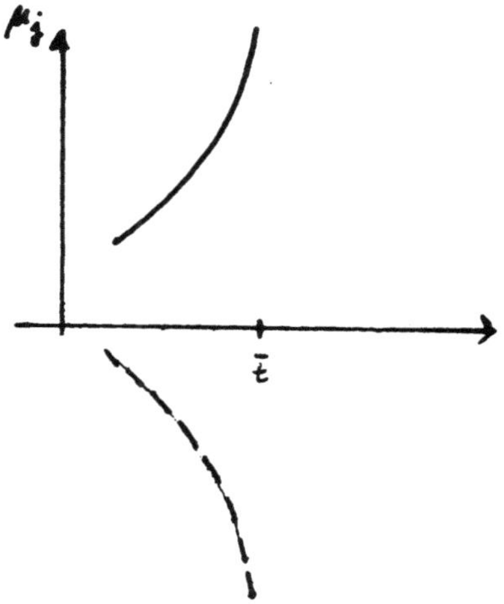

Fig. 2.3

In case we walk along branches of $\bar{\bar{\Sigma}}_{loc}$ for increasing values of t, the situation of Fig. 2.1.d),f),i) will not occur.
Let us consider Fig. 2.1 again. As long as we are in the situation of Fig. 2.1.a),b),g) we follow a branch of local minimizers which can be computed by pathfollowing methods (cf. e. g. H. Gfrerer et al. [2],[3], J. Guddat [4]) in an efficient manner. We can numerically trace also the KKT-curve corresponding to Fig. 2.1.c), in particular the turning point $\bar{z}$ can be computed (G. Pönisch and H. Schwetlick [13]).
The behaviour of the Lagrange parameters (cf. Fig. 2.3) indicates the approach of a point $\bar{z}$ of Type 4, but we cannot compute $\bar{z}$

sufficiently exactly using the KKT-system. In order to get $\bar{z}$, we have to compute across the turning point $\bar{z}$ of Σ. This can be done by using the system

$$
\begin{aligned}
&\lambda D_x f + \sum_{i \in I} \lambda_i D_x h_i + \sum_{j \in J} \mu_j D_x g_j = 0 \\
&h_i = 0,\ i \in I, \quad \mu_j g_j = 0,\ j \in J \qquad\qquad (2.3) \\
&g_j \le 0,\ j \in J, \quad \lambda^2 + \sum_{i \in I} \lambda_i^2 + \sum_{j \in J} \mu_j^2 = 1
\end{aligned}
$$

instead of the KKT-system. Formula (2.3) refers directly to the definition of a g.c. point. We mention that the system (2.3) is also basic for the study of A. B. Poore and C. A. Tiahrt [14].
In the last part of this chapter we characterize possible jumps at points of Type 3 and Type 4 (cf. J. Guddat et al. [6]).

(i) <u>The jump at a point of Type 3</u>

Let $\bar{z} = (\bar{x}, \bar{t})$ be a point of Type 3. The only degeneracy lies in the fact that the restricted Hessian $D_x^2 L(\bar{z})|_{T(\bar{z})}$ becomes singular. So, we can still compute across the turning point by using a pathfollowing method. But, as we pass the turning point, a local minimizer switches into a saddle point since (exactly) one eigenvalue of $D_x^2 L(z)|_{T(z)}$ changes from the positive to the negative sign; see also [11, Chapter 10] for a detailed discussion. Therefore, $\bar{x}$ cannot be a global minimizer for $P(\bar{t})$. Since the feasible set $M(\bar{t})$ was assumed to be compact, the existence of a global minimizer $\tilde{x}$ for $P(\bar{t})$ is assured. So, if we use a descent method starting at $\bar{x}$, we can arrive at a local minimizer $\hat{x}$, and the point $(\hat{x}, \bar{t})$ lies on another branch of $\bar{\Sigma}_{loc}$. Starting at $(\hat{x}, \bar{t})$, a pathfollowing procedure can again be exploited. The actual jump consists in the transition from $(\bar{x}, \bar{t})$ to $(\hat{x}, \bar{t})$. The problem now consists in finding a tangential direction of descent (cf. Fig. 2.4).

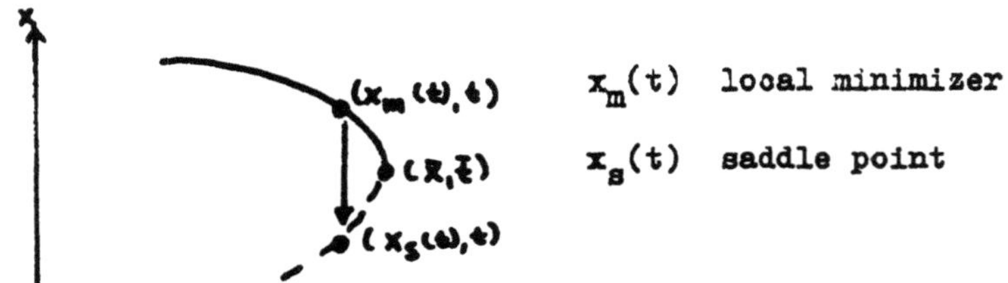

Fig. 2.4

Then, as t tends to $\bar{t}$, the vector
$v(t) := (x_s(t) - x_m(t))/\|x_s(t) - x_m(t)\|$ tends to a tangential vector $\bar{v}$, which is a direction of descent.

(ii) <u>The jump at a point of Type 4</u>

Let $\bar{z} = (\bar{x},\bar{t})$ be a point of Type 4. Then, when passing $\bar{z}$ along Σ, the local minimizer switches into a local maximizer. We know that $\bar{x}$ is not a stationary point for $P(\bar{t})$, but we cannot compute a direction of descent at $\bar{x}$ in the classical way. For a further insight into properties of a point of Type 4 it suffices to consider the case in which only one (in)equality constraint is present, i. e. we consider $M(t) = \{x \in \mathbb{R}^n \mid g(x,t) \leq 0\}$. For simplicity assume that g is a C^∞-function. Then, the existence of a local C^∞-coordinate transformation

$$(y,u) = \Phi(x,t) \tag{2.4}$$

having the structure

$$\Phi(x,t) = (\psi(x,t), \eta(t)), \quad \eta'(\bar{t})>0 \tag{2.5}$$

such that $\Phi(\bar{x},\bar{t}) = 0$ and g takes the following form in these new coordinates $x := y$ and $u := t$ (cf. [11, Chapter 10])

$$g(x,t) = -\sum_{i=1}^{k} x_i^2 + \sum_{j=k+1}^{n} x_j^2 - \delta t \tag{2.6}$$

where

$$\delta = \operatorname{sign}(-D_t g(\bar{x},\bar{t})) \in \{+1,-1\}.$$

We have to distinguish two cases, namely $\delta > 0$ and $\delta < 0$.

<u>Case I:</u> $\delta > 0$.

Then it holds:

$k = 1$, i. e. $-x_1^2 + \sum_{j=2}^{n} x_j^2 - t \leq 0$, and the corresponding local maximal value is less than the local minimal value (cf. Fig. 2.5a)).

<u>Case II:</u> $\delta < 0$.

Then it holds:

$k = 0$, i. e. $\sum_{i=1}^{n} x_i^2 + t \leq 0$, and the corresponding local maximal value is greater than the local minimal value (cf. Fig. 2.5b)).

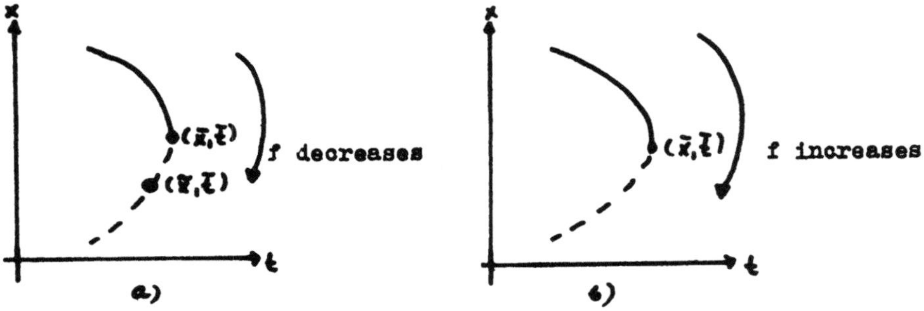

Fig. 2.5

Suppose that we walk on a branch of local minimizers and that we approach a point of Type 4 as t increases. If the functional value of f has decreased, meanwhile then start a descent procedure at a point $\tilde{\tilde{x}}$ with $(\tilde{\tilde{x}},\tilde{t}) \in \Sigma$ on the other side of the turning point (cf. Fig. 2.5.a)). Such a point can be computed by using the system (2.3). If we are in the situation of Fig. 2.5b), then the corresponding component of the feasible set $M(t)$ shrinks to a point and gets empty for increasing t. In this case we do not have any appropriate proposal yet.

3. Properties of the parametric optimization problems

For simplicity we consider the problem of global optimization (1.1), i. e.

$$\text{glob min } \{F(x) \mid x \in \mathbb{R}^n\} , \tag{3.1}$$

and we assume

(D1) F has compact level sets, i. e. $\{x \in \mathbb{R}^n \mid F(x) \leq \alpha\}$ is compact for all $\alpha \in \mathbb{R}$.

Of course, (D1) restricts the class of functions considered. But, if we follow the approach proposed in Chapter 1, we have only one constraint in the problems

$$(P): \quad \min \{f(x) \mid x \in M\} \tag{3.2}$$

where f is an arbitrarily chosen function with the property (A1) and

$$M := \{x \in \mathbb{R}^n \mid g(x) \leq 0\} , \tag{3.3}$$

$$g(x) := F(x) - F(\hat{x}) + \varepsilon , \tag{3.4}$$

and (A2) is satisfied, and

$$\hat{P}(t): \quad \min\{f(x) \mid x \in \hat{M}(t)\}, \quad t \in [0,1] \tag{3.5}$$

where

$$\hat{M}(t) := \{x \in \mathbb{R}^n \mid g(x,t) \leq 0\}, \tag{3.6}$$

$g(x,t)$ is defined by (1.10) and (A3) is satisfied.

Then, the analysis will be simplified. Of course, $\hat{P}(t)$ has the properties (C1), (C2) and (B3) (follows from (D1)) and is useful as an embedding from this point of view.

Now we consider the problem $\hat{P}(t)$ under the assumption (B1), i. e. $(f,g) \in \mathcal{F}$.

We will discuss this condition later.

First, we ask the question: What kind of singularities can appear? The answer will be given by the statements (i) and (ii) of the following

<u>Theorem 3.1.:</u> Let (A1), (A2), (A3), (B1) and (D1) be fulfilled. Then we have

(i) If $\bar{z} \in (\bar{x},\bar{t}) \in \bar{\Sigma}_{loc} \setminus \Sigma^1$ with $\bar{t} \in (0,1]$, then $\bar{z} \in \Sigma^3 \cup \Sigma^4$.

(ii) $(x^0,0) \in \Sigma_{loc} \cap \Sigma^2$

(iii) If $F \in C^\infty(\mathbb{R}^n,\mathbb{R})$ and $\bar{z} \in \bar{\Sigma}_{loc} \cap \Sigma^4$, then there exists a local C^∞-coordinate transformation (2.4) having the structure (2.5), sending z onto the origin, such that g takes the form

$$g(x,t) = \sum_{i=1}^{n} x_i^2 + t$$

(in these new coordinates). □

<u>Proof:</u> (i) First, we consider $\bar{z} \in \bar{\Sigma}_{loc}$ and suppose that $\bar{z} \in \Sigma^i$, $i \in \{2,\dots,5\}$. Let $\bar{z} \in \Sigma^2$ be. We know that x^0 is a global minimizer and the only stationary point for $\min\{f(x) \mid x \in \mathbb{R}^n\}$ (cf. (A1)) and that

$$g(x^0,0) = 0 \tag{3.7}$$

and $g(x^0,t) > 0$ for all $t \in (0,1]$ (cf. (A3)).

As we have only one constraint, $\bar{z} \in \Sigma^2$ does not exist for $t \in (0,1]$. Let $\bar{z} \in \Sigma^5$. Then, the property $|I| + |J_0(\bar{z})| = n + 1$ for the general problem $P(t)$ defined by (1.3) (cf. [8]) cannot be fulfilled for an arbitrary $n \geq 1$. These points of Type 5 cannot appear.

(ii) By application of (1.10) and (A3) it is easy to see that

$$g(x^0,t) < 0 \quad \text{for} \quad t < 0. \tag{3.8}$$

Then, we have that x^0 is a global minimizer for $\hat{P}(t)$, $t \leq 0$. By using (3.7) and (3.8) we obtain (ii).

(iii) From Chapter 2 we know that there exists a local C^∞-coordinate transformation, sending $\bar{z}$ onto the origin, which can be constructed such that g takes the form (2.6) in these new coordinates. Using (1.10) we obtain $\delta := \text{sign}(-D_t g(\bar{x},\bar{t}) = \text{sign}(-g(x^0))$ (cf. (2.5)). As $g(x^0) > 0$ holds (cf. (A3)), we have $\delta = -1$ and (iii) is proved. □

We introduce the following problem

$$\tilde{P}(t): \quad \min\{f(x) \mid g(x,t) = 0\} \tag{3.9}$$

and denote by $\tilde{\psi}(t)$ the set of all global minimizers for $\tilde{P}(t)$. From the proof of Theorem 3.1 we derive the following two consequences:

1) $\hat{\psi}(t) = \tilde{\psi}(t)$, for all $t \in [0,1]$. (3.10)

2) If $\bar{z} \in \bar{\Sigma}_{loc}$ is a point of Type 4, then $\bar{z}$ fulfils the condition of Case II ($\delta < 0$) in Chapter 2.

We remind the reader of the fact, that Case II was the difficult one (cf. Fig. 6.3.3b), namely, if f increases the corresponding connected component of the feasible set $\hat{M}(t)$ will shrink to one point and get empty for increasing t. Fig. 6.5.1 shows that this situation is typical of the application of the embedding $\hat{P}(t)$ defined by (3.6) to the problem of global optimization.

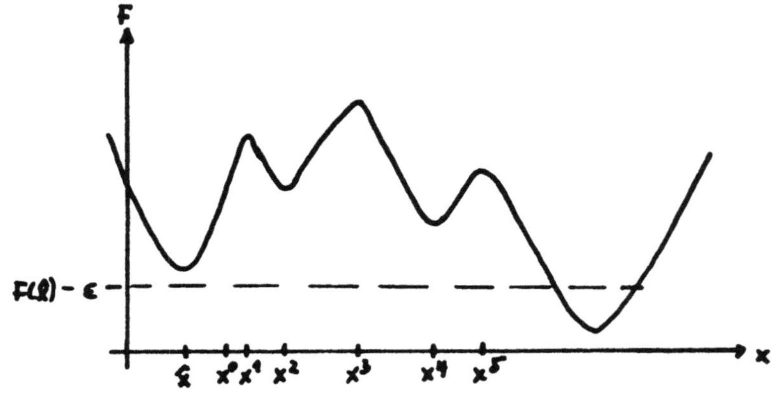

Fig. 3.1

$\hat{x}$ is the local minimizer computed by Step 1. x^0 is the starting point for Step 2. Then we see that the MFCQ is not fulfilled at the parameter values $t \in [0,1]$ where $g(x^i,t) = 0$, $i = 1,2,3,4,5$ and $g(\hat{x},t) = 0$. As noted in Chapter 3 up to now, we do not have a proposal for a possible jump which works.

This is the reason why we have the "forward" strategy (increasing t, pathfollowing with jumps) and follow the curve of g.c. points for $\tilde{P}(t)$ (cf. (3.10)), on which $(x^0,0)$ lies, numerically in the case that a point $\bar{z}\in\Sigma^4_{go}$ appears. Therefore, we have to consider the following system of equations corresponding to $\tilde{P}(t)$ defined by (3.9) (cf. (2.3))

$$\begin{aligned} \lambda_0 Df(x) + \lambda_1 DF(x) &= 0 , \\ g(x,t) &= 0 , \\ \lambda_0^2 + \lambda_1^2 &= 1 , \end{aligned} \tag{3.11}$$

instead of the KKT-system of $\tilde{P}(t)$. We can use standard continuation methods for a parameter-dependent equality system where turning points are passable (cf. e. g. H. Schwetlick [14]). A point $\bar{z}$ of Type 4 corresponds to a quadratic turning point $(\bar{x},\bar{t},\bar{\lambda}_0,\bar{\lambda}_1)$ satisfying (3.11). Now we have to distinguish between the following two cases:

Case 1: If $(x^0,0)$ lies on a curve of g.c. points that leads to $t = 1$, then Step 2 can be realized.

Case 2: If $(x^0,0)$ does not lie on such a curve of g.c. points, then we have to choose another starting point $x^0\in\mathbb{R}^n$ and restart the procedure.

How can we choose another starting point $x^0\in\mathbb{R}^n$? We propose the well-known strategy in considering a lattice on M. Of course, if one lattice point x^0 fulfils $g(x^0)\leq 0$, Step 2 is realized. In the other case we consider the embedding (3.5) and choose a lattice point x^0 as a starting point. Of course, if the chosen lattice is sufficiently fine, the proposed procedure is successful. It is easy to see (cf. the example which follows from Fig. 3.1) that x^0 does not necessarily belong to the radius of convergence of a stationary point of $\tilde{P}(1)$ if $(x^0,0)$ lies on a curve that leads to $t = 1$.
As noted before, Assumption (B2) is a restrictive one. But there is a similar analysis if we consider the problem glob min $\{F(x) \mid x\in K\}$ (1.2) where K is a given compact subset of $\mathbb{R}^n$, e. g. $K = K(q_0) := \{x\in\mathbb{R}^n \mid \|x\|^2\leq q_0\}$, $q_0>0$ and the corresponding parametric optimization problem

$$\hat{P}(t):\quad \min\{f(x) \mid x\in\hat{M}(t)\},\quad t\in[0,1] \tag{3.12}$$

where

$$\hat{M}(t) = \{x\in K \mid g(x,t)\leq 0\} \text{ (cf. 1.7)} \tag{3.13}$$

Of course, (B2) can be cancelled in this case. If we choose $K = \{x\in\mathbb{R}^n \mid \|x\|^2\leq q_0\}$ then we have two constraints in $\hat{M}(t)$ defined by (3.13).

4. Concluding remarks

(i) The analysis included in this paper allows a deeper insight into the essential difficulties to find an ever successful deterministic algorithm for the problem of global optimization on the one hand and to develop a globally convergent algorithm for the general nonlinear optimization problem without any additional information on the other hand.
Of course, if we can find a jump for Case II ($\delta < 0$), a theoretical foundation for solving these problems is given for the class $\mathcal{F}$.

(ii) Since we do not have a proposal for a possible jump for a point of Type 4 (Case II), the proposed approach (to follow numerically a curve of g.c. points with different starting points x^0 and (x^0,v^0), respectively) can be helpful, in particular, in combination with other techniques, e. g. stochastic search procedures. However, there are examples where the chosen starting point x^0 and (x^0,v^0), respectively, does not lie in a radius of convergence (for a given locally convergent algorithm) for any stationary points of $\hat{P}(1)$ and $\bar{P}(1)$ respectively, but $(x^0,0)$ and $(x^0,v^0,0)$, respectively, lies on a curve of g.c. points leading to a g.c. point for $\hat{P}(1)$ and $\bar{P}(1)$, respectively. This is an advantage, without doubt.

References

[1] Diener, I.: Trajectory nets connecting all critical points of a smooth function. Math. Programming 36 (1986), 340-352.

[2] Gfrerer, H., J. Guddat, Hj. Wacker and W. Zulehner: Path-following methods for Kuhn-Tucker curves by an active index set strategy. In: Systems and Optimization (A. Bagchi, H. Th. Jongen, eds.), Lecture Notes in Control and Information Sciences, Vol. 66, Springer-Verlag, Berlin-Heidelberg-New York 1985, 111-132.

[3] Gfrerer, H., J. Guddat, H. Th. Jongen, Hj. Wacker and W. Zulehner: Parametric optimization and continuation method (forthcoming).

[4] Guddat, J.: Parametric optimization: pivoting and predictor-corrector continuation, a survey. In: [5], 125-162.

[5] Guddat, J., H. Th. Jongen, B. Kummer and F. Nožička (eds): Parametric optimization and relatic topics. Akademie-Verlag, Berlin 1987.

[6] Guddat, J., H. Th. Jongen and D. Nowack: Parametric Optimization: Pathfollowing with jumps, Memorandum Nr. 607, Department of Applied Mathematics, Twente University of Technology, Febr. 1987, to appear in: Proceedings of the Seminar "Approximation and Optimization", Jan. 1987, Havana, Cuba.

[7] Hirsch, M. W.: Differential topology, Springer Verlag, Berlin-Heidelberg-New York 1976.

[8] Jongen, H. Th., P. Jonker and F. Twilt: One-parameter families of optimization problems: equality constraints. JOTA 48 (1986), 141-161.

[9] Jongen, H. Th., P. Jonker and F. Twilt: Critical sets in parametric optimization. Math. Programming 34 (1986), 333-353.

[10] Jongen, H. Th., P. Jonker and F. Twilt: Nonlinear optimization in $\mathbb{R}^n$, I. Morse theory, Chebyshev approximation. Peter Lang Verlag, Frankfurt a.M., Bern, New York 1983.

[11] Jongen, H. Th., P. Jonker and F. Twilt: Nonlinear optimization in $\mathbb{R}^n$, II. Transversality, flows, parametric aspects. Peter Lang Verlag, Frankfurt a.M., Bern, New York 1986.

[12] Kojima, M. and R. Hirabayashi: Continuation Deformation of Nonlinear Programs. Math. Programming Study 21 (1984), 150-198.

[13] Pönisch, G. and H. Schwetlick: Computing turning points of curves implicitly defined by nonlinear equations depending on a paramater. Computing 26 (1981), 107-121.

[14] Poore, A. B. and C. A. Tiahrt: Bifurcation problems in nonlinear parametric programming, Preprint Colorado State University 1986.

[15] Schwetlick, H.: Numerische Lösungen nichtlinearer Gleichungen. VEB Deutscher Verlag der Wissenschaften, Berlin 1979.

Extended subdifferential
and minimization of nondifferentiable functions

Horst Hollatz[1])

Let us study the minimization of a real function f on the n-dimensional euclidean space

$$\min \{f(x) \mid x \in R^n\}, \qquad (1)$$

where we do not require f to be smooth, but we assume that the directional derivative

$$f'(x,h) = \lim_{\alpha \downarrow 0} \frac{1}{\alpha}[f(x+\alpha h) - f(x)]$$

exists for all $h \in R^n$; for simplifying the presentation we restrict the development to the convex case. It is well-kown that convex functions are of this kind.
The subdifferential of f is defined as

$$\partial f(x) = \{ s \mid s'(y-x) \leq f(y) - f(x) \text{ for all } y \in R^n\}.$$

The set X^* of optimal points for (1) is characterized by

$$X^* = \{ x^* \mid o \in \partial f(x^*)\}.$$

On the other hand the directional derivative $f'(x,.)$ is the support function of $\partial f(x)$:

$$f'(x,h) = \max_{s \in \partial f(x)} s'h .$$

For the iterative process

$$x_{k+1} = x_k - \alpha_k s_k/||s_k||, \quad s_k \in \partial f(x_k), \qquad (2)$$

$$\alpha_k > 0, \quad \alpha_k \to 0, \quad \sum_k \alpha_k = \infty$$

----

[1]) Techn. Univ. Otto von Guericke, Dep. of Math., PSF 124, DDR-3010 Magdeburg

it was proved by POLJAK [1967] that for arbitrary starting point a subsequence of $\{x_k\}$, provided by (2), converges to some $x^* \in X^*$ (if X^* is nonempty and bounded). Under the additional assumption

$$\sum_k \alpha_k^2 < \infty$$

it was proved by EHLERT [1972] that

$$\lim_{k \to \infty} d(x_k, X^*) = 0,$$

if X^* is nonempty and bounded; here $d(.,.)$ denotes

$$d(x, X^*) = \inf_{y \in X^*} \|x - y\|^2 .$$

Using the assumption that $\sum_k \alpha_k = \infty$ it is easy to prove the rate of convergence for the generated sequence $\{x_k\}$ of is less than geometric type.

By EHLERT [1972] was introduced the concept of the extended linear supporting functions. Let be $c \geq 0$. A vector $w \in R^n$ defines an extended linear supporting function at the point y of a convex set X if

$$w'y \leq w'x + c\|w\|^2 \qquad \text{for all } x \in X.$$

We denote with $W^c(X,y)$ the set of all extended linear supporting functions at the point y for fixed $c \geq 0$. In the case $c = 0$ the set $W^0(X,y)$ will be the set of all linear supporting functions at y.

For $0 \leq c_1 \leq c_2$ we have

$$W^{c_1}(X,y) \subseteq W^{c_2}(X,y).$$

Let be $W_\alpha^c(X,y)$ the set of all extended linear supporting functions at y on X with a given norm $\alpha > 0$.

We consider the following iterative process.

$$c > 0,\ x_0\ R^n,\ \alpha_k > 0,\ \alpha_k \to 0,\ \sum_k \alpha_k = \infty,\ \sum_k \alpha_k^2 < \infty,$$

$$X_k = \{ x \mid f(x) \leq \min_{0 \leq i \leq k} f(x_i)\},\ w_k\ W_{\alpha_k}^c(X_k, x_k),$$

$$x_{k+1} = x_k + w_k.$$

It is proved by EHLERT [1972] that

$$\lim_{k \to \infty} d(x_k, X^*) = 0,$$

if X^* is a nonempty set.

The typical behaviour of the subgradient methods is an oscillating around the minimal point: the stepsizes α_k are chosen independently of the special situation. The behaviour of the EHLERT-method is smoother than the other one.

The ε-subdifferential is defined as follows (ZOWE [1985]):

$$\partial_\varepsilon f(x) = \{ s \mid s'(y-x) \leq f(y) - f(x) + \varepsilon \quad \text{for all } y \in R^n\} \quad (\varepsilon \geq 0).$$

The ε-directional derivative

$$f'_\varepsilon(x,h) = \inf_{\alpha > 0} \frac{1}{\alpha}[f(x+\alpha h)-f(x) + \varepsilon]$$

is the support function of the ε-subdifferential:

$$f'_\varepsilon(x,h) = \max_{s \in \partial_\varepsilon f(x)} s'h.$$

Using this fact it holds: If $o \notin \partial_\varepsilon f(x)$, then there are an $\bar{\alpha} > 0$ with

$$f(x - \alpha \bar{s}) < f(x) - \varepsilon \qquad \text{for all } \alpha \in (0, \bar{\alpha}),$$

where

$$\bar{s} = \arg \min_{s \in \partial_\varepsilon f(x)} ||s||.$$

If $o \in \partial_\varepsilon f(x)$, then

$$f(x) \leq f(y) + \varepsilon \qquad \text{for all } y \in R^n.$$

These properties imply the following iteration.

$$\varepsilon > 0,\ x_0 \in R^n; \quad s_k = \arg \min_{s \in \partial_\varepsilon f(x_k)} ||s||, \quad \bar{s}_k = -s_k/||s_k||$$

$$x_{k+1} = x_k + \alpha_k \bar{s}_k, \text{ where}$$

$$\alpha_k = \arg \min_{\alpha \geq 0} f(x_k + \alpha \bar{s}_k).$$

This algorithm finds in finitely many steps an ε-optimal point, i. e. if

$$f^* = \inf f(x) > -\infty \quad ,$$

then there exists k such that

$$f(x_k) \leq f^* + \varepsilon$$

(comp. f. e. ZOWE [1985]).
In general it is not possible to know the whole ε-subdifferential at an iteration point. The bundle concept overcomes this situation. The basic idea of a bundle-type-algorithm consists in replacing the ε-subdifferential $\partial_\varepsilon f(x_k)$ by an inner approximating convex polyeder P_k (forming of a bundle) and the direction s_k will be determined as

$$s_k = \arg\min_{s \in P_k} ||s||.$$

Provided P_k is a good approximation of the current ε-subdifferential $\partial_\varepsilon f(x_k)$, then we will find a direction along a line search provides some x_{k+1} with a decrease of almost ε. In case P_k is a bad approximation then we stay at x_k and add a further subgradient to P_k. These two items are the typical steps of all bundle methods (WOLFE [1975], MIFFLIN [1977], LEMARECHAL [1980]). In LEMARECHAL ...[1981] an implementation of the bundle concept is discussed in details.
Now let us define the extended subdifferetial of the function f.
For fixed $c \geq 0$ and $g \neq o$ let be

$$\partial_g^c f(x) = \{ s \mid s'(y-x) \leq f(y) - f(x) + c||g||^2 \text{ for all } y \in R^n\}.$$

This so-called extended subdifferential is a nonempty convex compact set (same proof as for c = 0).
The extended directional derivative is defined as follows:

$$f'_{c,g}(x,h) = \inf_{\alpha>0} \frac{1}{\alpha} [f(x+\alpha h)-f(x) + c||g||^2] \qquad (3)$$

This extended directional derivative is the support function of the extended subdifferential (same proof as for c = 0):

$$f'_{c,g}(x,h) = \max_{s \in \partial_g^c f(x)} s'h.$$

Now we consider the direction finding problem

$$\min_{||h|| \leq 1} \max_{s \in \partial_g^c f(x)} s'h. \qquad (4)$$

By a well-known minimax theorem (4) is equivalent to

$$\max_{s \in \partial_g^c f(x)} \min_{||h|| \leq 1} s'h. \qquad (5)$$

Hence, (5) becomes

$$\max_{s\in\partial_g^c f(x)} \min_{\|h\|\le 1} s'h = - \min_{s\in\partial_g^c f(x)} \|s\|.$$

In words: if $\bar{s}$ solves

$$\min_{s\in\partial_g^c f(x)} \|s\|, \tag{6}$$

then

$$\tilde{s} = - \bar{s}/\|\bar{s}\|$$

minimizes $f'_{c,g}(x,.)$ and $f'_{c,g}(x,s) = - \|\bar{s}\|$.
If $o \notin \partial_g^c f(x)$, then the optimal $\bar{s}$ in (6) is not the zero-vector and thus $f'_{c,g}(x,s) < 0$. Definition (3) implies that there exists a $\bar{t} > 0$ with

$$f(x + t.s) < f(x) - t\|g\|^2 \text{ for all } t\in (0,\bar{t}). \tag{7}$$

If $o\in \partial_g^c f(x)$, then it holds

$$f(x) \le f(y) + c\|g\|^2 \quad \text{for all } y\in R^n. \tag{8}$$

(7) together with (8) motivates the following iteration process.

$c_0 > 0$, $x_0 \in R^n$, $\tilde{s}_{-1} \in R^n$, $\tilde{s}_{-1} \neq o$;

$\bar{s}_k = \arg \min_{s\in\partial_{g_{k-1}}^{c_k} f(x_k)} \|s\|$

if $\bar{s}_k \neq o$, then $\tilde{s}_k = - \bar{s}_k/\|\bar{s}_k\|$,

$x_{k+1} = x_k + t_k.\tilde{s}_k$, $t_k = \arg \min_{t\ge 0} f(x_k + t.\tilde{s}_k)$,

$c_{k+1} = c_k$;

if $\bar{s}_k = o$, then $c_{k+1} = c/2$, $x_{k+1} = x_k$, $\tilde{s}_{k+1} = \tilde{s}_k$.

Let $\{x_k\}$ be the generated sequence; let us assume that

$$f^* = \inf f(x) > -\infty \tag{9}$$

$$X^* = \{ x^* \mid o\in \partial f(x^*)\} \neq \emptyset, \text{ bounded.} \tag{10}$$

There are two cases.
Case 1. If $\bar{s}_k \neq o$ holds for all k, then it is easy to prove using standard arguments that

$$\lim d(x_k, X^*) = 0.$$

Case 2. If there exists k such that $\bar{s}_k = o$, then either it holds $o \in \partial f(x_k)$ or after finitely many steps (reductions of c_k) it will be found a nonzero vector $\bar{s}_k$.

We summarize:

THEOREM. If inf $f(x) = -\infty$, then lim $f(x_k) = -\infty$. If the assumptions (9), (10) are fullfilled, then every accumulation point x^* of the sequence $\{x_k\}$ minimizes the function f over R .

For an implementation it is possible to apply a bundle concept nearly to the ε-subdifferential method.

References:

EHLERT, J., Ueber Erweiterungen eines allgemeinen Verfahrens von POLJAK zur Loesung von Extremalaufgaben mit Nebenbedingungen und seine Anwendung in der nichlinearen Optimierung, Diss. Humboldt-Univ. Berlin 1972
Math. Operationsforsch. und Statist. 6 (1975) 1, 91 - 105

LEMARECHAL, C., Nondifferentiable optimization, in: Dixon, Spedicato and Szegoe (eds.), Nonlinear Optimization, Theory and Algorithms (1980)
-, J. J. STRODIOT and A. BIHAIN, On the bundle algorithm for nonsmooth optimization, in: Mangasarian, Meyer and Robinson (eds.), Nonlinear Programming 4 (1981)

MIFFLIN, R., An algorithm for constrained optimization with semi-smooth functions, Mathematics of OR 2 (1977), 191-207

POLJAK, B. T., A general method for solving extremal problems, Soviet Math. Dokl. 174, 33 - 36 (1967)

WOLFE, P., A method of conjugate subgradients for minimizing non-differentiable functions, Math. Progr. Study 3 (1975), 145 - 173

ZOWE, J., Nondifferentiable optimization - a motivation and a short introduction into the subgradient and the bundle concept, NATO ASI Series, VOL. F15, Computational Math. Progr., Ed. by K. Schittkowski, 321 - 356 (1985)

STOCHASTIC PROGRAMMING WITH RECOURSE: UPPER BOUNDS AND MOMENT PROBLEMS - A REVIEW

P. Kall [1)]

I. Introduction

Given a set $X \subset \mathbb{R}^n$, a random vector ξ with range $\Xi \subset \mathbb{R}^k$, endowed with the Borel σ-algebra and a (induced) probability measure $\tilde{P}$, and a vector $c \in \mathbb{R}^n$, the stochastic program with recourse is usually stated as

$$\min_{x \in X} \{ f(x) + E_{\tilde{P}} Q(x,\xi) \} , \tag{1}$$

where $Q : X \times \Xi \to \mathbb{R}$ is the socalled recourse function and $E_{\tilde{P}}$ stands for the expectation with respect to the measure $\tilde{P}$.

In general the recourse function is assumed to be not only measurable with respect to ξ for any $x \in X$, but also convex in ξ for each $x \in X$, which implies Ξ to be assumed as a convex set. Throughout this paper we shall make this convexity assumption, except for one particular convex-concave case treated recently by K. Frauendorfer [8].

In particular in stochastic linear programming X is assumed to be a convex polyhedral set and the recourse function is defined by

$$Q(x,\xi) = \min \{ q'y | Wy = h(\xi) - T(\xi)x, y \geq 0 \} , \tag{2}$$

where $q \in \mathbb{R}^{n_1}$, W is a $m_1 \times n_1$ - matrix, and $h : \Xi \to \mathbb{R}^{m_1}$, $T : \Xi \to \mathbb{R}^{m_1 \times n}$ are assumed to be linear affine in ξ.

Assuming that W is a complete recourse matrix, i.e.

$$\left\{ \begin{array}{l} \{ t | t = Wy, y \geq 0 \} = \mathbb{R}^{m_1} \\ \text{and} \\ \{ u | W^T u \leq q \} \neq \emptyset , \end{array} \right. \tag{3}$$

it is well known that $Q(x,\cdot)$ is not only convex for all $x \in X$, but even proper, piecewise linear and convex, more precisely it is polyhedral and sublinear in $h(\xi) - T(\xi)x$ (see e.g. [12], [20]).

From the convexity of $Q(x,\cdot)$ we get for

$$\bar{\xi} = E_{\tilde{P}}\xi$$

by Jensen's inequality that

$$Q(x,\bar{\xi}) \leq E_{\tilde{P}} Q(x,\xi). \tag{4}$$

Hence we have a lower bound which requires only the evaluation of Q in $(x,\bar{\xi})$ -

1) University of Zurich, Institut für Operations Research und mathematische Methoden der Wirtschaftswissenschaften, Moussonstrasse 15, CH-8044 Zurich

being in general a very simple task compared to the evaluation of the multiple integral $E_{\tilde{P}}Q(x,\xi)$, the latter being usually too complicated to be carried out repeatedly in some iterative solution method for (1). Therefore, there is also a need for upper bounds being computable easily compared to the evaluation of $E_{\tilde{P}}Q(x,\xi)$. It turns out that most of the upper bounds proposed so far are derived by considering various (generalized) moment problems. In this paper we shall give a survey on these approaches. To make them more accessible, the underlying theoretical results on semi-infinite programming and generalized moment problems are reviewed first.

The important question for strategies how to improve the lower and upper bounds and to drive simultaneously x towards a solution of (1) goes beyond the aim of this paper and is not answered in general yet. The reader may find however some special proposals in this direction for instance in [14], [9], [7] and [1].

II. Semi-Infinite Programming and the Generalized Moment Problem

As reference for semi-infinite programming we have chosen the textbook of K. Glashoff and S.A. Gustafson [11] because it is very well written and does - for the statements of our interest - not need more than elementary real analysis and linear algebra.

For

S an arbitrary index set (in general an infinite set)
$c \in \mathbb{R}^n$
$a : S \to \mathbb{R}^n$, $b : S \to \mathbb{R}$ arbitrary functions

we may consider as primal problem the semi-infinite program

$$(P) \quad \min \{ c^T y \mid a(s)^T y \geq b(s) \quad \forall s \in S \} \tag{5}$$

and its dual problem, the generalized moment problem to find a positive finite discrete measure with values x_i at points $s_i \in S$ solving

$$(D) \quad \max \{ \sum_{i=1}^{q} b(s_i)x_i \mid \sum_{i=1}^{q} a(s_i)x_i = c,\ x_i \geq 0,\ s_i \in S,\ q \geq 1 \} . \tag{6}$$

For any pair of feasible solutions y of (P) and $\{ s_1,\dots,s_q; x_1,\dots,x_q \}$ of (D) we get immediately

$$y^T c = y^T \sum_{i=1}^{q} a(s_i)x_i = \sum_{i=1}^{q} [y^T a(s_i)]x_i \geq \sum_{i=1}^{q} b(s_i)x_i \tag{7}$$

and hence for the values v(P) and v(D) of (P) and (D), respectively, defined by

$$v(P) := \inf \{ c^T y \mid a(s)^T y \geq b(s) \quad \forall s \in S \} \tag{8}$$

$$v(D) := \sup \{ \sum_{i=1}^{q} b(s_i)x_i \mid \sum_{i=1}^{q} a(s_i)x_i = c,\ x_i \geq 0,\ s_i \in S,\ q \geq 1 \} , \tag{9}$$

the weak duality theorem

Lemma 1 [11] : $v(D) \le v(P)$.

The following statement allows in some sense for minimal representations of solutions of (D) and is a trivial extension of the socalled reduction theorem in [11].

Theorem 1

If $\{ s_1,\dots,s_q; x_1,\dots,x_q \}$ is feasible in (D), then there exist a subset $\{s_{i_1},\dots,s_{i_m}\} \subset \{s_1,\dots,s_q\}$ and scalars $\bar{x}_{i_1},\dots,\bar{x}_{i_m}$ such that $\{ s_{i_1},\dots,s_{i_m}; \bar{x}_{i_1},\dots,\bar{x}_{i_m} \}$ is feasible in (D), the set of vectors $\{ a(s_{i_j}) \mid \bar{x}_{i_j} > 0 \}$ is linearly independent, and, if $v(D) < +\infty$,

$$\sum_{j=1}^{m} b(s_{i_j})\bar{x}_{i_j} \ge \sum_{i=1}^{q} b(s_i)x_i \; .$$

Proof

Just rephrase the well-known fact in linear programming that feasibility implies the existence of basic solutions and boundedness yields the existence of an optimal basic solution. ■

Defining the set of vectors $a(S) := \{a(s) \mid s \in S\}$, the constraints of (D) involve the positive hull of $a(S)$, i.e. the set

$$M_n := \text{pos}\; a(S) = \{ z \mid z = \sum_{i=1}^{q} a(s_i)x_i ,\; x_i \ge 0,\; s_i \in S,\; q \ge 1 \} ,$$

a convex cone called the moment cone. Obviously holds

Lemma 2 [11] : (D) is feasible iff $c \in M_n$.

We may have a situation where (P) is solvable, (D) is feasible but not solvable, but at least the values of (P) and (D) coincide. There are even cases where $v(D) < v(P)$ (duality gap), as can be seen in [11]. To state conditions under which the solvability of (D) and (P), respectively, follow, it is convenient to reformulate (D) as follows:

Define

$$M_{n+1} := \text{pos}\; \tilde{a}(S)$$

where

$$\tilde{a}(s) = \begin{pmatrix} b(s) \\ a(s) \end{pmatrix} \in \mathbb{R}^{n+1} .$$

Then - for c fixed as in (6) - our generalized moment problem is obviously equivalent to

$$\text{(D)} \max \{ c_0 \mid \begin{pmatrix} c_0 \\ c \end{pmatrix} \in M_{n+1} \} . \qquad (10)$$

For (D) being solvable we trivially need that (D) is feasible and v(D) is bounded above. But these properties are not sufficient. Since we want to achieve that the value $v(D) = \sup \{ c_0 \mid \begin{pmatrix} c_0 \\ c \end{pmatrix} \in M_{n+1} \}$ is really attained in M_{n+1}, we need at least that the section $\begin{pmatrix} \zeta \\ c \end{pmatrix} \in M_{n+1}$ is closed to the right with respect to ζ. To facilitate the formulation we therefore may state immediately

Theorem 2 [11]

If (D) is feasible, $v(D) < +\infty$ and M_{n+1} closed, then (D) is solvable.

To apply this theorem we are left with the question when M_{n+1} can be guaranteed to be closed. We may expect that $a(\cdot)$, $b(\cdot)$ should be continuous implying that S is some topological space which we assume for simplicity to be a subset of $\mathbb{R}^k$. But this does not yet imply M_{n+1} to be closed.

Therefore we often find in the literature the

Assumption C : $S \subset \mathbb{R}^k$ is compact; $a(\cdot)$, $b(\cdot)$ are continuous.

As mentioned above we choose $S \subset \mathbb{R}^k$ just for simplicity; instead, the assumptions "S a compact metric space" or "S a compact Hausdorff space" are quite common, too. However, assumption C would also not yet yield M_{n+1} closed in general. Reminding to nonlinear programming and the duality statements (and Kuhn-Tucker conditions) discussed there, we may hardly expect to get through without any regularity assumption in semi-infinite programming. Hence we may try the natural extension of the familiar Slater condition for convex programs, which is called in [11]

Superconsistency of (P)

(P) is superconsistent if $\exists\, \tilde{y} : a(s)^T \tilde{y} > b(s) \quad \forall\, s \in S$.

Theorem 3 [11]

If (P) is superconsistent and assumption C holds, then M_{n+1} is closed.

Proof

$M_{n+1} = \text{pos}\, \tilde{a}\,(S) = \bigcup_{\lambda \geq 0} \lambda \cdot \text{co}\, \tilde{a}\,(S)$, where $\text{co}\, \tilde{a}\,(S)$ is the convex hull of $\tilde{a}\,(S)$. By C obviously $\tilde{a}\,(S)$ and hence $\text{co}\, \tilde{a}\,(S)$ are compact. With $z_\nu \in M_{n+1}$ and $z_\nu = \lambda_\nu y_\nu$, $y_\nu \in \text{co}\, \tilde{a}(S)$, $\lambda_\nu \geq 0$, let $\bar{z} = \lim_{\nu\to\infty} z_\nu = \lim_{\nu\to\infty} \lambda_\nu y_\nu$.

Then either $\{\lambda_\nu\}$ is bounded implying $\bar{z} \in M_{n+1}$, or $\{\lambda_\nu\}$ is unbounded and hence $\left\{\frac{1}{\lambda_\nu}\cdot(\lambda_\nu y_\nu)\right\} \subset \text{co}\, \tilde{a}\,(S)$ accumulates to 0, i.e. $0 \in \text{co}\, \tilde{a}\,(S)$.

This implies the existence of $s_i \in S$, $d_i \geq 0$, $i = 1,\dots,n+2$, such that $\sum_{i=1}^{n+2} \alpha_i = 1$ and $\sum_{i=1}^{n+2} \alpha_i (y^T a(s_i) - b(s_i)) = 0 \;\forall\, y \in \mathbb{R}^n$ in contradiction to the Slater condition. ■

Theorem 4 [11]

If

(i) assumption C holds,

(ii) (D) is feasible,

(iii) (P) is superconsistent,

then (D) is solvable and $v(P) = v(D)$.

Proof

By Theorem 3, Lemma 1 and Theorem 2, (D) is solvable with $\begin{pmatrix} c_0 \\ c \end{pmatrix} \in M_{n+1}$, where $c_0 = v(D)$. Hence $\begin{pmatrix} c_0+\varepsilon \\ c \end{pmatrix} \notin M_{n+1}$ and $\exists\, y \in \mathbb{R}^{n+1}$, $y \neq 0$, such that

$$y^T x \leq 0 < y^T \begin{pmatrix} c_0+\varepsilon \\ c \end{pmatrix} \quad \forall\, x \in M_{n+1}, \forall\, \varepsilon > 0 \,, \tag{11}$$

where $y_0 > 0$, since $\begin{pmatrix} c_0 \\ c \end{pmatrix} \in M_{n+1}$. Writing $y = \begin{pmatrix} y_0 \\ \hat{y} \end{pmatrix}$, it follows, observing $(b(s), a(s)^T)^T \in M_{n+1} \quad \forall\, s \in S$, that $\left(\frac{-\hat{y}}{y_0}\right)$ is feasible in (P) with

$$v(P) \leq -\frac{c^T\hat{y}}{y_0} < c_0 + \varepsilon = v(D) + \varepsilon \leq v(P) + \varepsilon \quad \forall\, \varepsilon > 0 \,. \quad ∎$$

From Theorem 2 and the construction in the proof of Theorem 4 follows

Corollary

If (D) is feasible, $v(D) < +\infty$ and M_{n+1} closed, then (P) is feasible and $v(D) = v(P)$.

Instead of (P) one can as regularity condition for (D) introduce

Superconsistency of (D)

(D) is superconsistent if $c \in \operatorname{int} M_n$ ($\operatorname{int} M_n$ the interior of M_n).

Then we get

Theorem 5 [11]

If

(i) (P) is feasible,

(ii) (D) is superconsistent,

then (P) is solvable and $v(D) = v(P)$.

Proof

By Lemma 1 we have $-\infty < v(D) \leq v(P) < +\infty$. Hence for $c_0 = v(D)$ $\begin{pmatrix} c_0 \\ c \end{pmatrix} \in \operatorname{cl} M_{n+1}$ (the closure of M_{n+1}). Since $\begin{pmatrix} c_0+\lambda \\ c \end{pmatrix} \notin M_{n+1}$ $\forall \lambda > 0$, there exist $(y_0, y^T) \neq 0^T$ such that

$$y_0(c_0+\lambda) + y^T c \leq 0 = y_0 c_0 + y^T c \leq y_0 z_0 + y^T z \quad \forall (z_0, z^T)^T \in M_{n+1} \,. \tag{12}$$

Hence $y_0 \cdot \lambda \leq 0$ $\forall \lambda > 0$. Since $y_0 = 0$ would yield a supporting hyperplane to M_n at $c \in \operatorname{int} M_n$, we have

$$y_0 < 0 \,. \tag{13}$$

By (12),(13) we get for $\tilde{y} = \frac{y}{-y_0}$ that $\tilde{y}^T a(s) \geq b(s)$ $\forall\, s \in S$ and hence $v(D) \leq v(P) \leq \tilde{y}^T c = c_0 = v(D)$. ∎

As we have seen so far, in semi-infinite programming the generalized moment problem comes up as a dual problem and is discussed in the frame of duality statements. However, the (generalized) moment problem has been considered much earlier in probability theory (see e.g. [19]) and in some other areas of applied mathematics (see e.g. [16]). The most general setting of the problem as discussed by J.H.B. Kemperman [18] reads as follows:

For

(S,Σ) a measurable space
$c \in \mathbb{R}^n$
$a : S \to \mathbb{R}^n$, $b : S \to \mathbb{R}$ measurable
$\mathcal{P} = \{ P \mid P$ a probability measure on Σ such that $\int |a_i(s)| P(ds) < \infty$,
$i=1,\dots,n$, and $\int |b(s)| P(ds) < \infty \}$

consider the problem

$(\tilde{D})$ $\quad v(\tilde{D}) := \sup \{ \int b(s)\, P(ds) \mid P \in \mathcal{P}, \int a(s)\, P(ds) = c \}$. (14)

The restriction to probability measures can easily be incorporated into our former formulation (6) of the moment problem by adding the function $a_0(s) \equiv 1$ with the moment $c_0 = 1$. But contrary to (6) the problem $(\tilde{D})$ is not restricted to measures with finite support and seems therefore to be more general. However according to Kemperman [18], Richter (1957) [21] and Rogosinski (1958) [23] have found

Theorem 6 [18, 23]
Let $f_1,\dots,f_N$ be integrable on the probability space (Ω,F,P). Then there exists a probability measure $\tilde{P}$ with finite support in Ω such that
$\int f_i(\omega)\, P(d\omega) = \int f_i(\omega)\, \tilde{P}(d\omega)$, $i=1,\dots,N$.
Even card (supp $\tilde{P}$) $\leq N+1$ may be achieved.

Proof
According to [23] the proof may be given as follows: For
$f(\omega) := (f_1(\omega),\dots,f_N(\omega))^T$
$\mathcal{C} := \{ c \in \mathbb{R}^N \mid \exists P \in \mathcal{P} : \int f(\omega)\, P(d\omega) = c \}$
$K := \operatorname{co} f(\Omega)$,
the statement $\mathcal{C} = K$ is shown by induction.
For N=1 it is obvious.
Using the assumption that for some

$N \geq 1 : \mathcal{C} = K$,

it is shown that for any supporting hyperplane
$H = \{ x \mid h^T x = h_0 \}$ to $K \subset \mathbb{R}^{N+1}$ follows

$K \cap H = \mathcal{C} \cap H$. (15)

Then it is easy to see that for $c \in \mathcal{C} \subset \mathbb{R}^{N+1}$ the case $c \notin \operatorname{cl} K$ is impossible, and for $c \in \operatorname{cl} K \setminus K$ there is a supporting hyperplane to K at c such that by (15) $c \in \mathcal{C} \cap H = K \cap H$. The second part of the theorem then follows by Carathéodory's Theorem. ∎

According to this theorem (D) turns out not to be more general than (D), i.e. we can restate $(\tilde{D})$ as

$(\tilde{D}) \qquad v(\tilde{D}) := \sup \{ \sum_{i=1}^{q} b(s_i)p_i \mid \sum_{i=1}^{q} a(s_i)p_i = c, \sum_{i=1}^{q} p_i = 1, p_i \geq 0, s_i \in S, q \geq 1 \} .$ (16)

Due to the additional constraints $\sum_{i=1}^{q} p_i = 1$ we get instead of Lemma 2 obviously.

Lemma 3 [18]

$(\tilde{D})$ is feasible iff $c \in K_n := \text{co } a(S)$.

Instead of Theorem 2 we get, defining

$K_{n+1} := \text{co } \tilde{a}(S)$

(where $\tilde{a}(s) = \begin{pmatrix} b(s) \\ a(s) \end{pmatrix} \in \mathbb{R}^{n+1}$),

and restating $(\tilde{D})$ as

$(\tilde{D}) \qquad v(\tilde{D}) := \sup \{ c_0 \mid \begin{pmatrix} c_0 \\ c \end{pmatrix} \in K_{n+1} \} ,$ (17)

obviously the formulation of

Theorem 7

If $(\tilde{D})$ is feasible, $v(\tilde{D}) < +\infty$ and K_{n+1} closed, then $(\tilde{D})$ is solvable.

However, compared to Theorem 3, the closedness of K_{n+1} does not require an assumption on the primal problem $(\tilde{P})$ corresponding to $(\tilde{D})$ since we have the obvious

Theorem 8

If C holds, then K_{n+1} is compact.

Proof

From S compact and $a(\cdot)$, $b(\cdot)$ continuous follows that $\tilde{a}(S)$ is compact, and hence $K_{n+1} = \text{co } \tilde{a}(S)$ is compact. $\blacksquare$

Remark

Since for the solvability of $(\tilde{D})$, given $-\infty < v(D) < +\infty$, it would be sufficient that $\{ \begin{pmatrix} \zeta \\ c \end{pmatrix} \} \subset K_{n+1}$ is closed to the right with respect to ζ, assumption C can be weakened such that S is compact and $b(\cdot)$ is only required to be upper semi-continuous (u.s.c.), as was remarked in [18].

The primal problem $(\tilde{P})$ corresponding to $(\tilde{D})$ reads now as

$(\tilde{P}) \qquad v(\tilde{P}) := \inf \{ y_0 + c^T y \mid y_0 + a(s)^T y \geq b(s) \quad \forall s \in S \} .$ (18)

Observe that now, if $(\tilde{P})$ is feasible at all, it is superconsistent; to see this, for any feasible (y_0, y) just increase y_0 by $\varepsilon > 0$.

By weak duality we have

$- \infty \leq v(\tilde{D}) \leq v(\tilde{P}) \leq + \infty .$

If $(\tilde{D})$ is consistent, i.e. $c \in K_n$, then we get

$-\infty < v(\tilde{D}) \leq v(\tilde{P}) \leq + \infty .$

In case that $v(\tilde{D}) = +\infty$, it follows trivially that $v(\tilde{D}) = v(\tilde{P})$, i.e. $(\tilde{P})$ is inconsistent. The question is whether for $-\infty < v(\tilde{D}) < +\infty$ also $v(\tilde{P}) = v(\tilde{D})$ holds. A positive answer can be given at least for $(\tilde{D})$ being superconsistent, which is closely related to Theorem 5, but more general.

<u>Theorem 9</u> [18]

For $c \in \text{int } K_n$ holds $v(\tilde{D}) = v(\tilde{P})$, i.e.

$$v(\tilde{D}) = \inf \{ y_0 + c^T y \mid y_0 + a(s)^T y \geq b(s) \quad \forall s \in S \} . \tag{19}$$

If $v(\tilde{D})<+\infty$, then the infimum in (19) is attained for some $(y_0,y^T)^T$, i.e. $(\tilde{P})$ is solvable. If $v(\tilde{D})<+\infty$ for all $c \in \text{int } K_n$, then $(\tilde{P})$ has for almost all $c \in \text{int } K_n$ a unique solution.

<u>Proof</u>

For $v(\tilde{D}) = +\infty$ Lemma 1 yields $v(\tilde{D}) = v(\tilde{P})$. For $v(\tilde{D})<+\infty$, we have again for $c_0 = v(\tilde{D})$ that $\begin{pmatrix} c_0 \\ c \end{pmatrix} \in \text{cl } K_{n+1}$ and that there exists a supporting hyperplane $H = \{ (x_0,x^T)^T \mid y_0x_0+y^Tx = \alpha \}$, $(y_0,y^T)\neq 0^T$, to K_{n+1} at $(c_0,c^T)^T$, i.e.

$$y_0(c_0+\lambda) + y^Tc \leq \alpha = y_0c_0 + y^Tc \leq y_0z_0 + y^Tz \quad \forall \begin{pmatrix} z_0 \\ z \end{pmatrix} \in K_{n+1} \tag{20}$$

for $\lambda>0$, and that, as in Theorem 5, $y_0<0$.

Then by (20) $(\eta_0 := \frac{\alpha}{y_0}, \eta^T := \frac{y^T}{-y_0})^T$ is feasible in $(\tilde{P})$,

$$\text{i.e. } \eta_0 + \eta^T a(s) \geq b(s) \quad \forall s \in S , \tag{21}$$

$$\text{and } v(\tilde{P}) \leq \eta_0 + c^T\eta = c_0 = v(\tilde{D}) \leq v(\tilde{P}) , \tag{22}$$

proving the first two parts of the theorem.

For $c \in \text{int } K_n$ with $v(\tilde{D})<+\infty$ denote $\psi(c) := v(\tilde{D}) = v(\tilde{P})$.

Obviously $\psi(c)$ is concave in c and by (21), (22)

$$\psi(c) = \eta_0 + \eta^T c , \tag{23}$$

$b(s) \leq \eta_0 + \eta^T a(s) \quad \forall s \in S$, implying $z_0 \leq \eta_0 + \eta^T z \quad \forall \begin{pmatrix} z_0 \\ z \end{pmatrix} \in K_{n+1}$

and hence for $\psi(z) := \sup \{\zeta \mid \begin{pmatrix} \zeta \\ z \end{pmatrix} \in K_{n+1}\}$

$$\psi(z) \leq \eta_0 + \eta^T z . \tag{24}$$

From (23),(24) follows $\psi(z) - \psi(c) \leq \eta^T(z-c)$, i.e. η is a subgradient of the concave function ψ in c. The last part of the theorem now follows from the fact that a concave function in $\mathbb{R}^n$ is differentiable almost everywhere. $\blacksquare$

<u>Theorem 10</u> [18]

For $c \in \text{int } K_n$ the conditions

i) $v(\tilde{D})$ is attained, i.e. $(\tilde{D})$ is solvable, and

ii) $\exists (\hat{y}_0,\hat{y}^T)^T$ feasible in $(\tilde{P})$ such that there exist $s_i \in S$, $i=1,\dots,q$ $(q\geq 1)$ for which $\hat{y}_0 + a(s_i)^T\hat{y} = b(s_i)$, $i=1,\dots,q$ and $c \in \text{co } \{a(s_i) \; ; \; i=1,\dots,q\}$

are equivalent.

<u>Proof</u>

<u>(i) $\Rightarrow$ (ii)</u>

Let $\{s_1,\dots,s_q\}$ be the support of the distribution solving $(\tilde{D})$, i.e. we have

$$P_i > 0 \; ; \; i=1,\dots,q \; ; \; \sum_{i=1}^{q} P_i = 1 \; , \; \sum_{i=1}^{q} a(s_i)P_i = c \; , \; v(\tilde{D}) = \sum_{i=1}^{q} b(s_i)P_i \; .$$

Since $c \in \text{int}\, K_n$, due to Theorem 9 $\exists (y_0, y^T)^T$ solving $(\tilde{P})$ and satisfying $y^0 + c^T y = v(\tilde{D})$. Hence

$$v(\tilde{P}) = y^0 + c^T y = y_0 + \sum_{i=1}^{q} P_i \; a(s_i)^T y$$

$$= \sum_{i=1}^{q} P_i (y_0 + a(s_i)^T y) \geq \sum_{i=1}^{q} P_i \; b(s_i) = v(\tilde{D})$$

which, due to $P_i > 0$, $i=1,\dots,q$ and $v(\tilde{P}) = v(\tilde{D})$, implies

$y_0 + a(s_i)^T y = b(s_i)$, $i=1,\dots,q$.

(ii) $\Rightarrow$ (i)

According to the assumptions $\exists \; p_i \geq 0$, $\sum_{i=1}^{q} p_i = 1$ such that

$c = \sum_{i=1}^{q} p_i \; a(s_i)$, i.e. $\{s_1,\dots,s_q \; ; \; p_1,\dots,p_q\}$ is feasible in $(\tilde{D})$. Hence

$v(\tilde{D}) \geq \sum_{i=1}^{q} p_i \; b(s_i) = \hat{y}_0 + \sum_{i=1}^{q} p_i \; a(s_i)^T \hat{y} = \hat{y}_0 + c^T \hat{y} \geq v(\tilde{P})$ and therefore, using weak duality, $(\hat{y}_0, \hat{y}^T)^T$ and $\{s_1,\dots,s_q \; ; \; p_1,\dots p_q\}$ solve $(\tilde{P})$ and $(\tilde{D})$ respectively. Observe that in this second part of the proof the assumption $c \in \text{int}\, K_n$ is not needed. ∎

III. Stochastic Programs with Recourse: Upper Bounds and Moment Problems

As stated in Chapter I the recourse function $Q : X \times \Xi \to \mathbb{R}$ is in many cases convex on Ξ for any $x \in X$. Since we focus on the problem to get upper bounds for $E_{\tilde{P}}Q(x,\xi)$ for any fixed $x \in X$, we may simplify the notation by considering a convex function $\psi : \Xi \to \mathbb{R}$ instead of $Q(\cdot,\cdot)$. There have been derived various upper bounds as solutions of moment problems depending on the assumptions on Ξ and on some special properties of ψ, on the probability distributions involved (dependence vs. independence) and on the required moment conditions.

III.1 Linear Moment Conditions

As mentioned in Chapter I we assume $S := \Xi \subset \mathbb{R}^k$ endowed with the σ-algebra Σ of Borel sets.

Case a)

$\Xi := [a,b] \subset \mathbb{R}^1$, moment condition $\int_a^b \xi P(d\xi) = \mu$, where $\mu = E_{\tilde{P}}\xi$ and $\tilde{P}$ the original measure in (1).

To get an upper bound for $E_{\tilde{P}}\psi(\xi)$, we may try to solve the moment problem (compare (14))

$$v(\tilde{D}) = \sup \{ \int_a^b \psi(\xi)P(d\xi) \mid P \in \mathcal{P}, \int_a^b \xi \, P(d\xi) = \mu \} \tag{25}$$

with the associated semi-infinite program (see (18))

$$\begin{aligned} v(\tilde{P}) &= \inf \{ y_0 + \mu y_1 \mid y_0 + \xi y_1 \geq \psi(\xi) \quad \forall \, \xi \in [a,b] \} \\ &= \inf \{ y_0 + \mu y_1 \mid y_0 + a y_1 \geq \psi(a), \, y_0 + b y_1 \geq \psi(b) \} , \end{aligned} \tag{26}$$

making use of the convexity of ψ. Since $a \leq \mu \leq b$, the solution of this latter LP may be found by solving the linear equations

$$\begin{aligned} y_0 + a y_1 &= \psi(a) \\ y_0 + b y_1 &= \psi(b) \end{aligned}$$

yielding $y_1 = \frac{\psi(b)-\psi(a)}{b-a}$, $y_0 = \psi(a) - a \frac{\psi(b)-\psi(a)}{b-a}$,

and hence, applying Theorem 10, $v(\tilde{P}) = v(\tilde{D}) = \frac{b-\mu}{b-a} \psi(a) + \frac{\mu-a}{b-a} \psi(b)$,

which shows that $\{\xi_1=a, \, \xi_2=b; \; P_1 = \frac{b-\mu}{b-a}, \, P_2 = \frac{\mu-a}{b-a}\}$ solves $(\tilde{D})$ independently of the particular choice of ψ. This socalled Edmundson-Madansky inequality

$$\int_a^b \psi(\xi)\tilde{P}(d\xi) \leq \frac{b-\mu}{b-a} \psi(a) + \frac{\mu-a}{b-a} \psi(b) \tag{27}$$

may be extended immediately to the case $\Xi := \mathop{X}_{\nu=1}^{k} [a_{0\nu}, a_{1\nu}]$, $k > 1$,

if $\tilde{P}$ is the product measure of probability measures $\tilde{P}_\nu$ on $[a_{0\nu}, a_{1\nu}]$, $\nu=1,\dots,k$, i.e. if the components of the random vector ξ are stochastically independent. However this leads to a moment problem with nonlinear moment conditions, namely

$$v(\tilde{D}) = \sup \{ \int_\Xi \psi(\xi)P(d\xi) \mid P \in \mathcal{P}, \int_\Xi (\prod_{i \in \Lambda} \xi_i) P(d\xi) = m_\Lambda, \, \Lambda \in \mathcal{B} \} , \tag{28}$$

where $\mathcal{B} := \{ \text{all subsets of } \{1,2,\dots,k\} \}$, $m_\Lambda := \int_\Xi (\prod_{i \in \Lambda} \xi_i) \tilde{P}(d\xi)$.

Problem (28) is solved by the distribution P^* on the vertices of Ξ defined as the product of the one-dimensional distributions P^*_ν solving (25) on $[a_{0\nu}, a_{1\nu}]$, $\nu=1,\dots,k$, as can be seen e.g. in [13] and [15].

Case b)

$\Xi := \text{co} \{v_1, v_2, \dots, v_r\}$, $v_i \in \mathbb{R}^k$, moment conditions $\int_\Xi \xi_i P(d\xi) = \mu_i \; (=E_{\tilde{P}} \xi_i)$.

In this case we get the moment problem

$$(\tilde{D}) \qquad \sup \{ \int_\Xi \psi(\xi)P(d\xi) \mid P \in \mathcal{P}, \int_\Xi \xi_i P(d\xi) = \mu_i, \, i=1,\dots,k \}$$

with the associated semi-infinite program.

$$(\tilde{P}) \qquad \inf \{ y_0 + \mu^T y \mid y_0 + \xi^T y \geq \psi(\xi) \quad \forall \, \xi \in \Xi \} .$$

Since ψ is convex, the constraints in $(\tilde{P})$ are satisfied for all ξ in the convex polyhedron Ξ iff they are satisfied for all $\xi \in \text{ext} \, \Xi$, the set of extreme points of Ξ. Hence $(\tilde{P})$ can be restated equivalently as

$(\tilde{P})\qquad \inf \{ y_0 + \mu^T y \mid y_0 + v_i^T y \geq \psi(v_i),\ i=1,\dots,r \}$.

Furthermore, ψ is u.s.c. on Ξ as a convex real-valued function on the (compact) polyhedron Ξ and hence, according to the remark after Theorem 8 and to Theorem 9, using LP duality we have

$$v(\tilde{D}) = \min \{ y_0 + \mu^T y \mid y_0 + v_i^T y \geq \psi(v_i),\ i=1,\dots,r \}$$

$$= \max \{ \sum_{i=1}^{r} \psi(v_i)p_i \mid \sum_{i=1}^{r} p_i = 1,\ \sum_{i=1}^{r} p_i v_i = \mu,\ p_i \geq 0 \},$$

i.e. $(\tilde{D})$ can be solved by a probability measure P^* with supp $P^* \subset$ ext Ξ, as was pointed out in [4].

In [10] and [2] - considering unbounded convex polyhedral sets and arbitrary closed convex sets, respectively - the assumptions have been generalized as follows:

Case c)

$\Xi \subset \mathbb{R}^k$ convex and closed, ψ l.s.c. proper convex, moment conditions $\int_\Xi \xi_i\, P(d\xi) = \mu_i\ (= E_{\tilde{P}}\, \xi_i)$. Hence we have the moment problem

$(\tilde{D})\qquad \sup \{ \int_\Xi \psi(\xi)P(d\xi) \mid P \in \mathcal{P},\ \int_\Xi \xi P(d\xi) = \mu \}$

and the associated semi-infinite program

$(\tilde{P})\qquad \inf \{ y_0 + \mu^T y \mid y_0 + \xi^T y \geq \psi(\xi)\quad \forall\, \xi \in \Xi \}$.

Since Ξ may be represented as Ξ = co(ext Ξ) + pos(ext-rcΞ), where rcΞ is the recession cone of Ξ, i.e. the largest closed convex cone such that Ξ + rc$\Xi \subset \Xi$, which may be represented as generated by a compact set ext-rcΞ (e.g. $\{0\} \cup [\mathrm{rc}\,\Xi \cap \{\xi \mid \|\xi\| = 1\}]$, it is easy to see that - observing the convexity of ψ - the constraints in $(\tilde{P})$ are equivalent to

$$y_0 + \xi^T y \geq \psi(\xi)\qquad \forall\, \xi \in \text{ext}\ \Xi \tag{29}$$

and

$$y_0 + (\hat{\xi} + \lambda\zeta)^T y \geq \psi(\hat{\xi} + \lambda\zeta)\quad \forall\, \hat{\xi} \in \text{ext}\ \Xi,\ \forall \lambda > 0, \text{ and } \forall\, \zeta \in \text{ext-rc}\Xi,$$

the latter being equivalent to

$$\zeta^T y \geq \frac{\psi(\hat{\xi} + \lambda\zeta) - \psi(\hat{\xi})}{\lambda} + \frac{\psi(\hat{\xi}) - [y_0 + \hat{\xi}^T y]}{\lambda}$$

for all $\zeta \in$ ext-rcΞ, $\lambda>0$ and any $\hat{\xi} \in$ ext Ξ. By the convexity of ψ and (29), the right-hand side is increasing in λ to the socalled recession function (see [2], [22])

$$\mathrm{rc}\ \psi(\zeta) = \sup_{\lambda>0} \frac{\psi(\hat{\xi} + \lambda\zeta) - \psi(\hat{\xi})}{\lambda},$$

which is known to be an extended real-valued sublinear function. Therefore we may state $(\tilde{P})$ equivalently as

$$(\tilde{P}) \quad \begin{cases} \inf\ [y_0 + \mu^T y\ \} \\ \text{s.t.} \\ \qquad y_0 + \xi^T y \geq \psi(\xi) \quad \forall\ \xi \in \text{ext}\ \Xi \\ \qquad\qquad \zeta^T y \geq \text{rc}\ \psi(\zeta) \quad \forall\ \zeta \in \text{ext-rc}\Xi\ . \end{cases}$$

Following (5) and (6) the associated moment problem is

$$(\bar{D}) \quad \begin{cases} \sup\ \{ \sum_{i=1}^{q} \psi(\xi_i) P_i + \sum_{j=1}^{r} \text{rc}\ \psi(\zeta_j) x_j\ \} \\ \text{s.t.} \quad \sum_{i=1}^{q} \xi_i P_i + \sum_{j=1}^{r} \zeta_j x_j = \mu \\ \qquad \sum_{i=1}^{q} P_i = 1 \\ \qquad P_i \geq 0,\ x_i \geq 0,\ \xi_i \in \text{ext}\ \Xi,\ \zeta_j \in \text{ext-rc}\Xi\ , \end{cases}$$

which amounts to determining a probability measure P on ext Ξ and a nonnegative measure ν on ext-rcΞ such that $\int_{\text{ext}\Xi} \xi\ P(d\xi) + \int_{\text{ext-rc}\Xi} \zeta\ \nu(d\zeta) = \mu$ and

$$\int_{\text{ext}\Xi} \psi(\zeta)\ P(d\zeta) + \int_{\text{ext-rc}\Xi} \text{rc}\ \psi(\zeta)\ \nu(d\zeta)$$

is maximized, yielding an upper bound for $\int_{\Xi} \psi(\xi)\ \tilde{P}(d\xi)$. This bound was developed in [2], where we find the remark that the bound is nontrivial only if $\text{rc}\ \psi(\zeta) < +\infty \quad \forall\ \zeta \in \text{ext-rc}\Xi$, which corresponds to the growth condition given in [10] for polyhedral Ξ.

However, the upper bound given by $(\bar{D})$ may also be worthless if ext Ξ is unbounded because the constraints in $(\tilde{P})$

$$y_0 + \xi^T y \geq \psi(\xi) \quad \forall\ \xi \in \text{ext}\ \Xi$$

may turn out to be infeasible, which at least for $\mu \in \text{int}\ \Xi$ means that $v(\bar{D}) = +\infty$, which follows from Theorem 9 and the obvious fact that $v(\tilde{D}) \leq v(\bar{D}) \leq v(\tilde{P})$. A simple example of this kind is

$$\Xi := \{\ \xi \in \mathbb{R}^2 \mid \xi_1 \geq 1,\ \xi_2 \geq \frac{1}{\xi_1}\ \}\ ,\ \psi(\xi) := \xi_1 + \frac{1}{\xi_2^2}\ .$$

Then,

$$\text{ext}\ \Xi = \{\ \xi \in \mathbb{R}^2 \mid \xi_1 \geq 1,\ \xi_2 = \frac{1}{\xi_1}\ \} \text{ is unbounded,}$$

$$\text{rc}\ \ \Xi = \{\ \zeta \in \mathbb{R}^2 \mid \zeta\ \ \geq 0\ \}\ ,\ \text{i.e. ext-rc}\ \Xi = \{(1,0),(0,1)\}.$$

$$\text{rc}\ \psi(\zeta) = \zeta_1 \quad \forall\ \zeta \in \text{rc}\ \Xi\ .$$

Hence the recession function is bounded on ext-rcΞ, but the constraints of $(\tilde{P})$

$$\begin{aligned} y_0 + \xi_1 y_1 + \frac{1}{\xi_1} y_2 &\geq \xi_1 + \xi_1^2 \quad \forall\ \xi_1 \geq 1 \\ y_1 &\geq 1 \\ y_2 &\geq 0 \end{aligned}$$

obviously cannot be satisfied. So for $\mu \in \text{int}\ \Xi$, e.g. $\mu = (2,2)$, we only get the upper bound $v(\bar{D}) = +\infty$.

III.2 Nonlinear Moment Conditions

As we have mentioned in Case a) above, the Edmundson-Madansky inequality can easily be generalized to an interval in $\mathbb{R}^k$ (which may be assumed to be the unit cube without loss of generality) if the components of ξ are stochastically independent. In [7] Frauendorfer has shown that a similar result also exists for the dependent case.

Case d)

$$\Xi = \mathop{X}_{i=1}^{k} [0,1], \text{ moment conditions}$$

$$\int_\Xi (\prod_{i\in\Lambda} \xi_i) P(d\xi) = m_\Lambda, \ \Lambda \in B,$$

where B := { all subsets of {1,2,...,k} }

$$m_\Lambda := \int_\Xi (\prod_{i\in\Lambda} \xi_i)\tilde{P}(d\xi) .$$

Here we have the moment problem

$$(\tilde{D}) \quad v(\tilde{D}) = \sup \{ \int_\Xi \psi(\xi)P(d\xi) \mid \int_\Xi (\prod_{i\in\Lambda} \xi_i)P(d\xi) = m_\Lambda, \ \Lambda \in B \}$$

with the associated semi-infinite program

$$(\tilde{P}) \quad v(\tilde{P}) = \inf \{ y_0 + \sum_{\Lambda\in B} m_\Lambda y_\Lambda \mid y_0 + \sum_{\Lambda\in B} (\prod_{i\in\Lambda} \xi_i) y_\Lambda \geq \psi(\xi), \ \xi \in \Xi \} .$$

Requiring the constraints in $(\tilde{P})$ only to be satisfied in the vertices of Ξ yields the LP

$$(\bar{P}) \quad v(\bar{P}) = \inf \{ y_0 + \sum_{\Lambda\in B} m_\Lambda y_\Lambda \mid y_0 + \sum_{\Lambda\in B} (\prod_{i\in\Lambda} \hat{\xi}_i) y_\Lambda \geq \psi(\hat{\xi}), \ \hat{\xi} \in \text{ext } \Xi \}$$

with its dual LP

$$(\bar{D}) \quad v(\bar{D}) = \sup \{ \int_{\text{ext}\Xi} \psi(\hat{\xi})P(d\hat{\xi}) \mid \int_{\text{ext}\Xi} (\prod_{i\in\Lambda} \hat{\xi}_i)P(d\hat{\xi}) = m_\Lambda, \ \Lambda \in B \} .$$

Obviously then holds

$$v(\bar{P}) = v(\bar{D}) \leq v(\tilde{D}) .$$

For any $\xi \in \Xi$ determine a probability measure $Q(\xi,\cdot)$: ext $\Xi \to [0,1]$ satisfying

$$\int_{\text{ext}\Xi} (\prod_{i\in\Lambda} \hat{\xi}_i) Q(\xi,d\hat{\xi}) = \prod_{i\in\Lambda} \xi_i, \ \Lambda \in B . \tag{30}$$

This system of linear equations has a regular matrix, as shown in [13], and hence a unique solution $Q(\xi,\cdot)$ which then obviously is continuous in $\xi \in \Xi$. Therefore for any feasible P in $(\tilde{D})$ we get from (30)

$$m_\Lambda = \int_\Xi (\prod_{i\in\Lambda} \xi_i) P(d\xi)$$

$$= \int_\Xi (\int_{ext\Xi} (\prod_{i\in\Lambda} \hat{\xi}_i) Q(\xi, d\hat{\xi})) P(d\xi)$$

$$= \int_{ext\Xi} (\prod_{i\in\Lambda} \hat{\xi}_i) \hat{P}(d\hat{\xi}) \text{ ; where}$$

$$\hat{P}(\cdot) = \int_\Xi Q(\xi, \cdot) P(d\xi). \tag{31}$$

This shows that $\hat{P}$ is a probability measure on ext Ξ feasible in $(\tilde{D})$. For its objective value we get by the convexity of ψ

$$\int_{ext\Xi} \psi(\hat{\xi}) \hat{P}(d\hat{\xi}) = \int_\Xi \int_{ext\Xi} \psi(\hat{\xi}) Q(\xi, d\hat{\xi}) P(d\xi)$$

$$\geq \int_\Xi \psi(\xi) P(d\xi) ,$$

since (30) also includes $\int_{ext\Xi} \hat{\xi} Q(\xi, d\hat{\xi}) = \xi$.

Therefore,

$$v(\bar{D}) \geq v(\tilde{D}) \text{ , implying } v(\bar{D}) = v(\tilde{D}) .$$

Since the equality constraints of $(\bar{D})$ are uniquely solvable (for feasible moment conditions), the solution of $(\bar{D})$ is independent of ψ and is determined explicitly in [7].

Recently upper and lower bounds have been derived in [8] for a convex-concave function ψ as coming up in stochastic linear programs with random objectives $q(\eta)^T y$ in (2), $q : \Theta \to \mathbb{R}^{n_1}$ assumed to be linear affine in $\eta \in \Theta$. More precisely: Let Ξ and Θ be cubes in $\mathbb{R}^k$ and $\mathbb{R}^l$, respectively. Assume that

$$\psi : \Xi \times \Theta \to \mathbb{R} \text{ is convex in } \xi \text{ and concave in } \eta ,$$

and that with probability measures $\tilde{P}$ on Ξ and $\tilde{Q}$ on Θ we have the measure $\tilde{P} \times \tilde{Q}$ on $\Xi \times \Theta$, i.e. η is stochastically independent of ξ.

Then by using (31) we get

$$\int_\Xi \psi(\xi,\eta) \tilde{P}(d\xi) \leq \int_{ext\Xi} \psi(\hat{\xi},\eta) \hat{P}(d\hat{\xi}) \quad \forall \eta \in \Theta .$$

For any $\hat{\xi} \in$ ext Ξ and an arbitrary fixed $\tilde{\eta} \in \Theta$ from the concavity of ψ in η we have

$$\psi(\hat{\xi},\eta) \leq \psi(\hat{\xi},\tilde{\eta}) + (\eta-\tilde{\eta})^T w(\hat{\xi},\tilde{\eta}) ,$$

where $w(\hat{\xi},\tilde{\eta})$ is some subgradient of $\psi(\hat{\xi},\cdot)$ at $\tilde{\eta}$. Hence we get

$$\int_\Theta \int_\Xi \psi(\xi,\eta) \tilde{P}(d\xi) \tilde{Q}(d\eta) \leq \int_\Theta \int_{ext\Xi} \psi(\hat{\xi},\eta) \hat{P}(d\hat{\xi}) \tilde{Q}(d\eta)$$

$$\leq \int_{ext\Xi} \psi(\hat{\xi},\tilde{\eta}) \hat{P}(d\hat{\xi}) + (\bar{\eta}-\tilde{\eta})^T \int_{ext\Xi} w(\hat{\xi},\tilde{\eta}) \hat{P}(d\hat{\xi}) ,$$

Due to the special geometrical shape of epi ψ and the epigraphs of

$$h_y(\xi) = y_0 + y^T \begin{pmatrix}\xi \\ \phi(\xi)\end{pmatrix}, \quad (y_0, y^T)^T \text{ feasible in } (\tilde{P}),$$

$(\tilde{P})$ is solved explicitly in [3] yielding $\{ \xi^1, \dots, \xi^{k+1}; P_1, \dots, P_{k+1} \}$ feasible in $(\tilde{D})$, $(\hat{y}_0, \hat{y}^T)^T$ feasible in $\tilde{P}$ such that applying Theorem 10

$$v(\tilde{D}) = \sum_{i=1}^{k+1} P_i \, \psi(\xi^i) = \hat{y}_0 + q \, \hat{y}_{k+1} \, .$$

A similar situation will arise in

Case f)

$\Xi := \mathbb{R}^k$, $\psi : \Xi \to \mathbb{R}$ positive conical,

moment conditions $\int_\Xi \xi \, P(d\xi) = \mu \; (= E_{\tilde{P}}\xi)$,

$$\int_\Xi \xi_i^2 \, P(d\xi) = v_i \quad (= \int_\Xi \xi_i^2 \, P(d\xi)) \, .$$

By ψ being positive conical we understand that epi ψ is a convex polyhedral cone in $\mathbb{R}^{k+1}$ pointed at the origin, and $\psi(\xi) > 0 \;\; \forall \, \xi \neq 0$. To get an upper bound for $E_{\tilde{P}}\psi(\xi)$, we consider the moment problem

$$(\tilde{D}) \quad v(\tilde{D}) = \sup \{ \int_\Xi \psi(\xi) \, P(d\xi) \mid \int_\Xi \xi \, P(d\xi) = \mu, \; \int_\Xi \xi_i^2 \, P(d\xi) = v_i, \; i=1,\dots,k \}$$

with the associated semi-infinite program

$$(\tilde{P}) \quad v(\tilde{P}) = \inf \{ y_0 + y^T\mu + \eta^T v \mid y_0 + y^T\xi + \sum_{i=1}^{k} \eta_i \xi_i^2 \geq \psi(\xi), \; \xi \in \Xi \} \, .$$

The assumptions on ψ imply obviously that $\eta_i > 0 \;\; \forall i$ for any feasible solution of $(\tilde{P})$. Hence we may restate the constraints of $(\tilde{P})$ by using

$$\psi(\xi) = \max_{1 \leq j \leq J} \beta^{(j)T}\xi, \quad \beta^{(j)} \in \mathbb{R}^k$$

as explicit representation of ψ, in the form

$$y_0 + \sum_{i=1}^{k} \eta_i \left(\xi_i^2 + \frac{y_i - \beta_i^{(j)}}{\eta_i} \, \xi_i \right) \geq 0 \quad \forall \, \xi \in \Xi \, , \; 1 \leq j \leq J$$

$$\eta_i > 0 \quad \forall i \quad \text{which is equivalent to}$$

$$\sum_{i=1}^{k} \left[\eta_i \left(\xi_i + \frac{y_i - \beta_i^{(j)}}{2\eta_i} \right)^2 - \frac{1}{4} \, \frac{(y_i - \beta_i^{(j)})^2}{\eta_i} \right] \geq -y_0, \quad \forall \, \xi \in \Xi \, , \; 1 \leq j \leq J$$

$$\eta_i > 0 \quad \forall i \, ;$$

obviously these constraints are satisfied for all $\xi \in \Xi$ iff

$$\sum_{i=1}^{k} \frac{1}{4} \, \frac{(y_i - \beta_i^{(j)})^2}{\eta_i} \leq y_0 \, , \quad 1 \leq j \leq J \, .$$

Hence we have to solve the fractional program

where $\bar{\eta} = E_{\tilde{Q}}\eta$. Obviously the best choice for $\tilde{\eta}$ (yielding the lowest upper bound) is $\tilde{\eta} := \bar{\eta}$, i.e. applying Jensen's inequality with respect to η. The lower bound is derived analogously.

In [3] Dulá considered the following situation:

Case e)

$\Xi := \mathbb{R}^k$, $\psi : \Xi \to \mathbb{R}$ a simplicial function,

moment conditions $\int_\Xi \xi \, P(d\xi) = \mu \; (= E_{\tilde{P}}\xi)$,

$$\int_\Xi \Big(\sum_{i=1}^k \xi_i^2\Big) P(d\xi) = q \;\Big(= E_{\tilde{P}} \Big(\sum_{i=1}^k \xi_i^2\Big)\Big) .$$

The motivation for the assumption on ψ to be simplicial, i.e. ψ is sublinear in ξ and the level sets of ψ are simplices in $\mathbb{R}^k$ containing the origin, stems from the fact mentioned in Chapter I, that the recourse function defined by (2) is sublinear and polyhedral in $h(\xi)-T(\xi)x$, considering the special case $h(\xi) \equiv \xi$, $T(\xi) \equiv T$ (constant) and bounding the recourse function from above by a simplicial function ψ. Under the assumptions

$$\mu = 0 \tag{32}$$

$$\tilde{P} \text{ nondegenerate, i.e. } E_{\tilde{P}}\, \xi_i^2 > \mu_i^2 \tag{33}$$

the moment problem

$$(\tilde{D}) \quad v(\tilde{D}) = \sup \Big\{ \int_\Xi \psi(\xi)\, P(d\xi) \;\Big|\; \int_\Xi \xi\, P(d\xi) = \mu = 0,\; \int_\Xi \sum_{i=1}^k \xi_i^2\, P(d\xi) = q \Big\}$$

is considered in [3] to yield an upper bound for $\int_\Xi \psi(\xi)\, \tilde{P}(d\xi)$ and hence for the expected recourse. The associated semi-infinite program is

$$(\tilde{P}) \quad v(\tilde{P}) = \inf \Big\{ y_0 + q \cdot y_{k+1} \;\Big|\; y_0 + \sum_{i=1}^k y_i\xi_i + y_{k+1} \sum_{i=1}^k \xi_i^2 \ge \psi(\xi),\; \xi \in \Xi \Big\} .$$

Since ψ is simplicial and hence represented as the maximum of n+1 linear homogenous functions, it may clearly be majorized by a quadratic function, i.e. $(\tilde{P})$ is feasible. Furthermore, $y_{k+1} > 0$ for any feasible solution of $(\tilde{P})$, since it is impossible to majorize a simplicial function by a linear or a concave quadratic function.

Finally $\begin{pmatrix}\mu\\ q\end{pmatrix} \in \operatorname{int} K_{k+1}$ by assumption (33), since $K_{k+1} = \operatorname{epi}\psi$, the epigraph of $\phi(\xi) = \sum_{i=1}^k \xi_i^2$, and $q = E_{\tilde{P}}\phi(\xi) > (E_{\tilde{P}}\xi)$ by (33), showing that

$$\begin{pmatrix}E_{\tilde{P}}\xi\\ q\end{pmatrix} = \begin{pmatrix}\mu\\ q\end{pmatrix} \in \operatorname{int}\operatorname{epi}\phi .$$

Hence by Theorem 9

$$v(\tilde{D}) = \min \Big\{ y_0 + q\, y_{k+1} \;\Big|\; y_0 + \sum_{i=1}^k y_i\xi_i + y_{k+1} \sum_{i=1}^k \xi_i^2 \ge \psi(\xi),\; \xi \in \Xi \Big\} .$$

$$(\tilde{P}) \quad v(\tilde{P}) = \inf \{ y_0 + y^T\mu + \eta^T v \mid \sum_{i=1}^{k} \frac{1}{4} \frac{(y_0-\beta_i^{(j)})^2}{\eta_i} \le y_0 \quad \forall j \ , \ \eta > 0 \} \ .$$

By the nondegeneracy assumption of Case (e) on $\tilde{P}$ we may ascert again the applicability of Theorem 9 such that then $v(\tilde{D}) = v(\tilde{P})$.

Problems of this kind were discussed by Dupačová, e.g. in [5], [6]. Their solution is in general apparently harder than the problem of Case e).

Finally we might point out first that the discrete distributions determined through moment problems to yield upper bounds for $E_{\tilde{P}}\psi(\xi)$ as described in this paper are distinct in the following respect:

Some of them are independent of the particular convex function ψ, as in Cases a) and d), others depend essentially on ψ, as in Cases b), c), e) and f).

This fact may be of interest in applying these variants in the course of a solution method for stochastic programs, because the recourse function is changing with x. And second, it seems not to be clear yet which moment conditions are in which particular stochastic programming situations the most effective to give tight or at least good upper bounds.

References

[1] Birge J.R. and Wets R.J.-B.: Designing Approximation Schemes for Stochastic Optimization Problems, in Particular for Stochastic Programs with Recourse. Mathematical Programming Study 27 (1986) 54-102

[2] Birge J.R. and Wets R.J.-B.: Computing Bounds for Stochastic Programming Problems by Means of a Generalized Moment Problem. Mathematics of OR 12 (1987) 149-162

[3] Dulá J.H.: An Upper Bound on the Expectation of Sublinear Functions of Multivariate Random Variables. Preprint, CORE (1987)

[4] Dupačová J: Minimax Stochastic Programs with Nonconvex Nonseparable Penalty Functions. In: Prékopa A. (ed.): Colloquia Mathematica Societatis Janós Bolyai 12, Progress in Operations Research, Eger 1974. North-Holland (1976) 303-316

[5] Dupačová J.: Minimax Stochastic Programs with Nonseparable Penalties. In: Iracki K., Malanowski K., Walukiewicz S. (eds.): Optimization Techniques, Part I. LN Control and Information Sci. 22 (1980) 157-163

[6] Dupačová J.: The Minimax Approach to Stochastic Programming and an Illustrative Application. Stochastics 20 (1987) 73-88

[7] Frauendorfer K.: Solving SLP Recourse Problems with Arbitrary Multivariate Distributions - The Dependent Case. To appear in Mathematics of OR

[8] Frauendorfer K.: Solving SLP Recourse Problems: The Case of Stochastic Technology Matrix, RHS & Objective. Manuscript, Institut für OR, University of Zurich (1987)

[9] Frauendorfer K. and Kall P.: A Solution Method for SLP Recourse Problems with Arbitrary Multivariate Distributions - The Independent Case. To appear in Problems of Control and Information Theory (1988)

[10] Gassmann H. and Ziemba W.T.: A Tight Upper Bound for the Expectation of a Convex Function of a Multivariate Random Variable. Mathematical Programming Study 27 (1986) 39-53

[11] Glashoff K. and Gustafson S.A.: Einführung in die Lineare Optimierung. Wissenschaftliche Buchgesellschaft Darmstadt (1978)

[12] Kall P.: Stochastic Linear Programming. Springer (1976)

[13] Kall P.: Stochastic Programs with Recourse: An Upper Bound and the Related Moment Problem. ZOR 31 (1987) A 119 - A 141

[14] Kall P., Frauendorfer K. and Ruszczynski A.: Approximation Techniques in Stochastic Programming. Manuscript, Institut für OR, University of Zurich (1984); to appear: Ermoliev Y.M. and Wets R.J.-B. (eds.): Stochastic Programming, Springer

[15] Kall P. and Stoyan D.: Solving Stochastic Programming Problems with Recourse Including Error Bounds. Mathematische Operationsforschung und Statistik, Series Optimization 13 (1982) 431-447

[16] Karlin S. and Studden W.J.: Tschebycheff Systems: With Applications in Analysis and Statistics. Interscience Publishers (1966)

[17] Karr A.F.: Extreme Points of Certain Sets of Probability Measures, with Applications. Mathematics of OR 8 (1983) 74-85

[18] Kemperman J.H.B.: The General Moment Problem, a Geometric Approach. Ann. Math. Statist. 39 (1968) 93-122

[19] Kreĭn M.G. and Nudel'man A.A.: The Markov Moment Problem and Extremal Problems. Translations of Math. Mon., Vol. 50, AMS (1977)

[20] Marti K.: Entscheidungsprobleme mit linearem Aktionen- und Ergebnisraum. Z. Wahrsch. verw. Geb. 23 (1972) 133-147

[21] Richter H.: Parameterfreie Abschätzung und Realisierung von Erwartungswerten. Blätter Dt. Ges. Versicherungsmath. 3 (1957) 147-161

[22] Rockafellar R.T.: Convex Analysis. Princeton University Press (1970)

[23] Rogosinski W.W.: Moments of Nonnegative Mass. Proc. Roy. Soc. London A 245 (1958) 1-27

ON STRONGLY STABLE LOCAL MINIMIZERS IN NONLINEAR PROGRAMS

Diethard Klatte [1]

Abstract: We give sufficient conditions for strong stability (in Kojima's sense) of local minimizers for nonlinear programs. It turns out that several results which are known for programs with twice differentiable data may be naturally extended to the case that the data are differentiable functions having a locally Lipschitzian gradient mapping, without being of the class C^2. The local minimizers being under consideration have to satisfy the Mangasarian-Fromovitz constraint qualification.

1. Introduction

Given a metric space T and functions $f_i: R^n \times T \longrightarrow R$ $(i=0,1,\dots,m)$, we consider the following optimization problem that depends on a parameter $t \in T$:

$$P(t): \qquad \min_x \left\{ f_o(x,t) \;/\; x \in M(t) \right\} \quad , \; t \in T,$$

where the multifunction $M: T \rightrightarrows R^n$ is defined by

$$M(t) := \left\{ x \in R^n \;/\; f_i(x,t)=0,\; i=1,\dots,p;\; f_j(x,t) \leqq 0,\; j=p+1,\dots,m \right\}$$

for all $t \in T$. Throughout the paper we suppose that for some open set $D \subset R^n$, for each $i \in \{0,1,\dots,m\}$ and for each $t \in T$,

$f_i(\cdot,t)$ is Fréchet differentiable on D,
f_i and $\nabla_x f_i(\cdot,\cdot)$ are continuous on $D \times T$,
and $\nabla_x f_i(\cdot,t)$ is locally Lipschitzian on D,

where $\nabla_x f_i(x,t)$ denotes the gradient of $f_i(\cdot,t)$ at x.
For fixed $t \in T$, we shall say that $(x,u) \in D \times R^m$ is a Kuhn-Tucker point of P(t), if (x,u) belongs to the solution set K(t) of the system

$$\nabla_x f_o(x,t) + \sum_{i=1}^{m} u_i \; \nabla_x f_i(x,t) = 0 \;, \quad x \in M(t),$$

$$u_j f_j(x,t) = 0 \quad , \quad u_j \geq 0 \quad , \; (j=p+1,\dots,m) \quad .$$

Each point $x \in D$ satisfying $(x,u) \in K(t)$ for some $u \in R^m$ is said to be a stationary solution of P(t), in symbols: $x \in S(t)$. The set of Lagrange multiplier vectors associated with x and t is denoted by LM(x,t), i.e. $LM(x,t) := \{ u \in R^m \;/\; (x,u) \in K(t) \}$. A point $z \in M(t) \cap D$ is called a local minimizer (strict local minimizer) of P(t) if there is some

1) Pädagogische Hochschule Halle, Kröllwitzer Straße 44, DDR - 4050 Halle/Saale

neighborhood U of z such that $f_0(z,t) \leq f_0(x,t)$ for all $x \in M(t) \cap U$ ($f_0(z,t) < f_0(x,t)$ for all $x \in M(t) \cap U$, $x \neq z$).

In this paper, the notion of a strongly stable stationary solution plays a central role. Let $B(y,r)$ and $\mathring{B}(y,r)$ denote the closed and the open r-neighborhood of y, respectively, where we use the same notation no matter whether $y \in R^n$ or $y \in T$. Adapting Kojima's definition [10] to the parametric problem P(t), $t \in T$, we shall say that a stationary solution x^0 of the problem $P(t^0)$ for fixed $t=t^0$ is <u>strongly stable</u> if for some real number $r > 0$ and each $\varepsilon \in (0,r]$, there exists a real number $\delta = \delta(\varepsilon)$ such that whenever $t \in \mathring{B}(t^0,\delta)$, $B(x^0,\varepsilon)$ contains a stationary solution of the problem P(t) which is unique in $B(x^0,r)$. A local minimizer which is also a strongly stable stationary solution will briefly be called a strongly stable local minimizer.

The concept of strong stability has essentially been used in homotopy methods, multi-level methods and statements on local convergence in nonlinear optimization, cf., for example, [5, 7, 10, 13]. It has been introduced and developed by Kojima [10] for optimization problems with twice differentiable data. We note that, in this case, strong stability is closely related to the concept of strong regularity in Robinson's [14] sense, provided that the corresponding stationary solution satisfies the linear independence constraint qualification (cf. [7]).

In the case of non-C^2 or nondifferentiable data there are several approaches to the study of qualitative stability (persistence and continuity of Kuhn-Tucker points, stationary solutions and local minimizers under perturbations of the problem data) or quantitative stability (estimates for Kuhn-Tucker points and local minimizers of perturbed programs), but without aiming at the s t r o n g stability, we refer, e.g., to Auslender [1], Klatte [8], Kummer [11, 12], Robinson [16].

Theorem 1, the main statement of the present paper, will extend a result of Kojima on strong stability of local minimizers for C^2-programs (namely, the if-part of Th. 7.2 and some version of Th. 8.2 in [10]) to the problem P(t), $t \in T$, in which twice differentiability of the data is not supposed. We shall assume that the Mangasarian-Fromovitz constraint qualification (CQ) holds w.r. to the stationary point being under consideration. As a consequence of Theorem 1, we shall give a non-C^2 version of a theorem which characterizes local minimizers of two-level optimization problems, it generalizes Th. 3.1 in [7].

In what follows we use the symbols $C^2(D)$ ($C^{1,1}(D)$) to denote the class of functions $f: R^n \longrightarrow R$ which are twice differentiable (once differentiable with a locally Lipschitzian gradient mapping) on D.

By $\nabla^2 f(x)$ we denote the Hessian matrix of $f \in C^2(D)$ at $x \in D$. The linear space of (n,n)-matrices is identified with $R^{n \times n}$. Each $x \in R^k$ is considered to be a column vector, $x^T y$ is the scalar product of $x,y \in R^k$. If X,Y and X_i $(i \in I)$ are subsets of R^k then we write (with $\beta \in R$)

$$\beta X + Y := \{\beta x + y \,/\, x \in X,\, y \in Y\},\quad \sum_{i \in I} X_i := \Big\{\sum_{i \in I} x^i \,/\, x^i \in X_i,\, i \in I\Big\},$$

and conv X (ri X) denote the convex hull (the relative interior) of X.

2. Strong stability of local minimizers

By the general suppositions on P(t), $t \in T$, the functions $f_i(\cdot,t)$ $(i \in \{0,1,\dots,m\},\ t \in T)$ belong to the class $C^{1,1}(D)$. We recall that for each function $f \in C^{1,1}(D)$ the mapping ∇f is almost everywhere on D differentiable, by Rademacher's theorem. Here ∇f denotes the gradient of f. Hence, it is reasonable to define a generalized Hessian matrix of $f \in C^{1,1}(D)$ via the Clarke generalized derivative of the mapping ∇f, as it has been done by Hiriart-Urruty et al. [6]: The set of (n,n)-matrices

$$\partial^2 f(x^o) := \operatorname{conv}\{M \,/\, \exists\, x^k \to x^o:\ x^k \in E_f\ (\forall k) \text{ and } \nabla^2 f(x^k) \to M\}$$

is called the generalized Hessian matrix of $f \in C^{1,1}(D)$ at $x^o \in D$, where E_f is the set of all points x for which $\nabla^2 f(x)$ exists. $\partial^2 f(x^o)$ is a nonempty compact convex subset of $R^{n \times n}$. If f is twice continuously differentiable at x^o then $\partial^2 f(x^o) = \{\nabla^2 f(x^o)\}$. Now we summarize some further properties of generalized Hessian matrices (cf. Clarke [4, § 2.6], Hiriart-Urruty et al. [6, §2]). Let $x^o \in D$ and $f,g \in C^{1,1}(D)$. Then it holds:

(E 1) The multifunction $\partial^2 f(\cdot)$ is locally bounded and closed at x^o.

(E 2) For any $h \in R^n$, $\partial^2(f+g)(x^o)h \subset \partial^2 f(x^o)h + \partial^2 g(x^o)h$, where for $W \subset R^{n \times n}$, $h \in R^n$: $Wh := \{Mh \,/\, M \in W\}$.

(E 3) If $a,b \in D$ and $\operatorname{conv}\{a,b\} \subset D$, then there exist $c \in \operatorname{ri}\operatorname{conv}\{a,b\}$ and $M \in \partial^2 f(c)$ such that $f(b) = f(a) + (b-a)^T \nabla f(a) + \frac{1}{2}(b-a)^T M (b-a)$ (second-order Taylor expansion of f at a).

Given $(x,u,t) \in D \times R^m \times T$ we put

$$I_2(x,t) := \{j \in \{p+1,\dots,m\} / f_j(x,t) = 0\},\quad I(x,t) := \{1,\dots,p\} \cup I_2(x,t), \tag{2.1}$$

$$I_2^+(u) := \{j \in \{p+1,\dots,m\} /\ u_j > 0\},\quad I^+(u) := \{1,\dots,p\} \cup I_2^+(u), \tag{2.2}$$

$$W^+(x,u,t) := \{h \in R^n \,/\, h^T \nabla_x f_i(x,t) = 0,\ i \in I^+(u)\}, \tag{2.3}$$

$$l(x,u,t) := f_o(x,t) + \sum_{i \in I^+(u)} u_i f_i(x,t). \tag{2.4}$$

If (x,u) is a Kuhn-Tucker point of P(t), then we obviously have

$$I_2^+(u) \subset I_2(x,t) \qquad \text{and} \qquad I^+(u) \subset I(x,t).$$

In the following $\partial^2_x l(x,u,t)$ denotes the generalized Hessian matrix of the function $l(\cdot,u,t)$ at x for fixed $(u,t) \in R^m \times T$.
Now we present the main result of the paper. The use of the multifunction F which will occur in (ii) and (iii) is a more or less technical trick. In §3 we shall discuss how to specify these requirements.

<u>Theorem 1</u>. Consider the parametric program $P(t)$, $t \in T$. Let $t^o \in T$, and let x^o be a stationary solution of $P(t^o)$. Suppose that

(i) x^o satisfies the Mangasarian-Fromovitz CQ w.r. to $M(t^o)$, i.e., the gradients $\nabla_x f_1(x^o,t^o),\dots,\nabla_x f_p(x^o,t^o)$ are linearly independent, and there is some $z \in R^n$ such that $z^T \nabla_x f_i(x^o,t^o) = 0$ $(\forall i \in \{1,\dots,p\})$ and $z^T \nabla_x f_j(x^o,t^o) < 0$ $(\forall j \in I_2(x^o,t^o))$,

(ii) for some open set $Q \supset LM(x^o,t^o)$ and some multifunction $F: D \times Q \times T \rightrightarrows R^{n\times n}$, and for each $(x,u,t) \in D \times Q \times T$,
(ii)' F is locally bounded and closed at (x,u,t), and
(ii)" $\partial^2_x l(x,u,t)\,h \subset F(x,u,t)\,h$ $(\forall h \in R^n \setminus \{0\})$,

(iii) for each Lagrange multiplier vector $u^o \in LM(x^o,t^o)$, for each vector $h \in W^+(x^o,u^o,t^o) \setminus \{0\}$ and for each matrix $M \in F(x^o,u^o,t^o)$, it holds $h^T M h > 0$.

Then we have:

(a) x^o is a strict local minimizer of $P(t^o)$.

(b) For some real number $r > 0$ and each $\varepsilon \in (0,r]$, there exists a real number $\delta = \delta(\varepsilon)$ such that whenever $t \in T$ belongs to $\mathring{B}(t^o,\delta)$, then $B(x^o,\varepsilon)$ contains a local minimizer of $P(t)$ which is the unique stationary solution of $P(t)$ in $B(x^o,r)$.

(c) Let $x(\cdot)$ denote the mapping which assigns, by (b), to each $t \in \mathring{B}(t^o,\delta(r))$ the unique stationary solution $x(t)$ of $P(t)$ in $B(x^o,r)$. Then, in particular, $x(t^o) = x^o$, and for any $t \in \mathring{B}(t^o,\delta(r))$, the point $x(t)$ is a strongly stable local minimizer of $P(t)$.

<u>Remark 1</u>. In order to prove (a) it even suffices to suppose that the following 2nd-order condition, together with assumption (i), holds: For some (fixed) $u^o \in LM(x^o,t^o)$, for all $h \in W^+(x^o,u^o,t^o) \setminus \{0\}$ and all $M \in \partial^2_x l(x^o,u^o,t^o)$, one has $h^T M h > 0$, cf. [9, Th. 1 and Remark 1].

The proof of Theorem 1 is essentially based on

<u>Lemma 1</u>. Let the assumptions of Theorem 1 be fulfilled. Then for any three sequences $\{t^k\} \subset T$, $\{x^k\}$, $\{y^k\}$ satisfying $t^k \to t^o$, $x^k \to x^o$, $y^k \to x^o$, $x^k \in S(t^k)$ and $y^k \in S(t^k)$ for all k, one has for k sufficiently large, $x^k = y^k$.

Proof of Lemma 1: Assume, on the contrary, that the assertion is not true. We then can suppose without loss of generality that

$$x^k \neq y^k \qquad \text{and} \qquad f_o(x^k,t^k) \geq f_o(y^k,t^k) \qquad \text{for all } k. \tag{2.5}$$

Hence, the sequence $h^k := \| y^k - x^k \|^{-1}(y^k - x^k)$ has an accumulation point $h \in R^n$ with $\|h\| = 1$. Since $x^k \in S(t^k)$ for all k, there exist Lagrange multiplier vectors u^k such that $(x^k,u^k) \in K(t^k)$ for all k. Assumption (i) implies that the sequence $\{u^k\}$ has an accumulation point u^o such that $(x^o,u^o) \in K(t^o)$, cf. Robinson [15 , Th. 2.3]. Without loss of generality, we may assume

$$h^k \longrightarrow h \qquad \text{and} \qquad u^k \longrightarrow u^o. \tag{2.6}$$

Following an idea of Robinson [15 , Th. 2.4] , we shall show that $h \in W^+(x^o,u^o,t^o)$. Property (2.6), the definitions (2.1), (2.2) and the continuity of the functions $f_1,\dots,f_m$ imply that

$$I^+(u^o) \subset I^+(u^k) \subset I(x^k,t^k) \subset I(x^o,t^o) \quad \text{for k large.}$$

For $j \in I^+(u^o)$ and for sufficiently large k, we thus obtain

$$f_j(y^k,t^k) = (y^k-x^k)^T \nabla_x f_j(x^k,t^k) + o(\| y^k-x^k \|).$$

Hence, in virtue of the continuity of $\nabla_x f_j(\cdot,\cdot)$, of (2.6) and of $y^k \in M(t^k)$ (for all k), it follows that $h^T \nabla_x f_j(x^o,t^o) \leq 0$ for all $j \in I^+(u^o)$, but even $h^T \nabla_x f_i(x^o,t^o) = 0$ if $i \in \{1,\dots,p\}$.
As $(x^o,u^o) \in K(t^o)$, one then has, with $J := I_2{}^+(u^o)$,

$$\begin{aligned} 0 = h^T \nabla_x l(x^o,u^o,t^o) = \; & h^T \nabla_x f_o(x^o,t^o) + \sum_{j \in J} u_j{}^o h^T \nabla_x f_j(x^o,t^o) \\ & \leq h^T \nabla_x f_o(x^o,t^o). \end{aligned} \tag{2.7}$$

Further, by (2.5), we know that for all k,

$$f_o(x^k,t^k) \geq f_o(y^k,t^k) = f_o(x^k,t^k) + (y^k-x^k)^T \nabla_x f_o(x^k,t^k) + o(\| y^k-x^k \|),$$

which implies

$$h^T \nabla_x f_o(x^o,t^o) \leq 0,$$

where (2.6) and the continuity of $\nabla_x f_o(\cdot,\cdot)$ were taken into account. By (2.7), we then have $h^T \nabla_x f_j(x^o,t^o) = 0$ in the case $j \in I_2{}^+(u^o)$, too, and so

$$h \in W^+(x^o,u^o,t^o) \setminus \{0\}. \tag{2.8}$$

As for large k, conv $\{x^k,y^k\} \subset D$, property (E 3) above allows a 2nd-order Taylor expansion of $l(\cdot,u^k,t^k)$ at y^k. Taking (2.5), $y^k \in M(t^k)$ and $(x^k,u^k) \in K(t^k)$ (for all k) into account, we obtain that for k sufficiently large,

$$\begin{aligned} f_o(x^k,t^k) \geq f_o(y^k,t^k) \geq \; & l(y^k,u^k,t^k) \\ & = l(x^k,u^k,t^k) + (y^k-x^k)^T \nabla_x l(x^k,u^k,t^k) + \\ & \qquad + \tfrac{1}{2} (y^k-x^k)^T M_k (y^k-x^k) \\ & = f_o(x^k,t^k) + \tfrac{1}{2} (y^k-x^k)^T M_k (y^k-x^k), \end{aligned}$$

where M_k is some matrix belonging to $\partial^2_x l(c^k,u^k,t^k)$ with some $c^k \in \mathrm{ri}\ \mathrm{conv}\{x^k,y^k\}$. For sufficiently large k, $u^k \to u^o \in LM(x^o,t^o)$ implies $u^k \in Q$, and hence, by assumption (ii)",

$$M_k h^k \in F(c^k,u^k,t^k)\, h^k .$$

Thus, since $t^k \to t^o$, $c^k \to x^o$, $u^k \to u^o$ and $h^k \to h$ with $h \neq 0$, assumption (ii)' then yields the existence of some (n,n)-matrix M_o such that

$$M_o\, h \in F(x^o,u^o,t^o)\, h ,$$

and, by the above relations,

$$h^T M_o\, h \leq 0,$$

which contradicts, together with (2.8), assumption (iii). This completes the proof. //

Proof of Theorem 1: (a) cf. Remark 1 above.

(b) Applying Lemma 1 with $t^k \equiv t^o$, $x^k \equiv x^o$ and for some sequence $\{y^k\} \subset S(t^o)$ satisfying $y^k \to x^o$ and $f_o(y^k,t^o) \leq f_o(x^o,t^o)$ (for all k), we obtain that $y^k = x^o$ if k is large enough, i.e., there exists some real number $s > 0$ such that

$$S(t^o) \cap B(x^o,s) = \{x^o\} . \tag{2.9}$$

The continuity assumptions on f_i and $\nabla_x f_i(\cdot,\cdot)$ $(i=1,\dots,m)$ and assumption (i) of the theorem allow to find positive real numbers $r \leq s$ and q such that the Mangasarian-Fromovitz CQ remains satisfied for all $x \in B(x^o,r)$ w.r. to $M(t)$, where $t \in \mathring{B}(t^o,q)$, and, moreover, such that the mapping

$$t \rightrightarrows S(t) \cap B(x^o,r) \tag{2.10}$$

is upper semicontinuous in Berge's sense (u.s.c.) at t^o. For the u.s.c. property of mapping (2.10) we refer to Robinson [15, Th. 2.3]. In virtue of $S(t^o) \cap B(x^o,r) = \{x^o\}$ (by (2.9)) and of the u.s.c. of the mapping (2.10) at t^o, it follows immediately from Lemma 1 that there is some positive real number q' such that $S(t) \cap B(x^o,r)$ contains at most one point if $t \in \mathring{B}(t^o,q')$. On the other hand, the continuity of the functions $f_o, f_1, \dots, f_m$, assumption (i) and the fact that, by (a)), x^o is a strict local minimizer of $P(t^o)$ allow to apply a standard argument on persistence and continuity of local minimizers for perturbed programs, cf., for example, Robinson [16, Th. 4.3 and §5] or Klatte [8, Th. 1]:
For each $\varepsilon \in (0,r]$ there exists some $q'' = q''(\varepsilon) > 0$ such that $B(x^o,\varepsilon)$ contains at least one local minimizer $x(t)$ of $P(t)$ if $t \in \mathring{B}(t^o,q'')$. If $q'' \leq q$ then $t \in \mathring{B}(t^o,q'')$ implies that $x(t)$ satisfies the Mangasarian-Fromovitz CQ w.r. to $M(t)$, as shown above, and hence

belongs to S(t). Setting for any $\varepsilon \in (0,r]$, $\delta(\varepsilon) := \min\{q, q', q''(\varepsilon)\}$, we have the desired result.

(c) In (b), it was particularly proved that $x(t^o) = x^o$ and that x^o is a strongly stable local minimizer of $P(t^o)$. If $x(\cdot)$ is the mapping constructed by (b), then it is a standard exercise to show that for all t sufficiently close to t^o, the assumptions of the theorem remain true if x^o is replaced by x(t) (and that, hence, x(t) is a strongly stable local minimizer of P(t)). The details of the proof are omitted here. //

3. Consequences and special cases

First we present a consequence of Theorem 1, which is of interest for two-level optimization problems. Consider the parametric program introduced in §1:

$$P(t): \qquad \min_x \{f_o(x,t) \ / \ x \in M(t)\} \ , \ t \in T,$$

suppose $T = R^k$ and put $v(t) := \inf_x \{f_o(x,t) / x \in M(t)\}$, $t \in T$. Let $\operatorname{dom} v := \{t \in T \ / \ v(t) \in R\}$. If the optimization problem

$$\min_{(x,t)} \{f_o(x,t)/x \in M(t)\} \tag{3.1}$$

is intended to be solved by a decomposition method in two phases, the first phase is (usually) the parametric problem P(t), $t \in T$, while the second phase will be an optimization problem w.r. to the variable t:

$$\min_t \{ v(t) \ / \ t \in \operatorname{dom} v\} . \tag{3.2}$$

Now we give a sufficient condition for a point (x^o, t^o) to be a local minimizer of problem (3.1). In the case of C^2-data, the subsequent theorem becomes a version of Theorem 3.1 in Jongen, Möbert and Tammer [7].

__Theorem 2__. Consider the parametric program P(t), $t \in T$, and the problems (3.1) and (3.2) under the general assumptions of §1, and let $T = R^k$. Then a point (x^o, t^o) is a local minimizer of the optimization problem (3.1) if x^o is a local minimizer of $P(t^o)$ which satisfies the assumptions of Theorem 1, and if t^o is a local minimizer of the program (3.2).

__Proof__: The mapping $x(\cdot)$ occuring in proposition (c) of Theorem 1 possesses all properties of the function $\bar{x}(\cdot)$ considered in the proof of Theorem 3.1 in [7]. In order to prove the assertion of our theorem, one may exactly follow the arguments given in the mentioned proof in [7]. //

In the preceding theorem, the assumptions imposed on the local minimizer x^o ensure, evidently, its strong stability. The importance of the strong stability property of x^o is illustrated in [7] by an example of an unconstrained optimization problem with a polynomial objective function of two variables.

We finish the paper by discussing the hypotheses of Theorem 1 for several special cases.

First we note that assumption (iii) may be verified in a rather simple way if the functions $f_o(\cdot,t^o)$, $f_1(\cdot,t^o),\dots,$ $f_m(\cdot,t^o)$ have a particular form or belong to certain subclasses of $C^{1,1}(D)$, we refer to Klatte and Tammer [9].

Now, a more extensive discussion will concern the question how to simplify assumption (ii). We recall that Kojima [10, §8] studies strong stability of local minimizers under the additional assumptions (for $i = 0,1,\dots,m$)

$$f_i(\cdot,t) \text{ is twice differentiable on } D \quad (\forall t\in T), \tag{3.3}$$

$$\nabla^2{}_x f_i(\cdot,\cdot) \text{ is continuous on } D\times T. \tag{3.4}$$

Under these additional assumptions, (ii) is automatically fulfilled. Indeed, we have in this case (with $x\in D$, $u\in R^m$, $t\in T$):

$$\partial^2{}_x l(x,u,t) = \left\{\nabla^2{}_x f_o(x,t) + \sum_{i\in I^+(u)} u_i \nabla^2{}_x f_i(x,t)\right\},$$

and $F:= \partial^2{}_x l(\cdot,\cdot,\cdot)$ evidently guarantees that (ii) holds. The complicated form of condition (ii) in Theorem 1 is due to the fact that the mapping $\partial^2{}_x l(\cdot,\cdot,\cdot)$ is not u.s.c., in general. We can meet this difficulty even in the case that the mappings $\nabla_x f_i$ are Lipschitz in both variables. Now we show how to choose F in particular cases.

Remark 2. Consider the parametric problem $P(t)$, $t\in T$. We further suppose that T is an open subset of R^k, and that for all $i\in\{0,1,\dots,m\}$, $f_i\in C^{1,1}(D\times T)$. Let $t^o\in T$ and $x^o\in S(t^o)$. Choose any open set $Q\supset LM(x^o,t^o)$. Obviously, we then have

$$l\in C^{1,1}(D\times Q\times T).$$

Therefore, the generalized Hessian mapping $\partial^2 l(\cdot,\cdot,\cdot)$ is locally bounded and closed on $D\times Q\times T$ (by (E 1)). Moreover, we have that $\partial^2{}_x l(x,u,t)\subset \mathrm{proj}_x\, \partial^2 l(x,u,t)$ (for $(x,u,t)\in D\times Q\times T$), where proj_x stands for the projection $(M_x,M_u,M_t)\in \partial^2 l(x,u,t) \longmapsto M_x$. It is easy to verify that the multifunction $(x,u,t)\rightrightarrows F(x,u,t):= \mathrm{proj}_x \partial^2 l(x,u,t)$ is also locally bounded and closed on $D\times Q\times T$ and thus fulfils (ii).

Remark 3. Literature on decomposition methods pays a special attention to optimization problems in which the objective function is separable w.r. to two groups of variables, cf., for example, Bank et al. [2], Beer [3]. In this context, an interesting special case of $P(t)$, $t\in T$, is the following: Suppose $f_o(x,t) = g_o(x) + h_o(x)$, where $g_o\in C^{1,1}(D)$, and $h_o\colon T \longrightarrow R$ is continuous on T. Further, let $f_1,\dots,f_m$ satisfy

(3.3) and (3.4), together with the general assumptions of §1. Of course

$$\partial^2_x l(x,u,t) = \partial^2 g_o(x) + \sum_{i \in I^+(u)} \{u_i \nabla^2_x f_i(x,t)\}$$

holds for all $(x,u,t) \in D \times R^m \times T$. If t^o, x^o and Q are given as in Remark 2, assumption (ii) is fulfilled for $F := \partial^2_x l(\cdot,\cdot,\cdot)$. For a simple desciption of $\partial^2 g_o(x)$ in an important particular case, we refer to the subsequent remark.

<u>Remark 4</u>. Let $g_1,\dots,g_N$ be real-valued functions which are defined and continuous on $D \times T$, and suppose that for each $j \in J := \{1,\dots,N\}$, g_j satisfies the properties (3.3) and (3.4) and that $\nabla_x g_j(\cdot,\cdot)$ is continuous on $D \times T$, where D and T are the sets introduced in §1. Denote by $CS^o(g_1,\dots,g_N)$ the following class of continuous selections $f: D \times T \longrightarrow R$ from the set $\{g_1,\dots,g_N\}$:

(a) f is continuous on $D \times T$.

(b) For each $(x,t) \in D \times T$ there is some $j \in J$ such that $f(x,t) = g_j(x,t)$.

(c) For each pair $j,k \in J$ and each $(x,t) \in D_j \cap D_k$, one has $\nabla_x g_j(x,t) = \nabla_x g_k(x,t)$, where $D_j := \{(x,t) \in D \times T / f(x,t) = g_j(x,t)\}$.

Then, by Theorem 4 in [9], it holds with $J(f,x,t) := \{j \in J / f(x,t) = g_j(x,t)\}$

$$\partial^2_x f(x,t) \subset \text{conv}\{\nabla^2_x g_j(x,t) \,/\, j \in J(f,x,t)\} \quad (\text{for } (x,t) \in D \times T).$$

For any $(\bar{x},\bar{t}) \in D \times T$, we have $J(f,x,t) \subset J(f,\bar{x},\bar{t})$ for all (x,t) near $(\bar{x},\bar{t})$, since f and $g_1,\dots,g_N$ are continuous functions. Taking the continuity of $\nabla_x^2 g_j(\cdot,\cdot)$ (for all $j \in J$) into account, we can consequently conclude that for each function $f \in CS^o(g_1,\dots,g_N)$, the multifunction

$$(x,t) \rightrightarrows \text{conv}\{\nabla^2_x g_j(x,t) \,/\, j \in J(f,x,t)\} \tag{3.5}$$

is locally bounded and closed on $D \times T$.

Consider now the parametric problem $P(t)$, $t \in T$, and suppose there are no equality constraints (i.e., $p=0$). Further suppose that for all $i=0,1,\dots,m$, $f_i \in CS^o(g_1,\dots,g_N)$. Let t^o, x^o and Q are given as in Remark 2. Replacing in (3.5) f by f_i, we thus have that for each $i=0,1,\dots,m$, the multifunction F_i defined by

$$F_i(x,t) := \text{conv}\{\nabla^2_x g_j(x,t) \,/\, j \in J(f_i,x,t)\}, \quad (x,t) \in D \times T, \tag{3.6}$$

is locally bounded and closed on $D \times T$. Hence, the multifunction F defined by

$$F(x,u,t) := F_o(x,t) + \sum_{i \in I^+(u)} u_i F_i(x,t), \quad (x,u,t) \in D \times Q \times T,$$

is locally bounded and closed on $D \times Q \times T$. Further, for each $h \in R^n$,

$$\partial^2_x l(x,u,t)h \subset \partial^2_x f_o(x,t)h + \sum_{i \in I^+(u)} u_i \partial^2_x f_i(x,t)h \subset F(x,u,t)h$$

holds on $D \times Q \times T$ in virtue of property (E 2) of $C^{1,1}$-functions, and so, F satisfies (ii). As p=0, hence $I^+(u^o)=I_2^+(u^o)$ and definition (3.6) yield that the verification of assumption (iii) reduces, in this case, to checking the positive definiteness of $\nabla^2_x g_i(x^o,t^o)$ on $W^+(x^o,u^o,t^o)$ for all indices i belonging to some subset of J.

References

[1] Auslender, A.: Stability in mathematical programming with nondifferentiable data. SIAM J. Control and Optimization 22 (1984) 239-254.

[2] Bank, B., R. Mandel and K. Tammer: Parametrische Optimierung und Aufteilungsverfahren. In: Lommatzsch, K.(ed.): Anwendungen der linearen parametrischen Optimierung. Akademie-Verlag, Berlin, 1979.

[3] Beer, K.: Lösung großer linearer Optimierungsaufgaben. VEB Deutscher Verlag der Wissenschaften, Berlin, 1977.

[4] Clarke, F.H.: Optimization and Nonsmooth Analysis. Wiley, New York, 1983.

[5] Guddat, J., Hj. Wacker and W. Zulehner: On imbedding and parametric optimization. Math. Programming Study 21 (1984) 79-96.

[6] Hiriart-Urruty, J.-B., J.J. Strodiot and V. Hien Nguyen: Generalized Hessian matrix and second-order optimality conditions for problems with $C^{1,1}$ - data. Appl. Math. Optim. 11 (1984) 43-56.

[7] Jongen, H.Th., T. Möbert and K. Tammer: On iterated minimization in nonconvex optimization. Math. Ops. Res. 11 (1986) 679-691.

[8] Klatte, D.: On the stability of local and global optimal solutions in parametric problems of nonlinear programming. Seminarbericht Nr. 75, Sektion Mathematik, Humboldt-Universität, Berlin, 1985.

[9] Klatte, D. and K. Tammer: On second-order sufficient optimality conditions for $C^{1,1}$-optimization problems. optimization 19 (1988), to appear.

[10] Kojima, M.: Strongly stable stationary solutions in nonlinear programs. In: Robinson, S.M. (ed.): Analysis and Computation of Fixed Points. Academic Press, New York, 1980.

[11] Kummer, B.: Linearly and nonlinearly perturbed optimization problems. In: Guddat, J., H.Th. Jongen, B. Kummer, F. Nožička (eds.): Parametric Optimization and Related Topics. Akademie-Verlag, Berlin, 1987.

[12] Kummer, B.: Ein Zugang zur quantitativen Stabilität in der nichtlinearen Optimierung: Variation von Kakutani-Abbildungen. Manuskript, Sektion Mathematik, Humboldt-Universität, Berlin, 1987.

[13] Lehmann, R.: On the numerical feasibility of continuation methods for nonlinear programming problems. Math. Operationsforsch. Stat., Series Optimization 15 (1984) 517-530.

[14] Robinson, S.M.: Strongly regular generalized equations. Math. Ops. Res. 5 (1980) 43-62.

[15] Robinson, S.M.: Generalized equations and their solutions, Part II: Applications to nonlinear programming. Math. Programming Study 19 (1982) 200-221.

[16] Robinson, S.M.: Local epi-continuity and local optimization. Math. Programming 37 (1987) 208-222.

NEWTON's METHOD FOR NON-DIFFERENTIABLE FUNCTIONS

Bernd Kummer[1)]

Abstract

The paper discusses extensions of Newton's method for solving equations with non-differentiable functions. Sufficient conditions of convergence, estimations for the speed as well as the range of convergence, and partial classes of functions allowing such extension are given. An example shows the difficulties which may arise in relation to regular zeros of Lipschitz equations.

Keywords

Newton's method, multifunction, Lipschitz functions, regularity, Clarke Jacobian, conditions of convergence

0. Introduction

The main objective of this paper is the study of Newton's method for solving the equation

$$f(x) = 0 \tag{2}$$

where $f: X \rightarrow Y$ is a (not necessarily differentiable) function, and X as well as Y are real Banach spaces. In order to extend Newton's idea to this case let G be some upper semicontinuous multifunction which assigns to each $x \in X$ some non-empty bounded subset $G(x)$ of the space $L(X,Y)$ of all linear, continuous operators A from X into Y. Let, further, $\bar{x}$ be a solution of (1) such that for all $A \in G(\bar{x})$, $A^{-1} \in L(Y,X)$ exists and that $\sup_{A \in G(\bar{x})} \| A^{-1} \|$ is finite.

Then, the following algorithm may be considered

ALG 1

step 1 $\quad x^o \in X$ (near $\bar{x}$)

step k $\quad$ select any $A \in G(x^k)$ and solve $f(x^k) + A(x-x^k) = 0$ to obtain x^{k+1}, put $k = k+1$, go to step k.

Obviously, ALG 1 coincides with Newton's method if both f is continuously (Frechét-) differentiable and $G(x) = \{Df(x)\}$. The aim of our paper is twofuld. At the one hand some sufficient condition for ALG 1 being (locally, superlinearly) converging will be derived. This condition shows, particularly, the convergence of ALG 1 in the following cases:

1) Humboldt-Universität zu Berlin, Sekt. Math. PSF 1297, GDR Berlin 1086

I $f=f(x)$ is a continuous selection of finitely many functions $f^{\nu}: X \to Y$ being continuously differentiable (Frechét-), briefly $f=CS(f^1,\dots,f^m)$, $f^{\nu} \in C^1$. The mapping G is then defined as

$$G(x) = \{D\, f^{\nu}(x) \,/\, f^{\nu}(x) = f(x)\}.$$

II $f: R^n \to R^n$ is locally Lipschitz, $G(x) = \partial f(x)$, the Clarke Jacobian of f at x, and card $\partial f(\bar{x}) = 1$.

At the other hand we present some example of a Lipschitz function $f: R \to R$ with the following properties:

$\partial f(x) \subset [a,b]$, $a > 0$

$Df(\bar{x})$ exists

For $G = \partial f$, ALG 1 do not converge with almost all x^0 (though f is continuously differentiable at all x^k attained).

Concerning case I some remarks are appropriate. First, that case has been discussed in the paper M. Kojima and S. Susumu (1986). There, the following assumptions are imposed

(I.1) $X = Y = R^n$

(I.2) $X = cl(int\, X_{\nu})$ with $X_{\nu} = \{x \,/\, f^{\nu}(x) = f(x)\}$.

I.1 implies (by W.W. Hager's famous theorem 1979) that f is locally Lipschitz and that, consequently, the Clarke-Jacobians $\partial f(x)$ always exist. By I.2, then the inclusion $G(x) \subset \partial f(x)$ is ensured.
The main advantage of the present approach consists less in considering Banach spaces instead of R^n than in the fact that mappings G of a large class may be used in ALG 1 not supposing the existence of $\partial f(x)$. At the same time, the main difficulty is to find, for given f, such a mapping G which makes ALG 1 converging and allows to determine $A \in G(x^k)$ in an efficient way.
The paper is organized as follows.

In section 1 the basic results will be formulated. They allow to study the equation (1) as well as the complementory problem

$$0 \in f(x) + N(x) \tag{2}$$

where N is a given multifunction.

In section 2 we specify these results for (1) and (2) and discuss sufficient conditions of convergence.

Finally, in section 3, some numerical topics in view of concrete realizations are added.

<u>Notations:</u> For subsets A_1, A_2 of a Banach space X, for elements $x \in X$ and $r \in R$, we use the (Minkovski-) notations.

$$A_1 + rA_2 = \{a_1 + ra_2 \,/\, a_1 \in A_1,\ a_2 \in A_2\}$$
$$x + rA_2 = \{x + ra_2 \,/\, a_2 \in A_2\}$$
$$-A_1 = (-1)A_1.$$

Further, B_X denotes the closed unit ball in X.

1. Basic results

Let X and Y be (real) Banach spaces and suppose $F: X \times X \longrightarrow Y$ to be a multifunction assigning to each pair $(x,t) \in X \times X$ some subset $F(x,t)$ of Y. We consider t as being some parameter and study the sets $S(t)$ of all x satisfying

$$0_Y \in F(x,t). \tag{1.1}$$

If we fix any $x^0 \in X$ and solve, for given x^k, the inclusion

$$0_Y \in F(x,x^k)$$

to obtain x^{k+1} (if such solution exists) we will call this procedure ALG 2. Thus, ALG 2 is defined by

$$x^{k+1} \in S(x^k),\ x^0 \text{ given}$$

and covers ALG 1 where F is specified as

$$\begin{aligned} F(x,t) &= f(t) + G(t)(x-t) := \\ &:= \{f(t)+A(x-t) \;/\; A \in G(t)\}, \\ & G(t) \subset L(X,Y). \end{aligned} \tag{1.2}$$

The convergence of ALG 2 obviously depends on the behaviour of the solution sets $S(t)$ for t near to some parameter $\bar{t}$ of interest. We are going now to investigate this crucial fact in a more detailed manner.

Let $\bar{t} \in X$ be arbitrarily fixed and suppose M to be some non-empty subset of $S(\bar{t})$. Depending on $\bar{t}$ and M, we define a function $\tau = \tau(\varepsilon, t)$ as follows

$$\tau(\varepsilon,t) = \inf \left\{ \delta > 0 \;/\; \begin{array}{l} F(x,\bar{t}+t) \subset F(x,\bar{t}) + \delta B_Y \\ \text{for all } x \in M + \varepsilon B_X \end{array} \right\}$$

$(\varepsilon > 0)$.

For given $y \in Y$, let $L(y)$ denote the set of all x satisfying $y \in F(x,\bar{t})$.

We shall say that M is regular (related to $\bar{t}$) if there are positive reals $\bar{\varepsilon}$ and $\bar{l}$ such that

$$(M+\bar{\varepsilon} B_X) \cap L(y) \subset M + \bar{l} \cdot \|y\| \cdot B_X \quad \forall y \in \bar{\varepsilon} B_Y \tag{1.3}$$

and

$$\lim_{t \to 0} \tau(\bar{\varepsilon},t) = 0 \tag{1.4}$$

hold. In the case $M = \{\bar{t}\}$ we will simply say that $\bar{t}$ is regular.

Proposition 1. If M is regular (related to $\bar{t}$) then there is some $\alpha > 0$ such that both

$$x \in S(t+\bar{t}),\ \|t\| < \alpha \quad \text{and} \quad r \overset{Df}{=} \operatorname{dist}(x,M) < \bar{\varepsilon} \tag{1.5}$$

imply

$$r \leq \bar{l} \cdot \inf_{r' > r} \tau(r',t) \tag{1.6}$$

where $\bar{\varepsilon}$ and $\bar{l}$ are from (1.3) and (1.4). Moreover, such α is defined by the following conditions

$$0 < \alpha < \bar{\varepsilon}$$
$$\tau(\bar{\varepsilon},t) < \bar{\varepsilon} \quad \forall t \in \alpha B_X \tag{1.7}$$

Proof. Let (1.5) and (1.7) be fulfilled, and let β belong to the open interval $(0, \bar{\varepsilon} - \tau(\bar{\varepsilon},t))$. Since $0 \in F(x,t+\bar{t})$ and $\mathrm{dist}(x,M) = r$, for each $r' > r$, we have $x \in M + r'B_X$. By the definition of τ one finds some y such that $\|y\| < \tau(r',t) + \beta$ and $0 \in F(x,\bar{t}) + y$.
Obviously, we may assume that $r' < \bar{\varepsilon}$ what implies

$$\tau(r',t) + \beta < \bar{\varepsilon} .$$

In view of (1.3) we may then estimate

$$x \in M + \bar{l}\cdot(\tau(r',t) + \beta)\cdot B_X \qquad (\forall \beta > 0).$$

This establishes the inequality (1.6).
As a particular result of the proposition we obtain that the solution sets $S(\bar{t}+t)$, restricted to the neighbourhood $M + \frac{\bar{\varepsilon}}{2} B_X$ of M, form an upper semicontinuous (at $t = 0$) mappings (in Hausdorff's sense), briefly u.s.c. (H).
It is very clear that several simple assumptions allow to replace the estimation (1.6) by

$$r \leq \bar{l}\cdot\tau(r,t). \tag{1.6'}$$

Such assumptions are M compact or $\tau(.,t)$ continuous.
Our main hypothesis for what follows is the estimation

$$\tau(\varepsilon,t) \leq c\cdot\|t\| + a_0(t)\cdot\varepsilon + a_1(t) \tag{1.8}$$

where we suppose $c \geq 0$ and

$a_k = a_k(t)$ to be non-negative functions satisfying

$$\lim_{t\to 0,\, \|t\|\neq 0} a_k(t)\cdot\|t\|^{-k} = 0 \qquad (k=0, k=1).$$

Since we aim at local statements only we shall seen (1.8) to be true . this inequality holds for sufficiently small ε and $\|t\|$.
The motivation of the hypothesis (1.8) is given by

Proposition 2. Let F be defined via (1.2) where $G: X \longrightarrow L(X,Y)$ is u.s.c. (H) at $\bar{t}$ and $G(\bar{t})$ is bounded. Let $M = \{\bar{t}\}$. Then, the estimation (1.8) holds (locally) whenever f is locally Lipschitz near $\bar{t}$.

Proof. Since G is u.s.c. (H) at $\bar{t}$ there is (by definition) some positive functions $\delta = \delta(t)$ satisfying $G(\bar{t}+t) \subset G(\bar{t}) + \frac{1}{2}\delta(t)B_{L(X,Y)}$; $\delta(t) \longrightarrow 0$ as $t \longrightarrow 0$. Therefore, to each $A_t \in G(\bar{t}+t)$, corresponds some $A_t^o \in G(\bar{t})$ such that

$\|A_t - A_t^o\| < \delta(t)$.

Consider any

$z \in F(x, \bar{t}+t)$ where $x \in \bar{t} + \varepsilon B_X$.

In view of (1.2) we may write

$z = f(\bar{t}+t) + A_t(x-(\bar{t}+t))$.

Setting $y = f(\bar{t}) + A_t^o(x-\bar{t})$ we obtain

$y \in F(x, \bar{t})$

and

$\|z-y\| \leq \|f(\bar{t}+t)-f(t)-A_t t\| + \delta(t)\ \|x-\bar{t}\|$.

Consequently (with $a_o(t) = \delta(t)$), the estimation

$\tau(\varepsilon, t) \leq \|f(\bar{t}+t)-f(\bar{t})-A_t t\| + a_o(t)\cdot\varepsilon$

is true.

Finally, since $\|A_t\|$ is bounded for sufficiently small $\|t\|$ and since f is Lipschitzian near $\bar{t}$, there is some c such that (1.8) holds true (with $a_1 \equiv 0$).

Note that the assumption $M = \{\bar{t}\}$ is needed anyway in order to ensure the convergence of the sequence $\{x^k\}$ in ALG 2 to $\bar{t}$.

At the same time one sees that, if F is given by (1.2) and $\bar{t}$ is regular, the mappings $A \in G(\bar{t})$ are injective and $f(\bar{t}) = 0$. Let us, additionally, assume that $f(\bar{t}+\varepsilon B)$ contains some neighbourhood of 0_Y in Y (for each $\varepsilon > 0$). If nothing more is known about f and if we want to guarantee the existence of x^{k+1} in ALG 1 for x^k near $\bar{t}$ we have to impose that all $A \in G(t)$ (for t near $\bar{t}$) are also surjective.

Therefore, our assumptions

$$G(x) \text{ bounded, } G \text{ u.s.c. (H) at } \bar{x},\ A^{-1} \in L(Y,X) \text{ exists for all } A \in G(\bar{x}) \text{ and } \sup_{A \in G(\bar{x})} \|A^{-1}\| < \infty \qquad (*)$$

which have been imposed in relation to ALG 1 (see introduction) are natural and ensure that the assumptions of the Proposition 1 and 2 are satisfied.

We proceed now to study the convergence of ALG 2 under the following hypothesis.

$$\bar{t} \text{ is a regular solution of } 0_Y \in F(x,\bar{t}) \text{ (see 1.3, 1.4)} \qquad (1.9)$$

$$\text{There exists } \bar{\delta} > 0 \text{ such that } \emptyset \neq S(\bar{t}+t) \subset \bar{t} + \delta(t)\cdot B_X \text{ for } \|t\| < \bar{\delta} \text{ where } \delta(t) \to 0 \text{ as } t \to 0_X. \qquad (1.10)$$

<u>Theorem 1.</u> Let (1.9) and (1.10) be fulfilled, and let (1.8) be true for $\varepsilon \leq \bar{\delta}$, $\|t\| \leq \bar{\delta}$. Suppose, additionally that the numbers $\bar{\varepsilon}$ and $\bar{I}$ in (1.3) and (1.4) satisfy

$\varepsilon < \bar{\delta}$ and $c\bar{l} < 1$.

Then, there exists $\alpha > 0$ such that ALG 2 generates an infinite sequence x^k satisfying

$$\|x^{k+1}-\bar{t}\| \leq 2\frac{\bar{l}c}{1+\bar{l}c}\|x^k-\bar{t}\| + \frac{2\bar{l}}{1+\bar{l}c}a_1(x^k-\bar{t}) \qquad (**)$$

whenever $\|x^0-\bar{t}\| < \alpha$.

The number α shows the property above if

$$0 < \alpha < \bar{\varepsilon},$$

and if $\|t\| < \alpha$ implies

$$\delta(t) < \bar{\varepsilon} \qquad (1.11)$$

$$c\|t\| + a_0(t)\cdot\bar{\varepsilon} + a_1(t) < \bar{\varepsilon} \qquad (1.12)$$

$$\bar{l}\, a_0(t) < \frac{1}{2}(1-\bar{l}c). \qquad (1.13)$$

and

$$a_1(t) \leq \frac{1}{4}\cdot\frac{1-\bar{l}c}{\bar{l}}\cdot\|t\|. \qquad (1.14)$$

Proof. Since $c\bar{l} < 1$, the existence of $\alpha > 0$ such that $\|t\| < \alpha$ implies (1.11),...,(1.14) is ensured.

Let α be choosen in the given way, and let, for some fixed t, $\|t\| < \alpha$. We consider any solution x of $0 \in F(x,\bar{t}+t)$, i.e. $x \in S(\bar{t}+t)$, and put $r = \|x-\bar{t}\|$. Because of (1.10) and (1.11) such solution x exists, and $r \leq \delta(t) < \bar{\varepsilon}$.

By (1.8) and (1.12) we obtain $\tau(\bar{\varepsilon},t) < \bar{\varepsilon}$. Therefore, proposition 1 may be applied with $M = \{\bar{t}\}$. This leads to the inequality

$$r \leq \bar{l}\cdot\inf_{r'>r}\tau(r',t) \leq \bar{l}(c\|t\|+a_0(t)r+a_1(t)). \qquad (1.15)$$

Viewing (1.13) we may continue

$$r \leq \bar{l}c\|t\| + \frac{1}{2}(1-\bar{l}c)r + \bar{l}\,a_1(t)$$

$$r\cdot\left(\frac{1+\bar{l}c}{2}\right) \leq \bar{l}c\|t\| + \bar{l}a_1(t)$$

$$r \leq 2\frac{\bar{l}c}{1+\bar{l}c}\|t\| + \frac{2\bar{l}}{1+\bar{l}c}a_1(t). \qquad (1.16)$$

Finally, recalling (1.14), the right-hand side in (1.16) is bounded above by

$$\frac{2\bar{l}c + \frac{1}{2} - \frac{1}{2}\bar{l}c}{1+\bar{l}c}\cdot\|t\| =$$

$$\frac{1}{2}\cdot\frac{1+3\bar{l}c}{1+\bar{l}c}\cdot\|t\| = \gamma\cdot\|t\|$$

with $\gamma < 1$. Hence

$$r \leq \gamma\|t\|, \quad \gamma < 1. \qquad (1.17)$$

Setting $t = x^k-\bar{t}$ (k=0,1,2,...) and $x = x^{k+1}$ the inequality (1.17) shows both that ALG 2 is converging for $\|x^0-\bar{t}\| < \alpha$ and that $S(x^k)$ is non-empty. The inequality (1.16) gives the estimation (**).

To discuss the theorem let us assume (1.9) and (1.10) to be satisfied (These are, of course, no trivial assumptions). According to the theorem we have then to estimate

$$\tau(\varepsilon,t) = \inf\left\{\delta>0 \;/\; F(x,\bar{t}+t)\subset F(x,\bar{t})+\delta B_Y \text{ for all } x\in\bar{t}+\varepsilon B_X\right\}$$

in the form (1.8) where the constant c is as small as possible. If $c < \frac{1}{\bar{1}}$ the local convergence of ALG 2 is ensured.

In the case $0 < c < \frac{1}{\bar{1}}$ this convergence is a linear one (or better).

In the case $c = 0$ it is of the order a_1 and, consequently, better than linear,

Finally, if such estimation for τ is impossible or is only possible with $c \geq \frac{1}{\bar{1}}$, we cannot say anything.

As a simple consequence of the inequality (1.15) one observes, for the case $0 \leq c < \frac{1}{\bar{1}}$, that

$$\limsup_{\substack{k\to\infty \\ x^k\neq\bar{t}}} \frac{\|x^{k+1}-\bar{t}\|}{\|x^k-\bar{t}\|} \leq c\bar{1} \quad \text{whenever } \|x^o-\bar{t}\| < \alpha .$$

2. Special types of multifunctions F

2.1. F is given by (1.2)

Let us suppose the multifunction G in (1.2) to satisfy the conditions (*) with $\bar{x} = \bar{t}$ and $f(\bar{x}) = 0$. Let f be locally Lipschitzian.

Then, we already know that (1.9) is true and that (1.8) holds with some c, a_o and a_1. To verify (1.10) is an elementary task of functional analysis which we can omit here.

Thus, the number c in (1.8) becomes the key for applying Theorem 1.

Proposition 3: Let G satisfy the condition (*) with $\bar{x} = \bar{t}$, and let $f(\bar{x}) = 0$.

Then, the estimation (1.8) holds with $c = 0$ if

$$\limsup_{\substack{t\to 0,\, t\neq 0 \\ A_t\in G(\bar{t}+t)}} \frac{1}{\|t\|}\cdot \|f(\bar{t}+t)-f(\bar{t})-A_t t\| = 0. \tag{2.1}$$

Proof. The statement immediately follows from the estimation

$$\tau(\varepsilon,t) \leq \|f(\bar{t}+t)-f(t)-A_t t\| + a_o(t)\cdot\varepsilon$$

derived in the proof of Proposition 2.

Let us discuss the condition (2.1) for the cases I and II in the introduction.

I $\quad f = CS(f^1,\dots,f^m)$, $f^\nu \in C^1(X,Y)$

I.1 $\quad G(x) = \{Df^\nu(x) \ / \ f^\nu(x) = f(x)\}$.

Since, for each ν, f^ν is continuously Frechét differentiable there is some function $a_1^\nu = a_1^\nu(t)$ satisfying

$$\lim_{t\to 0,\, t\neq 0} a_1^\nu(t)\cdot \|t\|^{-1} = 0$$

and

$$\|f^\nu(\bar t+t)-f^\nu(\bar t)-Df^\nu(\bar t)t\| \leq a_1^\nu(t)$$

Defining the finite sets

$$I(X) = \{\nu / \ f^\nu(x) = f(x)\}$$

and taking into account that f is continuous we obtain the existence of some $\delta > 0$ such that

$$I(t+\bar t) \subset I(\bar t) \qquad \text{for } \|t\|<\delta.$$

Thus, if $\|t\|<\delta$, for each ν such that

$$Df^\nu(\bar t+t) \in G(\bar t+t)$$

we have $f^\nu(\bar t) = f(\bar t)$. This yields the estimation

$$\|f(\bar t+t)-f(\bar t)-A_t t\| = \|f^\nu(\bar t+t)-f^\nu(\bar t)-Df^\nu(\bar t+t)t\|$$

$$\leq a_1^\nu(t) + \|-Df^\nu(\bar t+t) + Df^\nu(\bar t)\| . \|t\|$$

and shows (2.1).

In order to fulfil the remaining hypotheses of Proposition 3 we need:

$f(\bar t) = 0$ and

$[Df^\nu(\bar t)]^{-1} : Y \to X$ exists and is bounded for each $\nu \in I(0)$.

We consider now the case I under the assumptions

I.2 $\quad X = Y = R^n$, $G(x) = \partial f(x)$

where $\partial f(x)$ denotes the Clarke-Jacobian of f at x (see F.H. Clarke 1983)

$$\partial f(x) = \text{conv}\,\{A \ / \ A = \lim Df(x^k),\ f \text{ is Frechét-differentiable at } x^k,\ x^k \to x\}.$$

Since f is locally Lipschitz (see W.W. Hager, 1979) the mapping $G = \partial f$ fulfils (*) if (at $\bar x = \bar t$) all matrices in $\partial f(\bar t)$ are regular. For proving (2.1) we need

<u>Proposition 4.</u> Let $f = CS(f^1,\dots,f^m)$, $f^\nu \in C^1(R^n,R^n)$. Then it holds

$$\partial f(x) = \text{conv}\,\{Df^\nu(x) \ / \ \nu \in J(x)\}$$

where $J(x)$ is the set of all ν such that some sequence $x^k \to x$ exists satisfying both

$$f^\nu(x^k) = f(x^k) \text{ and } Df^\nu(x^k) = Df(x^k).$$

Proof. Let C be the convex hull of these $Df^{\nu}(x)$. Since $f^{\nu}\in C^1$ the inclusion $C \subset \partial f(x)$ is trivial. To verify $\partial f(x) \subset C$ we firstly consider any point x where Df(x) exists. Without loss of generality, let Df(x)=0. We fix any $u \in R^n$, $u \neq 0$. Since $f = CS(f^1,\dots,f^m)$ there are some sequence of positive $\lambda_k \to 0$ and some ν such that
$f(x+\lambda_k u) = f^{\nu}(x+\lambda_k u)$ for all k.
One then obtains

$$0 = \lim \frac{1}{\lambda_k} (f(x+\lambda_k u)-f(x)) = \lim \frac{1}{\lambda_k} (f^{\nu}(x+\lambda_k u)-f^{\nu}(x)).$$

Hence $Df^{\nu}(x)u = 0$.

Setting $U^{\nu} = \{u \mid Df^{\nu}(x)u = 0\}$ this means

$$R^n \subset \bigcup_{\nu\in V(x)} U^{\nu} \tag{2.2}$$

with $V(x) = \{\nu \mid f^{\nu}(x^k) = f(x^k) \text{ for some sequence } x^k \to x\}$.
If $Df^{\nu}(x) \neq 0$ the subspace U^{ν} is a proper one. Since V(x) is finite the inclusion (2.2) yields

$$\dim U^{\nu} = n \quad \text{for at least one } \nu\in V(x)$$

and, consequently, $Df^{\nu}(x) = 0 = Df(x)$.
Now, let $A \in \partial f(x)$ where x is arbitrarily fixed and, A is any extremal point of the compact convex set $\partial f(x)$. Then, some sequence $x^k \to x$ exists such that f is differentiable at each x^k and

$$A = \lim Df(x^k).$$

Recalling the first part of the proof one can assign to each k some $\nu(k)$ such that

$$f^{\nu}(x^k) = f(x^k) \text{ and } Df^{\nu}(x^k) = Df(x^k)$$

for $\nu = \nu(k)$. After selecting some subsequence satisfying $\nu(k) \equiv \nu$ (for all k) this shows $A\in C$ and leads to $\partial f(x) \subset C$.

Knowing proposition 4 the proof of (2.1), in the case I.2, can be organized as in the case I.1. before. We have only to note that it is enough to consider extremal points $A_t \in G(\bar{t}+t)$ in (2.1) and we must replace the set I(x) by J(x).

Finally, let us look at the case II in the introduction, i.e.

$$f: R^n \to R^n \text{ locally Lipschitz, } G(x) = \partial f(x)$$
$$\operatorname{card} \partial f(\bar{x}) = 1.$$

The verification of (2.1), for $\bar{t} = \bar{x}$, then follows via

$$\|f(\bar{t}+t) - f(\bar{t})-Df(\bar{t})t\| \le a_1(t)$$

and

$$\limsup_{A_t\in\partial f(\bar{t}+t),\, t\to 0} \|A_t-Df(\bar{t})\| = 0 \tag{2.3}$$

by the standard estimation

$$\|f(\bar{t}+t)-f(\bar{t})-A_t t\| \leq a_1(t)+ \|A_t-Df(\bar{t})\| \cdot \|t\|.$$

The condition (*) means nothing else than that $Df(\bar{x})$ is a regular matrix.

It should be mentioned that, in each of these cases, ALG 1 locally superlinear converges where the estimations are given by the theorem.

2.2. Complementarity problems

Let F be given by the formula

$$F(x,t) = f(t)+G(t)(x-t) + N(x). \tag{2.4}$$

This mapping differs from (1.2) by the fixed multifunction N not depending on t, and it is related to the complementarity problem (2)

$$0 \in f(x) + N(x).$$

The mapping N, most often, is a normal cone mapping associated with some closed convex set. In this case we have $Y = X^*$.

If we modify ALG 1 by solving, in step k,

$$0 \in f(x^k) + A(x-x^k) + N(x), \tag{2.5}$$

then the results above may be applied. Particularly, the function τ will not depend on N. However, it becomes now more difficult to ensure regularity (see 1.3) as well as solvability of each inclusion (2.5) for x^k near to some solution $\bar{x}$ of (2) and for any $A \in G(x^k)$. In this context, we refer to B. Kummer 1987, 1984 and to the basic paper S.M. Robinson 1979. For the case that (2) describes a Kuhn-Tucker-system, the papers D. Klatte/K. Tammer (1987) and D. Klatte(1988; in the present volume) are very interesting.

2.3. A pathological example

In this section, we define a real Lipschitz function f having the following properties.

$f(0) = 0$, $Df(0) = 1$, $\partial f(0) = [\frac{1}{2}, 2]$

f^{-1} exists

ALG 1 (with $G = \partial f$) fails to converge for each starting point $x^0 \neq 0$ provided that $Df(x^0)$ exists.

For the equation $f(x) - \lambda = 0$, ALG 1 finds the solution $x(\lambda)$ by one step whenever $\lambda \neq 0$ and $\|x^0-x(\lambda)\| < \varepsilon(\lambda)$.

In order to construct f we fix any natural number $n > 1$ and consider the interval $I_n = [\frac{1}{n}, \frac{1}{n-1}]$. Let m and m' be the middle points of I_n and I_{2n}, respectively. Setting

$$a = \frac{2n}{4n-1} \quad , \quad b = \frac{8n-4}{4n-3}$$

we define two linear functions by

$$f_n^1(x) = a(x+m),\ f_n^2(x) = b(x-m').$$

They fulfil

$$f_n^1(\tfrac{1}{n-1}) = \tfrac{1}{n-1},\ f_n^2(\tfrac{1}{n}) = \tfrac{1}{n},$$

$$f_n^1(m) < f_n^2(m),\quad a < b.$$

The point z defined by $f_n^1(z) = f_n^2(z)$ then belongs to the open interval $(\frac{1}{n},m)$. Note that a,b,m,m' and z depend on n. Now, we define f:

(i) $f(0) = 0$

$(ii)_n$ $f(x) = \begin{cases} f_n^1(x) & z \le x \le \frac{1}{n-1} \\ f_n^2(x) & \frac{1}{n} \le x \le z \end{cases}$

(iii) $f(x) = f_2^1(x)$ $\quad x \ge 1$

(iv) $f(x) = -f(-x)$ $\quad x < 0$

For $x \in I_n$, we have

$$\frac{1}{n} \le f(x) \le \frac{1}{n-1}$$

and

$$\frac{n-1}{n} \le \frac{f(x)}{x} \le \frac{n}{n-1},$$

viewing (iv) that leads to $Df(0) = 1$.
Since, for $n \to \infty$, one obtains $\lim a = \frac{1}{2}$, $\lim b = 2$ the equation $\partial f(0) = [\frac{1}{2},2]$ is evident.
Let us apply ALG 1 ($G = \partial f$) with $x^0 \in (z,\frac{1}{n-1})$.
Then $x^1 = -m$, $x^2 = m$, $x^3 = -m, \ldots$
If $x^0 \in (\frac{1}{n},z)$, then the sequence $x^2 = m'$, $x^3 = -m'$, $x^4 = m', \ldots$ will be generated.
If x^0 belongs to the corresponding negative intervals or $|x^0| > 1$, then the same situation (alternating sequences) will appear.
Finally, the equation $f(x) = \lambda$ $(\lambda \neq 0)$ will be solved by ALG 1 after one step since f is linear between $x(\lambda)$ and x^0 (near to $x(\lambda)$).
It should be noted that, at $\bar{t} = 0$, f does not satisfy the condition (2.3).

3. Some numerical topics

3.1. Approximation of G(x)

It is well-known from the ordinary Newton method that, in the case of regularity, the matrices $A = Df(x^k)$ may approximatively be determined up to some error ε_k. If $\varepsilon_k \to 0$ $(k \to \infty)$ then the order of convergence remains better than linear; if $\varepsilon_k \le \delta$ and δ is sufficiently small, then this order is linear and the better as the upper bound δ is smaller. The same situation appears for ALG 1 provided that $A_k \in G(x^k) + \varepsilon_k \cdot B_{L(X,Y)}$ (in step k) and that the hypotheses of Proposition 3 including (2.1) are fulfilled. Indeed, after assigning to A_k some $A \in G(x^k)$ such that $\|A_k - A\| \le 2\,\varepsilon_k$ one easily may estimate the distance d between the solutions y^{k+1} and x^{k+1} of

$$f(x^k) + A_k(x-x^k) = 0$$

and

$$f(x^k) + A(x-x^k) = 0,$$

respectively, i.e.

$$d \le \|A_k^{-1} - A^{-1}\| \; \|f(x^k)\|.$$

Since (*) and (2.1) imply the existence of $L < \infty$ such that $\|f(\bar{t}+t)-f(\bar{t})\| \le L \|t\|$ for $\|t\| < \frac{1}{L}$ an estimation of the type

$$d \le \delta(\varepsilon_k)\cdot L\cdot\|x^k-\bar{t}\| \; ; \; \delta(\varepsilon) \to 0 \text{ as } \varepsilon \to 0$$

is true. Together with (**) this gives the desired (local) convergence of the sequence y^k.

3.2. Difference quotients

Let $X = Y = R^n$. In the cases I.1, I.2 and II being considered in 2.1 one can try to approximate some elements of $G(x^k)$ by difference quotients. As case II is concerned the only question remaining is that of the reasonable step-size. For case I, however, an additional problem arises which we discuss now in relation to I.1.

Let $f = CS(f^1,f^2)$ where $f^\nu: R^2 \to R^2$, $f^\nu \in C^1$. Assume $f = f^1$ for $x_1 \ge x_2$ and $f = f^2$ for $x_1 \le x_2$. Finally, let $x^k = 0$.

If A is some approximation of $Df^1(0)$ or $Df^2(0)$ then it is an approximation of $G(0)$, too.

But a direct difference approximation of f via

$$\frac{\partial f}{\partial x_i} \approx \frac{1}{\varepsilon} \; (f(\varepsilon e_i)-f(0)),$$

where e_i denotes the i-th unit vector, leads to a matrix D being near to

$$\left(\frac{\partial f^1}{\partial x_1}, \frac{\partial f^2}{\partial x_2}\right)$$

and, possible, far from $G(0)$.

References:

Clarke, F.H. Optimization and non-smooth analysis, Wiley Interscience, New York (1983)

Hager, W.W. Lipschitz continuity for constrained processes. SIAM. J. Control and Optimiz. 17(1979) 321-338

Kojima, M.; Susumu, S. Extensions of Newton and Quasinewton Methods to Systems of PC^1 Equations. Journ. of Oper. Research, Society of Japan Vol. 29, No. 4, Dez. 1986

Kummer, B. /1/ Generalized equations: solvability and regularity, Math. Progr. Study 21(1984), 199-212

/2/ Linearly and non-linearly perturbed optimization problems. In Guddat/Jongen/Kummer/Nožička (eds.): Parametric optimization and related topics. Akademie-Verlag Berlin (1987)

Robinson, S.M. Generalized equations and their solutions: Part I: Basic theory. Math. Progr. Study 10(1979), 128-141

Klatte, D.; Tammer, K. On second order sufficient optimality conditions for $C^{1,1}$-optimization problems. Optimization (1987, to appear).

A MINIMIZATION OF A STRICTLY CONVEX FUNCTION ON A SIMPLEX

Klaus Lommatzsch [1)]

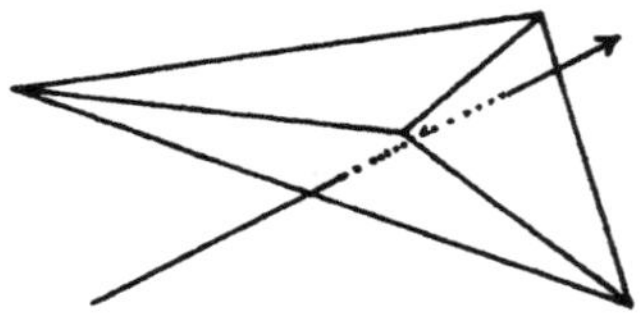

In environment consideration frequently arise optimization problems of the type

$$\sum_{i=1}^{n}(b_i - \sum_{j=1}^{m} a_{ij}p_j)^2$$

s. t.

$$\sum_{i=1}^{n} (b_i - \sum_{j=1}^{m} a_{ij}p_j) = 0 \ , \quad p_j \geqslant 0, \ j=1,\ldots,m.$$

The unknowns are $p_1,\ldots,p_m$. In real-life problems we obtained the coefficients a_{ij} and b_i were positive reals and $m \leqslant 20$ and $n \leqslant 1000$.

The given problem can be stated in the standard form of a quadratic optimization problem:

$$\min \{p^TA^TAp - 2b^TAp + b^Tb \mid p \in M\} \qquad (1)$$

where

$$M := \{p \in R^m \mid g^Tp = g_o, \ p \geqslant 0\} \ ,$$

$$A := (a_{ij})_{n,m} \ , \quad b^T := (b_1,\ldots,b_n),$$

$$p^T := (p_1,\ldots,p_m) \ , \quad g^T := (g_1,\ldots,g_m),$$

$$g_j := \sum_{i=1}^{n} a_{ij} \quad (j=1,\ldots,m) \quad \text{and} \quad g_o := \sum_{i=1}^{n} b_i .$$

The concrete algorithm we will describe takes into account the following two specifics:

- a serie of problems (1) has to be solved where the single problems arise by setting several unknowns p_j, a priori, equal to zero,
- the moduls of the coefficients g_j in the equation $g^Tp = g_o$

1) Humboldt-Universität zu Berlin, Sektion Mathematik
PSF 1297, 1086 Berlin, GDR

differ from the one of the right hand side g_o by 8 or 9 powers of 10.

Therefore a special algorithm for minimizing a strictly convex differentiable function $f(p)$ on a simplex M in the form given above is developed. Let us denote

$$\min \{f(p) \mid p \in M\}. \tag{2}$$

The idea of the algorithm is to solve a sequence of quadratic subproblems on linear manifolds in the Euclidian m-space R^m. Let $R^m_+ := \{p \in R^m \mid p \geqq 0\}$, $U^o := \{p \in R^m \mid g^T p = g_o\}$ and $U^j := \{p \in R^m \mid p_j = 0\}$, $j=1,\dots,m$. Then the simplex $M := R^m_+ \cap U^o$ is the feasible set of problem (2) and, for an arbitrary index set $J \subset \{1,\dots,m\}$ with $1 \leqq |J| \leqq m-1$ the set

$$S^J = R^m_+ \cap U^o \cap (\bigcap_{j\in J} U^j) \tag{3}$$

is a closed face of the simplex M. Its m vertices have the form $v^j = (0,\dots,0,{}^{g_o}/_{g_j},0,\dots,0)$. Since g_j and g_o are positive, every point in U^o has at least one positive coordinate.

Now, we can formulate the subproblems mentioned above:

$$\min \{f(p) \mid p \in U^o \cap (\bigcap_{j\in J} U^j)\} \tag{4}$$

Let p^J be the uniquely defined solution of (4) and $J_- := \{j \mid p^J_j < 0\}$. If $J_- = \emptyset$ then p^J is a feasible point of (2); otherwise in the case $J_- \neq \emptyset$ the optimal point of the problem $\min \{f(p) \mid p \in S^J\}$ belongs at least to one of the faces $S^J \cap U^j$, $j \in J_-$, of the simplex S^J (this immediately follows by the convexity of $f(p)$).

If the point p^J is feasible, then it is optimal for the original problem (2) if and only if there is a number u^J such that

$$\begin{aligned} u^J &= -\frac{1}{g_j}\frac{\partial f(p^J)}{\partial p_j}, \quad j \in J_+, \\ u^J &\geqq -\frac{1}{g_j}\frac{\partial f(p^J)}{\partial p_j}, \quad j \notin J_+, \end{aligned} \tag{5}$$

where $J_+ = \{j \mid p^J_j > 0\}$. Indeed, these are the well-known Kuhn-Tucker conditions for problem (2). If there is an index $j^* \notin J_+$ with $u^J < -\frac{1}{g_{j^*}}\frac{\partial f(p^J)}{\partial p_{j^*}}$, then

$$\frac{\partial f(p^J)}{\partial p}(v^{j^*} - p^J) < 0 \tag{6}$$

and, the problem $\min\{f(p) \mid p \in U^o \cap (\bigcap_{j \in J \setminus \{j^*\}} U^j)\}$ has points superior in the sense of optimality. Note, an index set $\hat{J} \subset \{1,\ldots,m\}$ exists for which the optimal point of the original problem (2) is a solution of the unbounded problem (4) where $J := \hat{J}$.

On account of these properties one can propose the following algorithm: Let the steps $1,2,\ldots,k-1$ be executed. By the step $k-1$ we obtained the index set J^k.

(a) Solve the unbounded problem (4) where $J := J^k$. If the resulting optimal point p^k has only non-negative coordinates, then go to (b); otherwise set $J^k := J^k \cup \{j \mid p_j^k < 0\}$ and repeat (a).

(b) Construct the set $J_-^k := \{j \mid u^{Jk} < -\frac{1}{g_j}\frac{\partial f(p^k)}{\partial p}\}$. If J_-^k is empty, then p^k is a solution of the given problem (2). Otherwise set $J^{k+1} := J^k \setminus J_-^k$ and go to step $k+1$.

As an initial index set J^1 we can choose any proper subset of $\{1,\ldots,m\}$.

The substep (a) of the general step can be repeated at most $m-1-|J^k|$ times. Not all indices dropped from index set J^k in substep (b) will return into the index set J^{k+1} constructed by the following substep (a). This is a consequence of the fact that $\frac{\partial f(p^k)}{\partial p}(p^{k+1} - p^k) < 0$ holds, where p^k and p^{k+1} are non-negative solutions of the unbounded problems (4) for $J := J^k$ and $J := J^{k+1}$ in the steps k and $k+1$, respectively. Namely rewriting this inequality we obtain (see the optimality conditions)

$$\sum_{j \in J_-^k} p_j^{k+1}(u^k + \frac{1}{g_j}\frac{\partial f(p^k)}{\partial p_j}) < 0 ,$$

i. e. at least one of the coordinates p_j^{k+1}, $j \in J_-^k$, is positive. By similar arguments it can be shown, that not all indices added in the substep (a) to the index set J^k can be dropped in the next substep (b). Of course, the values $f(p^k)$ corresponding to iteration points p^k not necessarily decrease.

A sequence p^k with decreasing values $f(p^k)$ one obtains by the following modified algorithm, where the substep (a) is replaced by the substep (a'): Solve the unbounded problem (4) where $J := J^k$. If the resulting optimal point p^k has only non-negative coordinates, then go to (b). If the set $J' = \{j \mid p_j^{k-1} = 0$ and $p_j^k < 0\}$ is non-empty, then put $J^k := J^k \cup J'$ and repeat (a') (the argument pointed out above ensure that not all coordinates dropped in (b) in the step k-1 can occur again in the new defined index set J^k). Otherwise, we compute an index j^* (and the corresponding number t^*) such that $p^{k-1} + t^*(p^k - p^{k-1}) \in U^{j^*} \cap M$. With the new index-set $J^k := J^k \cup \{j^*\}$ one repeats the substep (a').

By constructions this modified algorithm generates a sequence of iteration points $\{p^k\}$ with decreasing values f(p). But, of course, in general the number of repetitions of (a') has increased in comparation to the original algorithm.

Using the first algorithm by J. Goldstein a program for the quadratic problem described in the beginning was written and tested. The unbounded problems (4) were solved by a method of conjugate directions (the generation of such directions is the most time consuming procedure in the computation, but these directions, which belong to the linear manifolds occuring in the algorithm ensure the accuracy demanded by the user). In the first index set J^1 we collect indices j for which the values $f(v^j)$ at the vertices v^j of M are relatively large. In order to solve the given real-life problems the algorithm did not need more then two or three steps.

References

[1] Bernau, H.: Ein konjugiertes Richtungsverfahren zur Minimierung von quadratischen Funktionen mit beschränkten Variablen. Seminarbericht Nr. 94 der Sektion Mathematik der Humboldt-Universität zu Berlin, 1987.

[2] Goldstein, J.: Ein Verfahren zur Lösung streng konvexer quadratischer Optimierungsprobleme mit spezieller Struktur des Restriktionsbereiches. Diplomarbeit an der Sektion Mathematik der Humboldt-Universität zu Berlin, 1987.

[3] Hestenes, M. R.: Conjugate Direction Method in Optimization. Springer-Verlag, Berlin-Heidelberg-New York, 1980.

[4] Künzi, H. P. und W. Krelle: Nichtlineare Programmierung. Springer-Verlag, Berlin-Göttingen-Heidelberg, 1962.

[5] Lommatzsch, K.: Eine angewandte Aufgabe der quadratischen konvexen Optimierung. Seminarbericht Nr. 94 der Sektion Mathematik der Humboldt-Universität zu Berlin, 1987.

MONOTON-KONVEXE FUNKTIONEN - EINE BEMERKUNG ZUM SATZ VON BROWDER-MINTY

W. Oettli*)

Der Satz von Browder-Minty über die Lösbarkeit monotoner Operatorgleichungen gilt zu Recht als eindrucksvolles Beispiel für die Verbindung des Schönen mit dem Nützlichen in der Mathematik [7]. In der Literatur wird dieser Satz in der Regel mit Argumenten vom Fixpunkttyp bewiesen (Fixpunktsatz von Brouwer, KKM-Theorem). Hier soll ein Beweis gegeben werden, der als einziges strukturelles Hilfsmittel den Trennsatz für konvexe Mengen im $\mathbb{R}^n$ benutzt - genauer: eine fast offensichtliche Folgerung aus diesem Trennsatz, nämlich das sogenannte Lemma von Fan-Glicksberg-Hoffman [5]. Eine derartige Vorgehensweise kann einmal den Vorzug einer gewissen Ökonomie der eingesetzten Hilfsmittel geltend machen. Sie zeigt aber auch, daß das Existenzproblem für monotone Variationsungleichungen sich adäquat behandeln läßt durch Untersuchung einer geeignet gewählten Funktion $\varphi: B\times B \to \mathbb{R}$. Diese Funktion ist konvex im zweiten Argument und erfüllt ferner $\varphi(u,v) + \varphi(v,u) \leq 0 \ \ \forall u,v \in B$ sowie $\varphi(u,u) = 0 \ \ \forall u \in B$. Wir nennen derartige Funktionen *monoton-konvex*. Sie scheinen uns für das vorliegende Problem besser geeignet zu sein als die konkav-konvexen Funktionen, die in [4] zum gleichen Zweck herangezogen wurden.

Das bereits erwähnte Lemma von Fan-Glicksberg-Hoffman besagt:

<u>FGH-Lemma.</u> Sei Γ eine beliebige konvexe Menge, seien $f_i : \Gamma \to \mathbb{R}$ $(i=1,\dots,n)$ konvexe Funktionen mit der Eigenschaft, daß $\max_i f_i(x) \geq 0 \ \ \forall x \in \Gamma$. Dann gibt es reelle Zahlen $\mu_i \geq 0$ $(i=1,\dots,n)$ mit $\Sigma_i \mu_i = 1$, so daß $\Sigma_i \mu_i f_i(x) \geq 0 \ \ \forall x \in \Gamma$.

Der *Beweis* ergibt sich unmittelbar aus dem Trennsatz im $\mathbb{R}^n$: Nach Voraussetzung ist $0 \in \mathbb{R}^n$ kein innerer Punkt der konvexen Menge $D := \{ z \in \mathbb{R}^n \mid z_i \geq f_i(x) \ (i=1,\dots,n),\ x \in \Gamma \}$. Daher existiert im $\mathbb{R}^n$ ein lineares Funktional $\langle \mu, \cdot \rangle$ mit $\mu \neq 0$, so daß $\langle \mu, z \rangle \geq 0 \ \ \forall z \in D$. Wegen $D + \mathbb{R}^n_+ \subset D$ folgt $\mu \geq 0$. Man normiert auf $\Sigma_i \mu_i = 1$ und wählt als $z \in D$ die Punkte $z = (f_1(x),\dots,f_n(x))$ mit $x \in \Gamma$. Dies liefert die Aussage des Lemmas. □

Der Satz von Browder-Minty findet sich in der Literatur in verschiedenen Varianten. Wir legen uns hier auf die folgende mengenwertige Fassung fest:

<u>Satz 1</u> (Browder-Minty). Sei E ein reflexiver reeller Banach-Raum. Sei E*, sein topologischer Dualraum, mit der schwachen* Topologie versehen. Sei $B \subset E$ nichtleer, abgeschlossen und konvex. Sei $\Phi : B \rightrightarrows E^*$ eine mengenwertige Abbildung mit $\Phi(u) \neq \emptyset \ \ \forall u \in B$, die den folgenden Bedingungen genügt:

(i) Φ ist monoton, d.h. $\langle u^*-v^*, u-v \rangle \geq 0 \ \ \forall u,v \in B, \ \forall u^* \in \Phi(u), \ \forall v^* \in \Phi(v)$;

(ii) Φ ist hemistetig, d.h. die Abbildung $s \in [0,1] \rightrightarrows \Phi(u+s(v-u))$ ist oberhalbstetig in $s=0 \ \ \forall u,v \in B$, und $\Phi(u)$ ist konvex und kompakt $\forall u \in B$;

(iii) Φ ist koerzitiv, d.h. $\exists \bar{v} \in B, \ \bar{v}^* \in \Phi(\bar{v})$, so daß $\langle u^*-\bar{v}^*, u-\bar{v} \rangle \geq \gamma(\| u-\bar{v} \|) \cdot \| u-\bar{v} \|$ $\forall u \in B, \ \forall u^* \in \Phi(u)$, wobei $\gamma(s) \to \infty$ für $s \to \infty$.

Dann gibt es $u_o \in B$ und $u_o^* \in \Phi(u_o)$, so daß $\langle u_o^*, u-u_o \rangle \geq 0 \ \ \forall u \in B$.

---

*) Universität Mannheim, Fakultät für Mathematik und Informatik, Schloß, D-6800 Mannheim 1

Um den Satz zu beweisen, arbeitet man mit einer Funktion $\varphi: B\times B \to \mathbb{R}$; dabei spielen folgende Bedingungen eine Rolle:

(B1) φ ist konvex und unterhalbstetig im 2. Argument, und es gilt $\varphi(u,u) = 0 \quad \forall u \in B$.

(B2) φ ist monoton, d.h. $\varphi(u,v) + \varphi(v,u) \le 0 \quad \forall u,v \in B$.

(B3) φ ist radialstetig, d.h. die Abbildung $s \in [0,1] \mapsto \varphi(u+s(v-u),v)$ ist oberhalbstetig in $s = 0 \quad \forall u,v \in B$.

(B4) φ ist koerzitiv, d.h. $\exists C \subset B$ kompakt und konvex, mit $\mathrm{core}_B C \neq \emptyset$, so daß zu jedem $u \in C \setminus \mathrm{core}_B C$ ein $v \in \mathrm{core}_B C$ existiert mit $\varphi(u,v) \le 0$ (hierbei ist $v \in \mathrm{core}_B C :\Leftrightarrow \forall w \in B \; \exists \lambda_o > 0$, so daß $v + \lambda(w-v) \in C \quad \forall \lambda \in [0,\lambda_o]$).

(B5) Für beliebiges $u_o \in B$ gilt $\varphi(u_o,u) \ge 0 \quad \forall u \in B$ dann und nur dann, wenn ein Element $u_o^* \in \Phi(u_o)$ existiert, so daß $\langle u_o^*, u-u_o\rangle \ge 0 \quad \forall u \in B$.

Hilfssatz 1. Seien die Voraussetzungen von Satz 1 erfüllt. Sei E mit der schwachen Topologie versehen. Sei $\varphi: B\times B \to \mathbb{R}$ definiert durch

$$\varphi(u,v) := \sup_{u^* \in \Phi(u)} \langle u^*, v-u\rangle. \tag{1}$$

Dann erfüllt φ die Bedingungen (B1) bis (B5).

Beweis.

(B1): Für festes $u \in B$ ist $\varphi(u,v)$ als Supremum einer Familie von in v konvexen und unterhalbstetigen Funktionen selbst konvex und unterhalbstetig in v. Offensichtlich gilt auch $\varphi(u,u) = 0 \quad \forall u \in B$.

(B2): Wegen Eigenschaft (i) in Satz 1 gilt $0 \ge \langle u^*, v-u\rangle + \langle v^*, u-v\rangle \quad \forall u,v \in B, \; \forall u^* \in \Phi(u), \forall v^* \in \Phi(v)$, also auch bei Supremumsbildung $0 \ge \varphi(u,v) + \varphi(v,u) \quad \forall u,v \in B$. Insbesondere ist φ reellwertig.

(B3): Es gilt $\varphi(u+s(v-u),v) = (1-s) \cdot \sup_{z \in \Gamma(s)} \langle z, v-u\rangle$, wobei $\Gamma(s) := \Phi(u+s(v-u))$. Die Oberhalbstetigkeit folgt dann aus dem Maximumtheorem von Berge [1] wegen Eigenschaft (ii) in Satz 1.

(B4): Unter Verwendung von Eigenschaft (iii) in Satz 1 wähle man $R > 0$ so groß, daß $\gamma(R) \ge \| \bar{v}^* \|$, und hierauf $C := \{ u \in B \mid \| u-\bar{v} \| \le R \}$. Dann ist C konvex und schwach kompakt, und $\bar{v} \in \mathrm{core}_B C$. Sei $u \in C \setminus \mathrm{core}_B C$. Dann ist $\| u-\bar{v} \| = R$, und es gilt für alle $u^* \in \Phi(u)$, daß

$$\begin{aligned}\langle u^*, \bar{v}-u\rangle &= \langle u^*-\bar{v}^*, \bar{v}-u\rangle + \langle \bar{v}^*, \bar{v}-u\rangle \\ &\le -\gamma(\| u-\bar{v} \|) \cdot \| u-\bar{v} \| + \| \bar{v}^* \| \cdot \| u-\bar{v} \| \\ &= R \cdot (-\gamma(R) + \| \bar{v}^* \|) \le 0.\end{aligned}$$

Durch Supremumsbildung folgt $\varphi(u,\bar{v}) \le 0$. Der Punkt $\bar{v} \in \mathrm{core}_B C$ hat also für alle $u \in C \setminus \mathrm{core}_B C$ die geforderte Eigenschaft.

(B5): Aus $\langle u_o^*, u-u_o\rangle \ge 0$ mit $u_o^* \in \Phi(u_o)$ folgt trivialerweise $\varphi(u_o,u) \ge 0$. Sei nun umgekehrt $\varphi(u_o,u) \ge 0 \quad \forall u \in B$. Angenommen, es existiert kein $u_o^* \in \Phi(u_o)$ mit $\langle u_o^*, u-u_o\rangle \ge 0 \quad \forall u \in B$. Dann hat die Familie der abgeschlossenen Mengen

$$F(u,\varepsilon) := \{ z \in \Phi(u_o) \mid \langle z, u-u_o\rangle \ge -\varepsilon \} \quad (u \in B, \; \varepsilon > 0)$$

leeren Durchschnitt über $\Phi(u_o)$. Da $\Phi(u_o)$ als kompakt vorausgesetzt ist, gibt es dann endlich viele $u_i \in B$, $\varepsilon_i > 0$ $(i=1,\dots,n)$, so daß $\bigcap_i F(u_i,\varepsilon_i) = \emptyset$. Mit $\bar{\varepsilon} := \min_i \varepsilon_i$ folgt hieraus $\min_i \langle z, u_i-u_o\rangle \le -\bar{\varepsilon} \quad \forall z \in \Phi(u_o)$. Aus dem FGH-Lemma (übertragen auf konkave Funktionen) folgt dann die Existenz von $\mu_i \ge 0$ $(i=1,\dots,n)$ mit $\Sigma_i \mu_i = 1$, so daß $\Sigma_i \mu_i \langle z, u_i-u_o\rangle \le -\bar{\varepsilon} \quad \forall z \in \Phi(u_o)$. Für $\bar{u} := \Sigma_i \mu_i u_i \in B$ gilt also $\langle z, \bar{u}-u_o\rangle \le -\bar{\varepsilon} \quad \forall z \in \Phi(u_o)$, und damit auch $\varphi(u_o,\bar{u}) \leq -\bar{\varepsilon}$. Dies steht im Widerspruch zur Voraussetzung $\varphi(u_o,u) \geq 0 \quad \forall u \in B$. □

Bemerkung. Es sei

$$\partial_2\varphi(u,u) := \{\, f \in E^* \mid \varphi(u,v) - \varphi(u,u) \geq \langle f, v-u\rangle \quad \forall v \in B \,\} \qquad (u \in B),$$

$$N_B(u) := \{\, g \in E^* \mid \langle g, v-u\rangle \geq 0 \quad \forall v \in B \,\} \qquad (u \in B).$$

Ersetzt man Φ durch $\Phi - f$, wobei $f \in E^*$, so geht $\varphi(u,v)$ gemäß (1) über in $\varphi(u,v) - \langle f, v-u\rangle$. Die Voraussetzungen (i), (ii), (iii) von Satz 1 vererben sich auf $\Phi - f$. Daher erhält man aus der Gültigkeit von (B5) sofort die weitergehende Aussage, daß

$$(\Phi - N_B)(u) = \partial_2\varphi(u,u) \quad \forall u \in B. \tag{2}$$

Der *Beweis* von Satz 1 ergibt sich nun sofort aus dem folgenden Satz 2 in Verbindung mit Hilfssatz 1.

Satz 2. Sei B eine nichtleere, abgeschlossene, konvexe Menge in einem reellen topologischen Vektorraum E. Sei $\varphi: B \times B \to \mathbb{R}$ eine Funktion, die die Bedingungen (B1) - (B4) erfüllt. Dann gibt es ein $u_o \in B$, so daß $\varphi(u_o,u) \geq 0 \quad \forall u \in B$.

Beweis.

1. Sei C die Menge aus Voraussetzung (B4), und seien $x_i \in C$ $(i=1,\dots,n)$ beliebig gewählt. Wir zeigen die Existenz von $\xi \in C$, so daß $\varphi(x_i,\xi) \leq 0$ $(i=1,\dots,n)$. Sei $\Sigma^n := \{\, \lambda \in \mathbb{R}^n \mid \lambda \geq 0,\ \Sigma_i\ \lambda_i = 1 \,\}$. Für $i = 1,\dots,n$ sei $f_i(\lambda) := \varphi(x_i, \Sigma_{j=1}^n\ \lambda_j x_j)$ $(\lambda \in \Sigma^n)$. Die Funktionen $f_i(\cdot)$ sind konvex und unterhalbstetig. Die unterhalbstetige Funktion $\max_i\ f_i(\lambda)$ nimmt ihr Minimum auf der kompakten, konvexen Menge Σ^n an, etwa in $\bar{\lambda} \in \Sigma^n$. Sei dann $\alpha := \max_i\ f_i(\bar{\lambda}) = \min_{\lambda \in \Sigma^n}\ \max_i\ f_i(\lambda)$. Wegen $\max_i\ (f_i(\lambda) - \alpha) \geq 0 \quad \forall \lambda \in \Sigma^n$ folgt nach dem FGH-Lemma die Existenz von $\bar{\mu} \in \Sigma^n$, so daß $\Sigma_i\ \bar{\mu}_i f_i(\lambda) - \alpha \geq 0 \quad \forall \lambda \in \Sigma^n$, insbesondere also $\alpha \leq \Sigma_i\ \bar{\mu}_i f_i(\bar{\mu})$. Es gilt aber wegen der Konvexität von φ und wegen (B2) $\Sigma_i\ \bar{\mu}_i f_i(\bar{\mu}) = \Sigma_i\ \bar{\mu}_i \varphi(x_i, \Sigma_j\ \bar{\mu}_j x_j) \leq \Sigma_{i,j}\ \bar{\mu}_i \bar{\mu}_j \varphi(x_i,x_j) = \frac{1}{2}\ \Sigma_{i,j}\ \bar{\mu}_i \bar{\mu}_j (\varphi(x_i,x_j) + \varphi(x_j,x_i)) \leq 0$. Somit ist $\alpha \leq 0$. Es folgt wegen $\max_i\ f_i(\bar{\lambda}) = \alpha$, daß $f_i(\bar{\lambda}) \leq 0$ $(i=1,\dots,n)$, also $\varphi(x_i, \Sigma_j\ \bar{\lambda}_j x_j) \leq 0$ $(i=1,\dots,n)$. Setzt man $\xi := \Sigma_j\ \bar{\lambda}_j x_j$, so ist $\xi \in C$, und man hat $\varphi(x_i,\xi) \leq 0$ $(i=1,\dots,n)$.

2. Sei $S(x) := \{\, \xi \in C \mid \varphi(x,\xi) \leq 0 \,\}$ $(x \in C)$. Es wurde gezeigt, daß je endlich viele dieser Mengen nichtleeren Durchschnitt haben. Weil die Mengen $S(x)$ abgeschlossen in der kompakten Menge C sind, hat dann die gesamte Familie nichtleeren Durchschnitt. Es gibt also ein $u_o \in C$, so daß $\varphi(x,u_o) \leq 0 \quad \forall x \in C$.

3. Wir zeigen nun, daß mit diesem $u_o \in C$ auch gilt $\varphi(u_o,u) \geq 0 \quad \forall u \in C$. Sei $u \in C$ beliebig, $x(t) := u_o + t(u-u_o)$, $0 < t \leq 1$. Es ist $x(t) \in C$, und daher gilt, wie bereits gezeigt, $\varphi(x(t),u_o) \leq 0$. Aus (B1) folgt ferner $\varphi(x(t),x(t)) = 0$. Aus der Konvexität von φ im 2. Argument folgt dann $0 = \varphi(x(t),x(t)) \leq t\cdot\varphi(x(t),u) + (1-t)\cdot\varphi(x(t),u_o) \leq t\cdot\varphi(x(t),u)$. Division durch t und $t \to 0$ liefert wegen (B3) $0 \leq \varphi(u_o,u)$. Somit gilt $\varphi(u_o,u) \geq 0 \quad \forall u \in C$.

4. Nun zeigen wir $\varphi(u_o,u) \geq 0 \quad \forall u \in B$. Angenommen, es gäbe ein $u \in B \setminus C$, so daß $\varphi(u_o,u) < 0$. Dann gibt es ein $v_o \in \mathrm{core}_B C$, so daß $\varphi(u_o,v_o) \leq 0$ (ist $u_o \in \mathrm{core}_B C$, so wähle man $v_o := u_o$; ist $u_o \notin \mathrm{core}_B C$, so ergibt sich v_o aus (B4)). Es folgt, weil $\varphi(u_o,\cdot)$ konvex ist, $\varphi(u_o,\xi) < 0 \quad \forall \xi \in (v_o,u]$. Wegen $v_o \in \mathrm{core}_B C$ ist aber $C \cap (v_o,u] \neq \emptyset$. Somit gäbe es auch ein $\xi \in C$ mit $\varphi(u_o,\xi) < 0$. Dies ist ein Widerspruch zum bereits Bewiesenen. Es gilt also $\varphi(u_o,u) \geq 0 \quad \forall u \in B$. Damit ist der Beweis vollständig. □

Bemerkung. Seien B und E wie im Satz 2, und sei $\varphi: B\times B \to \mathbb{R}$ eine Funktion, die den Bedingungen (B1), (B2) genügt. Dann ist jede der folgenden Bedingungen (K1) bis (K4) hinreichend für die Gültigkeit der Koerzitivitätsbedingung (B4) (wobei wir bei (K1) voraussetzen wollen, daß $\bar{v}$ nicht bereits der Folgerung von Satz 2 genügt):

(K1) Es existiere $\bar{v}\in B$ derart, daß die Menge $K:=\{u\in B \mid \varphi(\bar{v},u)\le 0\}$ kompakt ist.

(K2) Es sei E ein reflexiver Banach-Raum, und es existiere $\bar{v}\in B$ derart, daß $\varphi(u,\bar{v})+\varphi(\bar{v},u)\le -\gamma(\|u-\bar{v}\|)\cdot\|u-\bar{v}\| \quad \forall u\in B$, wobei $\gamma(s)\to\infty$ für $s\to\infty$.

(K3) Es sei E ein reflexiver Banach-Raum, und es existiere $\bar{v}\in B$ derart, daß $\inf_{u\in B}\varphi(\bar{v},u)>-\infty$ und daß $\varphi(u,\bar{v})+\varphi(\bar{v},u)\le -\gamma\cdot\|u-\bar{v}\| \quad \forall u\in B$ mit $\|u-\bar{v}\|\ge 1$, wobei $\gamma>0$.

(K4) Es sei $E=\mathbb{R}^n$; φ sei stetig in jedem der beiden Argumente, und es gelte $\varphi(u,u+\lambda(v-u))=\lambda\varphi(u,v) \quad \forall u,v\in B, \ \forall\lambda\in[0,1]$; ferner existiere $\bar{v}\in B$ derart, daß $\inf_{u\in B}\varphi(\bar{v},u)>-\infty$ und daß $\varphi(u,\bar{v})+\varphi(\bar{v},u)<0 \quad \forall u\in B,\ u\ne\bar{v}$.

Beweis. Wir zeigen, daß jede dieser Bedingungen (B4) impliziert.

(K1): Falls $\varphi(\bar{v},u)\ge 0 \quad \forall u\in K$, so liefert die Definition von K, daß $\varphi(\bar{v},u)\ge 0 \quad \forall u\in B$. Schließt man diesen Fall aus, so gibt es also ein $u^*\in K$ mit $\varphi(\bar{v},u^*)\le -\varepsilon<0$. Man setzt $C:=\{u\in B \mid \varphi(\bar{v},u)\le 1\}$. Dann ist C konvex und abgeschlossen. Ferner gilt $C-u^*\subset\varepsilon^{-1}(1+\varepsilon)\cdot(K-u^*)$; daher ist C kompakt. Wegen $\varphi(\bar{v},\bar{v})=0$ ist $\bar{v}\in \mathrm{core}_B C$. Dann gilt $\varphi(u,\bar{v})\le-\varphi(\bar{v},u)\le 0$ für alle $u\in B$ mit $\varphi(\bar{v},u)=1$, mithin für alle $u\in C\setminus \mathrm{core}_B C$. (B4) ist erfüllt.

(K2): Wir können annehmen, daß E mit der schwachen Topologie versehen ist. Sei $\alpha:=\inf\{\varphi(\bar{v},u) \mid u\in B,\ \|u-\bar{v}\|\le 1\}$. Dann ist $-\infty<\alpha\le 0$, und es gilt für alle $u\in B$ mit $\|u-\bar{v}\|\ge 1$, daß $\varphi(\bar{v},u)\ge\alpha\cdot\|u-\bar{v}\|$ und damit auch $\varphi(u,\bar{v})\le\|u-\bar{v}\|\cdot(-\alpha-\gamma(\|u-\bar{v}\|))$. Man wählt $R\ge 1$ so groß, daß $-\alpha-\gamma(R)\le 0$, und setzt $C:=\{u\in B \mid \|u-\bar{v}\|\le R\}$. Es ist $\bar{v}\in\mathrm{core}_B C$, und es gilt $\varphi(u,\bar{v})\le 0$ für alle $u\in B$ mit $\|u-\bar{v}\|=R$, mithin für alle $u\in C\setminus\mathrm{core}_B C$. (B4) ist erfüllt.

(K3): Wegen $\varphi(\bar{v},u)\ge M \quad \forall u\in B$ folgt für alle $u\in B$ mit $\|u-\bar{v}\|\ge 1$, daß $\varphi(u,\bar{v})\le-\varphi(\bar{v},u)-\gamma\cdot\|u-\bar{v}\|\le -M-\gamma\cdot\|u-\bar{v}\|$. Man wählt $R\ge 1$ so groß, daß $-M-\gamma R\le 0$, und $C:=\{u\in B \mid \|u-\bar{v}\|\le R\}$. Dann ist wieder (B4) erfüllt.

(K4): Sei $\alpha:=\max\{\varphi(u,\bar{v})+\varphi(\bar{v},u) \mid u\in B,\ \|u-\bar{v}\|=1\}$. Es ist $\alpha<0$, und für alle $u\in B$ mit $\|u-\bar{v}\|\ge 1$ gilt dann $\varphi(u,\bar{v})+\varphi(\bar{v},u)\le\alpha\cdot\|u-\bar{v}\|$. Sei nämlich $u\in B$ mit $k:=\|u-\bar{v}\|\ge 1$. Setze $w:=\frac{1}{k}u+\frac{k-1}{k}\bar{v}$. Dann ist

$$0=\frac{k^2}{k-1}\varphi(w,w)\le\frac{k^2}{k-1}\left(\frac{1}{k}\varphi(w,u)+\frac{k-1}{k}\varphi(w,\bar{v})\right)=\frac{k}{k-1}\varphi(w,u)+k\varphi(w,\bar{v})=:L.$$

Wegen $\bar{v}=u+\frac{k}{k-1}(w-u)$, $u=\bar{v}+k(w-\bar{v})$, $\|w-\bar{v}\|=1$ folgt dann

$$\varphi(u,\bar{v})+\varphi(\bar{v},u)=\frac{k}{k-1}\varphi(u,w)+k\varphi(\bar{v},w)$$
$$\le\frac{k}{k-1}\varphi(u,w)+k\varphi(\bar{v},w)+L=\frac{k}{k-1}(\varphi(u,w)+\varphi(w,u))+k(\varphi(\bar{v},w)+\varphi(w,\bar{v}))\le k(\varphi(\bar{v},w)+\varphi(w,\bar{v}))$$
$$\le k\alpha.$$

Damit befindet man sich wieder in der Situation von (K3). □

Im Zusammenhang mit (K3) und (K4) verweisen wir auf den Fall, daß B ein konvexer Kegel ist mit zugehörigem Polarkegel $B^*:=\{u^*\in E^* \mid \langle u^*,u\rangle\ge 0 \ \forall u\in B\}$, und $\varphi(u,v):=\langle\Phi(u),v-u\rangle$ $(\Phi: B\to E^*)$. In diesem Fall ist die Bedingung $\bar{v}\in B$, $\inf_{u\in B}\varphi(\bar{v},u)>-\infty$ äquivalent zu $\bar{v}\in B$, $\Phi(\bar{v})\in B^*$, während die Aussage $u_o\in B$, $\varphi(u_o,u)\ge 0 \ \forall u\in B$ (die in der Konklusion von Satz 2 auftritt) äquivalent ist zu $u_o\in B$, $\Phi(u_o)\in B^*$, $\langle\Phi(u_o),u_o\rangle=0$ (Komplementaritätsbedingung).

Der folgende Satz kommt mit einer recht schwachen Koerzitivitätsbedingung aus.

Satz 3. Sei $E = E^*$ ein Hilbert-Raum, sei $B \subset E$ abgeschlossen, konvex, nichtleer. $\varphi: B \times B \to \mathbb{R}$ erfülle die Bedingungen (B1), (B2), (B3) sowie $\varphi(u,v) + \varphi(v,u) \le -p(\| u-v \|) \cdot \| u-v \|^2 \quad \forall u,v \in B$, wobei $p(s) > 0$ für $s > 0$ und monoton nicht wachsend ist. Dann gibt es ein $u_o \in B$ mit $\varphi(u_o,u) \ge 0 \quad \forall u \in B$.

Beweis. Sei $f \in E$ beliebig. Mit $\psi_f(u,v) := \varphi(u,v) + \langle u,v-u\rangle - \langle f,v-u\rangle$ gilt, daß ψ_f die Voraussetzungen (B1), (B2), (B3) erfüllt. Weiter gilt $\psi_f(u,v) + \psi_f(v,u) = \varphi(u,v) + \varphi(v,u) + \langle u-v,v-u\rangle \le -\| u-v \|^2$. Somit erfüllt ψ_f auch (K2) und damit (B4). Satz 2 ist auf ψ_f anwendbar und liefert ein $u \in B$ mit

$$\psi_f(u,v) \ge 0 \quad \forall v \in B. \tag{3}$$

u ist eindeutig. Sei $T: E \to B$ die Abbildung, die jedem $f \in E$ das $u \in B$ gemäß (3) zuordnet. Ist $u_1 = T(f_1)$, $u_2 = T(f_2)$, so gilt insbesondere

$0 \le \psi_{f_1}(u_1,u_2) + \psi_{f_2}(u_2,u_1) = \varphi(u_1,u_2) + \varphi(u_2,u_1) + \langle u_1-u_2,u_2-u_1\rangle + \langle f_1-f_2,u_1-u_2\rangle$
$\le (-p(\| u_1-u_2 \|) \cdot \| u_1-u_2 \| - \| u_1-u_2 \| + \| f_1-f_2 \|) \cdot \| u_1-u_2 \|$.

Hieraus folgt mit $q(\cdot) := (1+p(\cdot))^{-1}$, daß $\| u_1-u_2 \| \le q(\| u_1-u_2 \|) \cdot \| f_1-f_2 \|$. Wegen $q(\cdot) \le 1$ folgt hieraus $\| u_1-u_2 \| \le \| f_1-f_2 \|$, und weil $q(\cdot)$ monoton nicht fallend ist, folgt dann $\| u_1-u_2 \| \le q(\| f_1-f_2 \|) \cdot \| f_1-f_2 \|$. Insgesamt gilt dann $\| T(f_1)-T(f_2) \| \le q(\| f_1-f_2 \|) \cdot \| f_1-f_2 \|$ mit $q(s) < 1$ für $s > 0$. $T: B \to B$ genügt also der abgeschwächten Kontraktionsvoraussetzung des Fixpunktsatzes aus [2] und hat damit einen Fixpunkt $u_o = T(u_o)$. Mit $u := f := u_o$ besagt (3) aber gerade $\varphi(u_o,v) \ge 0 \quad \forall v \in B$. □

Im Zusammenhang mit dem Satz von Browder-Minty wird häufig anstelle der Hemistetigkeit von Φ maximale Monotonie vorausgesetzt. Wir gehen kurz auf diesen Fall ein.

Eine monotone Abbildung $\Phi: B \rightrightarrows E^*$ heißt bekanntlich maximal monoton über B, wenn aus $v \in B$, $v^* \in E^*$ und $\langle u^*-v^*,u-v\rangle \ge 0 \quad \forall u \in B, \ \forall u^* \in \Phi(u)$ folgt, daß $v^* \in \Phi(v)$. Dies legt für eine Funktion $\varphi: B \times B \to \mathbb{R}$, die den Bedingungen (B1), (B2) genügt, die folgende *Definition* nahe:

Die Funktion $\varphi: B \times B \to \mathbb{R}$ heißt maximal monoton, wenn für alle $v \in B$ und $v^* \in E^*$ aus $\varphi(u,v) + \langle v^*,u-v\rangle \le 0 \quad \forall u \in B$ folgt, daß $\varphi(v,u) + \langle v^*,v-u\rangle \ge 0 \quad \forall u \in B$.

Seien nun B und E wie im Satz 2, und $\varphi: B \times B \to \mathbb{R}$ genüge den Bedingungen (B1), (B2). Wir definieren $T_\varphi: B \rightrightarrows E^*$ als $T_\varphi(u) := \partial_2\varphi(u,u)$. Wegen $\varphi(u,v) \ge \langle u^*,v-u\rangle \quad \forall u \in B, \ \forall u^* \in T_\varphi(u)$ folgt sofort, daß T_φ monoton über B ist. Ist T_φ maximal monoton über B, so ist φ maximal monoton im Sinne der obigen Definition. In der Tat: Aus $\varphi(u,v) + \langle v^*,u-v\rangle \le 0 \quad \forall u \in B$ folgt $\langle u^*-v^*,u-v\rangle \ge 0 \quad \forall u \in B, \ \forall u^* \in T_\varphi(u)$, hieraus folgt dann $v^* \in T_\varphi(v)$, und hieraus $\varphi(v,u) + \langle v^*,v-u\rangle \ge 0 \quad \forall u \in B$. □

Ferner ist φ maximal monoton, wenn (B3) erfüllt ist. In der Tat: Die Bedingungen (B1), (B2), (B3) vererben sich auf die Funktion $\psi_{v^*}(u,v) := \varphi(u,v) + \langle v^*,u-v\rangle$, und aus $\psi_{v^*}(u,v) \le 0 \quad \forall u \in B$ folgt dann wie im Schritt 3 des Beweises von Satz 2, daß $\psi_{v^*}(v,u) \ge 0 \quad \forall u \in B$. □

Ist $E = E^*$ ein Hilbert-Raum, so sieht man auch leicht, daß aus (B1), (B2), (B3) die maximale Monotonie von T_φ folgt. In der Tat: Sei $v \in B$, $v^* \in E$ so, daß

(*) $\quad \langle u^*-v^*,u-v\rangle \ge 0 \quad \forall u \in B, \ \forall u^* \in T_\varphi(u)$.

Setze $f := v + v^*$ und $\psi_f(u,w) := \varphi(u,w) + \langle u,w-u\rangle - \langle f,w-u\rangle$ $(u,w \in B)$. Es folgt wie im Beweis zu Satz 3 - siehe (3) - die Existenz von $u_o \in B$, so daß $\psi_f(u_o,w) \geq 0 \;\; \forall w \in B$, also

(**) $\quad \varphi(u_o,w) + \langle u_o, w-u_o\rangle - \langle v+v^*, w-u_o\rangle \geq 0 \;\; \forall w \in B.$

Hieraus folgt $-u_o + v + v^* \in T_\varphi(u_o)$. Man wählt in (*) $u := u_o$ und $u^* := -u_o + v + v^* \in T_\varphi(u)$ und erhält $0 \leq \langle v-u_o, u_o-v\rangle = -\|u_o-v\|^2$. Es ist also $u_o = v$, und dies in (**) eingesetzt ergibt $\varphi(v,w) - \langle v^*, w-v\rangle \geq 0 \;\; \forall w \in B$, also $v^* \in T_\varphi(v)$. □

Es gilt nun die folgende Variante zu Satz 2.

Satz 4. Seien B und E wie im Satz 2. $\varphi: B\times B \to \mathbb{R}$ genüge den Bedingungen (B1), (B2), (K1), ferner sei φ maximal monoton. Dann gibt es ein $u_o \in B$, so daß $\varphi(u_o,u) \geq 0 \;\; \forall u \in B$.

Beweis. Wir erwähnen nur die Änderungen, die im Beweis zu Satz 2 vorgenommen werden müssen. Im Schritt 1 zeigt man jetzt, daß für je endlich viele Punkte $x_i \in B$ $(i=1,\ldots,n)$, unter denen der Punkt $\overline{v}$ aus Voraussetzung (K1) enthalten sei, ein $\xi \in B$ existiert mit $\varphi(x_i,\xi) \leq 0$ $(i=1,\ldots,n)$. Aus $\varphi(\overline{v},\xi) \leq 0$ folgt insbesondere $\xi \in K$, K aus Voraussetzung (K1). In Schritt 2 erhält man dann wegen der Kompaktheit von K ein $u_o \in K$, so daß $\varphi(x,u_o) \leq 0$ $\forall x \in B$. Hieraus folgt im Schritt 3 unmittelbar $\varphi(u_o,u) \geq 0 \;\; \forall u \in B$ aufgrund der maximalen Monotonie von φ. Schritt 4 entfällt. □

Seien schließlich B und E wie im Satz 1. Sei $\Phi: B \rightrightarrows E^*$ eine Abbildung mit $\Phi(u) \neq \emptyset$ $\forall u \in B$, die der Bedingung (i) aus Satz 1 genügt. Sei $\Phi_B(u) := (\Phi - N_B)(u)$ für $u \in B$, $\varphi: B\times B \to \mathbb{R}$ sei durch (1) definiert, T_φ wie zuvor. Man sieht unmittelbar, daß dann auch $\varphi(u,v) = \sup_{u^* \in \Phi_B(u)} \langle u^*, v-u\rangle$, und $\Phi_B(u) \subset T_\varphi(u)$. Ferner ist T_φ monoton, weil φ die Bedingungen (B1), (B2) erfüllt.

Es gilt nun: Wenn Φ der Bedingung (ii) aus Satz 1 genügt, oder wenn Φ_B maximal monoton über B ist, dann ist $\Phi_B(u) = T_\varphi(u) \;\; \forall u \in B$, und T_φ ist maximal monoton über B (insbesondere also φ maximal monoton). *Beweis.* Wenn Φ der Bedingung (ii) genügt, so genügt φ, wie im Beweis von Hilfssatz 1 gezeigt, der Bedingung (B3), und es gilt die Aussage (2), also $\Phi_B(u) = T_\varphi(u) \;\; \forall u \in B$. Um die maximale Monotonie von T_φ zu zeigen, sei $v \in B$, $v^* \in E^*$ mit $\langle u^*-v^*, u-v\rangle \geq 0 \;\; \forall u \in B, \; \forall u^* \in T_\varphi(u)$. Hieraus folgt, weil jetzt $\varphi(u,v) = \sup_{u^* \in T_\varphi(u)} \langle u^*, v-u\rangle$, daß $\psi_{v^*}(u,v) := \varphi(u,v) + \langle v^*, u-v\rangle \leq 0 \;\; \forall u \in B$. Weil ψ_{v^*} ebenfalls der Bedingung (B3) genügt, folgt wie im Schritt 3 des Beweises zu Satz 2, daß dann auch $\psi_{v^*}(v,u) = \varphi(v,u) + \langle v^*, v-u\rangle \geq 0 \;\; \forall u \in B$. Somit ist $v^* \in T_\varphi(v)$, und T_φ ist maximal monoton. Ist Φ_B maximal monoton, so folgt die Behauptung sofort aus $\Phi_B(u) \subset T_\varphi(u)$ und der Monotonie von T_φ. □

Minty [6] hat sein Resultat ursprünglich mittels eines Satzes von Kirszbraun über die Fortsetzbarkeit einer Lipschitz-stetigen Abbildung bewiesen. Der folgende Satz enthält den Satz von Kirszbraun als Sonderfall und läßt sich ebenfalls mittels des FGH-Lemmas beweisen. conv A bezeichne im folgenden die konvexe Hülle von A.

Satz 5. Sei X eine konvexe Untermenge eines reellen topologischen Vektorraums. Seien $x_i \in X$ $(i=1,\ldots,n)$. Seien $\varphi: X\times X \to \mathbb{R}$ und $\psi: X\times X \to \mathbb{R}$ Funktionen mit folgenden Eigenschaften: φ sei im 2. Argument konvex und unterhalbstetig; es gelte $\varphi(x_i,x_j) + \varphi(x_j,x_i) \leq 0$ $(i,j=1,\ldots,n)$; ψ sei im 2. Argument konvex und unterhalbstetig und im 1. Argument konkav; es gelte $\psi(x,x) \leq 0 \;\; \forall x \in X$. Dann existiert ein

$\xi \in \operatorname{conv}\{x_1,\dots,x_n\}$, so daß $\varphi(x_i,\xi) + \psi(x_i,\xi) \le 0$ $(i=1,\dots,n)$.

Beweis. Sei $\Sigma^n := \{\lambda \in \mathbb{R}^n \mid \lambda \ge 0,\ \Sigma_i \lambda_i = 1\}$. Für $i = 1,\dots,n$ definiert man auf Σ^n konvexe, unterhalbstetige Funktionen $f_i(\lambda) := \varphi(x_i, \Sigma_j \lambda_j x_j) + \psi(x_i, \Sigma_j \lambda_j x_j)$ $(\lambda \in \Sigma^n)$ und hat dann die Existenz von $\bar{\lambda} \in \Sigma^n$ zu zeigen, so daß $\max_i f_i(\bar{\lambda}) \le 0$. Dies folgt aus dem FGH-Lemma analog zum Schritt 1 des Beweises von Satz 2. Die hierbei benötigte Ungleichung $\Sigma_i \mu_i f_i(\mu) \le 0$ für beliebiges $\mu \in \Sigma^n$ ergibt sich jetzt folgendermaßen:

$$\Sigma_i \mu_i f_i(\mu) = \Sigma_i \mu_i \varphi(x_i, \Sigma_j \mu_j x_j) + \Sigma_i \mu_i \psi(x_i, \Sigma_j \mu_j x_j) \le$$

$$\le \Sigma_{i,j} \mu_i \mu_j \varphi(x_i, x_j) + \psi(\Sigma_i \mu_i x_i, \Sigma_j \mu_j x_j) =$$

$$= \frac{1}{2} \Sigma_{i,j} \mu_i \mu_j (\varphi(x_i,x_j) + \varphi(x_j,x_i)) + \psi(\Sigma_i \mu_i x_i, \Sigma_i \mu_i x_i) \le 0. \quad \square$$

Seien nun $Y := \mathbb{R}^k$, $Z := \mathbb{R}^\ell$, versehen mit der euklidischen Norm. Seien $(y_i, z_i) \in Y \times Z$ $(i=1,\dots,n)$ und $y_o \in Y$. Man setzt im vorangehenden Satz $X := Y \times Z$, $x_i := (y_i, z_i) \in X$ $(i=1,\dots,n)$ und definiert mit $x = (y,z) \in X$ und $\xi = (\eta,\zeta) \in X$ die Funktionen φ und ψ auf $X \times X$ folgendermaßen:

$$\varphi(x,\xi) := \| z \|^2 - \| y \|^2 - \langle z,\zeta\rangle + \langle y,\eta\rangle,$$

$$\psi(x,\xi) := \| \zeta \|^2 - \langle z,\zeta\rangle - \langle y,\eta\rangle + 2\langle y,y_o\rangle - \| y_o \|^2.$$

Dann ist $\varphi(x,\xi) + \varphi(\xi,x) = \| z-\zeta \|^2 - \| y-\eta \|^2$, $\varphi(x,\xi) + \psi(x,\xi) = \| z-\zeta \|^2 - \| y-y_o \|^2$, insbesondere also $\varphi(x_i,x_j) + \varphi(x_j,x_i) = \| z_i-z_j \|^2 - \| y_i-y_j \|^2$, $\varphi(x_i,\xi) + \psi(x_i,\xi) = \| z_i-\zeta \|^2 - \| y_i-y_o \|^2$. Damit erhält man aus Satz 5 das folgende Resultat:

> Für $(y_i,z_i) \in Y \times Z$ $(i=1,\dots,n)$ gelte $\| z_i-z_j \| \le \| y_i-y_j \|$ $(i,j=1,\dots,n)$. Dann gibt es zu jedem $y_o \in Y$ ein $\zeta \in \operatorname{conv}\{z_1,\dots,z_n\}$, so daß $\| z_i-\zeta \| \le \| y_i-y_o \|$ $(i=1,\dots,n)$.

Dies ist der *Satz von Kirszbraun* über die Fortsetzbarkeit einer Lipschitz-stetigen Abbildung. Man vergleiche hierzu auch [3].

Ist $Z = Y^*$, und setzt man mit den gleichen Verabredungen wie zuvor $\varphi(x,\xi) := -\langle z-\zeta,y\rangle$, $\psi(x,\xi) := \langle z-\zeta,y_o\rangle$, so ist $\varphi(x,\xi) + \varphi(\xi,x) = -\langle z-\zeta,y-\eta\rangle$ und $\varphi(x,\xi) + \psi(x,\xi) = -\langle z-\zeta,y-y_o\rangle$. Damit liefert Satz 5 das folgende Resultat:

> Für $(y_i,z_i) \in Y \times Y^*$ $(i=1,\dots,n)$ gelte $\langle z_i-z_j,y_i-y_j\rangle \ge 0$ $(i,j=1,\dots,n)$. Dann gibt es zu jedem $y_o \in Y$ ein $\zeta \in \operatorname{conv}\{z_1,\dots,z_n\}$, so daß $\langle z_i-\zeta,y_i-y_o\rangle \ge 0$ $(i=1,\dots,n)$.

Dies ist der *Satz von Debrunner-Flor* über die Fortsetzbarkeit einer monotonen Abbildung.

[1] C. Berge: Espaces topologiques. Fonctions multivoques. Dunod, Paris, 1966^2.

[2] D.W. Boyd, J.S.W. Wong: On nonlinear contractions. Proc. Amer. Math. Soc. 20 (1969), 458-464.

[3] S. Karamardian: A further generalization of Kirszbraun's theorem. In: Inequalities III, edited by O. Shisha. Academic Press, New York, 1972.

[4] E. Krauss: A representation of maximal monotone operators by saddle functions. Rev. Roumaine Math. Pures Appl. 30 (1985), 823-837.

[5] O.L. Mangasarian: Nonlinear Programming. McGraw-Hill, New York, 1969.

[6] G.J. Minty: Monotone (nonlinear) operators in Hilbert space. Duke Math. J. 29 (1962), 341-346.

[7] J. Naas, W. Tutschke: Große Sätze und schöne Beweise der Mathematik. Identität des Schönen, Allgemeinen, Anwendbaren. Akademie-Verlag, Berlin, 1986.

QUASI-DIFFERENTIABLE FUNCTIONS IN NON-DIFFERENTIABLE OPTIMIZATION THEORY

Diethard Pallaschke [1)]

0. Introduction

In this paper we give a report on quasi-differentiable functions, which have been introduced by V. F. Demyanov, A. N. Rubinov and L. V. Vasilev (see [1], [2]), and their applications in non-differentiable optimization theory. The essential property of quasi-differentiable functions is the existence of a full differential calculus. With this powerfull tool relatively complicated non-differentiable optimization problems can be solved.

1. Quasi-Differentiable Functions

Let $U \subseteq \mathbb{R}^n$ be an open set and $x_0 \in U$. A continuous function

$$f: U \longrightarrow \mathbb{R}$$

is said to be "quasi-differentiable" in x_0 if and only if the following two conditions are satisfied:

i) for every vector $g \in \mathbb{R}^n$ the directional derivative

$$\left.\frac{df}{dg}\right|_{x_0} = \lim_{\substack{\alpha \to 0 \\ \alpha > 0}} \frac{f(x_0+\alpha g) - f(x_0)}{\alpha}$$

exists, and

ii) there exist two compact convex sets $\underline{\partial} f|_{x_0}, \overline{\partial} f|_{x_0} \subseteq \mathbb{R}^n$, such that

$$\left.\frac{df}{dg}\right|_{x_0} = \sup_{v \in \underline{\partial} f|_{x_0}} \langle v,g\rangle + \inf_{w \in \overline{\partial} f|_{x_0}} \langle w,g\rangle .$$

The sets $\underline{\partial} f|_{x_0}$ resp. $\overline{\partial} f|_{x_0}$ are called a "sub-" resp. a "superdifferential" of f in x_0. The pair $Df|_{x_0} := (\underline{\partial} f|_{x_0}, \overline{\partial} f|_{x_0})$ is said to be a "quasi-differential" of f in x_0. It is uniquely determined in the lattice $\mathcal{M}$ of all pairs of compact convex subsets of $\mathbb{R}^n$ (see [1], chapter 6). The set of quasi-differentiable functions is very large. It is an algebra with respect to addition and multiplication and is closed with respect to finite maxima and minima. The differential calculus comprehends the classical differentiation rules, such as the rules for differentiation of sums, products and quotients, and contains in extension also rules for the differentiation of finite maxima and minima of quasi-differentiable functions. For a detailed presentation of this calculus we refer to [1], and to [2], chapter 2.

1) Institut für Statistik und Mathematische Wirtschaftstheorie, Universität Karlsruhe, Kaiserstr. 12, D-7500 Karlsruhe 1

Here we will discuss only a very special fact of this calculus, which follows from the isomorphy of the lattice $\mathcal{M}$ of pairs of compact convex subsets of $\mathbb{R}^n$ to the lattice of differences of sublinear functions defined on $\mathbb{R}^n$. This isomorphy is described in [1], chapter 6. We will now use the following approach:

Let $U \subseteq \mathbb{R}^n$ be an open set, $f: U \longrightarrow \mathbb{R}$ a continuous function, which is quasi-differentiable at $x_0 \in U$. Then from the definition of quasi-differentiability follows, that the directional derivative, as a function of the direction, is given by

$$\mathbb{R}^n \longrightarrow \mathbb{R}$$

$$g \longrightarrow \left.\frac{df}{dg}\right|_{x_0} = \sup_{v \in \underline{\partial} f|_{x_0}} \langle v,g\rangle + \inf_{w \in \overline{\partial} f|_{x_0}} \langle w,g\rangle .$$

In other words: The directional derivative of f at $x_0 \in U$, i.e.

$$\mathbb{R}^n \longrightarrow \mathbb{R}$$

$$g \longrightarrow \left.\frac{df}{dg}\right|_{x_0}$$

can be represented as a difference of two sublinear functions, defined on $\mathbb{R}^n$.

Let us now introduce the lattice

$$\mathcal{D}(\mathbb{R}^n) := \{T := p-q \mid p,q: \mathbb{R}^n \longrightarrow \mathbb{R} \text{ sublinear}\},$$

endowed with the partial order of functions. In this context, quasi-differentiability of f in $x_0 \in U$ can be expressed by

$$\left(g \longrightarrow \left.\frac{df}{dg}\right|_{x_0}\right) \in \mathcal{D}(\mathbb{R}^n).$$

Now the following result holds:

<u>Theorem:</u>
Let $U \subseteq \mathbb{R}^n$ be an open set, $x_0 \in U$, and $f: U \longrightarrow \mathbb{R}$ a locally-lipschitz function. Then the following properties are equivalent:

i) f is quasi-differentiable in x_0

ii) there exists a $T \in \mathcal{D}(\mathbb{R}^n)$ which satisfies the following condition: For every $\varepsilon > 0$ there exists a $\delta > 0$, such that for all $h \in \mathbb{R}^n$ with $\|h\| \le \delta$ and $x_0 + h \in U$ the inequality

$$|f(\bar{x}_o+h)-f(x_o)-T(h)|\leq\varepsilon\ \|h\|$$

holds.

For the proof of this result see [6].

This theorem shows, that for locally-lipschitz functions the quasi-differentiability is very close to the Fréchet-differentiability, namely the condition ii) of the above theorem is identically to the condition for Fréchet-differentiability, except of the fact, that $T\in\mathcal{D}(\mathbb{R}^n)$ and not a linear map.

Let us remark, that there exist quasi-differentiable functions, which are not locally-lipschitz. Namely take

$$f:\mathbb{R}^2\longrightarrow\mathbb{R}$$

with

$$f(x_1,x_2):=\begin{cases}\dfrac{|x_1^3\cdot x_2|}{x_1^4+x_2^2} & (x_1,x_2)\neq(0,0)\\ 0 & (x_1,x_2)=(0,0)\end{cases}$$

This function is not locally-lipschitz in a neighbourhood of $0\in\mathbb{R}^2$ but quasi-differentiable in this point (see [6]).
Although this result gives a nice theoretical understanding of quasi-differentiability, it is not very usefull for the practical work of solving optimization problems.

2. Quasi-Differentiable Optimization Problems

In this part, we will discuss a type of quasi-differentiable optimization problems, which have been considered by V. F. Demyanov, S. Gamidov and T. J. Sivelina [3].
The authors considered the class $\mathcal{F}$ of all functions given by

$$\mathcal{F}:=\{f:\mathbb{R}^n\longrightarrow\mathbb{R}\ \ f(x)=F(x,y_1(x),\dots,y_m(x))\},$$

where

$$y_i:\mathbb{R}^n\longrightarrow\mathbb{R}$$

is defined by

$$y_i(x):=\max_{j\in I_i}\phi_{ij}(x)$$

with

$$I_i:=\{1,\dots,N_i\}; \quad i\in\{1,\dots,m\}$$

and

$$\phi_{ij}:\mathbb{R}^n \longrightarrow \mathbb{R}$$

for all $j\in I_i$ and all $i\in\{1,\dots,m\}$.
The functions F and ϕ_{ij} under cons .deration are assumed to belong to the classes of continuous differentiable functions $C^1(\mathbb{R}^{n+m})$ and $C^1(\mathbb{R}^n)$ respectively.
The optimization problem is:

$$\min f(x)$$

under

$$g_i(x)\le 0, \quad i\in\{1,\dots,k\},$$

where

$$f,g_1,\dots,g_k\in \mathcal{F}.$$

For this type of problems a modified steepest descent algorithm is proposed in [3].

Let us shortly recall this procedure: For an arbitrary quasi-differentiable function

$$f:U\longrightarrow \mathbb{R},$$

defined on an open set $U\subseteq \mathbb{R}^n$, a steepest descent direction in $\hat{x}\in U$ is given as follows:

Let $Df|_{\hat{x}}=(\underline{\partial}f|_{\hat{x}},\bar{\partial}f|_{\hat{x}})$ be a quasi-differential of f in $\hat{x}$ and $v_o\in\underline{\partial}f|_{\hat{x}}, w_o\in\bar{\partial}f|_{\hat{x}}$ such that

$$(*) \qquad \|v_o+w_o\|=\max_{w\in\bar{\partial}f|_{\hat{x}}}\left(\min_{v\in\underline{\partial}f|_{\hat{x}}}\|v+w\|\right).$$

Here $\|\ \|$ denotes the Euklidean norm on $\mathbb{R}^n$. Then a steepest descent direction of the function f in $\hat{x}$ is given by:

$$g|_{\hat{x}} := -\frac{v_o+w_o}{\|v_o+w_o\|_2}.$$

The method of steepest descent is then given by the iteration procedure

$$x_{k+1}=x_k+\alpha_k g|_{x_k}$$

where α_k is determined by a line-search in the direction $g|_{x_k}$.

The algorithm in [3] has been implemented by H. P. Jandl (see [4]) using the fact, that for elements of the class $\mathcal{F}$ the sub- and superdifferential can be chosen as polyhedras. In this case, the quadratic optimization problem (*) for determining a steepest descent direction can be solved by a modification of the Frank-Wolfe algorithm. The algorithm is implemented for an IBM/XT personal computer with 640 KB in PASCAL.

As a further application of quasi-differentiable functions in non-differentiable optimization, let us consider the continuous set covering problem of a compact subset $\Omega \subseteq \mathbb{R}^2$.

Let $k \in \mathbb{N}$ be an integer and let us cover Ω by k balls with centers $x_1, \ldots, x_k \in \Omega$. Then the maximal radius R of these balls is given by

$$R = \max_{x \in \Omega} \min_{1 \le i \le k} \| x - x_i \| ,$$

where $\| \ \|$ denotes the Euklidean norm in $\mathbb{R}^2$.
A minimal covering of Ω with $k \in \mathbb{N}$ balls leads to the optimization problem

$$\min \phi(x_1, \ldots, x_k)$$

under

$$x_1, \ldots, x_k \in \Omega^k ,$$

where

$$\phi(x_1, \ldots, x_k) = \max_{x \in \Omega} \min_{1 \le i \le k} \| x - x_i \| .$$

Using the theory of quasi-differentiable functions, K. Wieder constructed an algorithm to solve this problem for convex decomposable subsets $\Omega \subseteq \mathbb{R}^2$ (see [5]).

References

[1] Demyanov, V. F., Rubinov, A. M.: Quasidifferential Calculus. Optimization Software, Inc. Springer-Verlag, New York 1986.

[2] Demyanov, V. F., Vasilev, L. V.: Nondifferentiable Optimization. Optimization Software, Inc. Springer-Verlag, New York 1985.

[3] Demyanov, V. F., Gamidov, S., Sivelina, T. I.: An Algorithm for Minimizing a Certain Class of Quasidifferentiable Functions. Math. Prog. Study 29 (1986), pp. 74-84.

[4] Jandl, H. P., Wieder, K.: An Implementation for a Class of Quasidifferentiable Optimization Problems. Mathematical Methods in Operations Research. Sommer School Lovech '86, Sofia 1986.

[5] Jandl, H. P., Wieder, K.: A Continuous Set Covering Problem as a Quasidifferentiable Optimization Problem. To appear in: Optimization (1988).

[6] Pallaschke, D., Recht, P., Urbański, R.: On Locally-Lipschitz Quasi-Differentiable Functions in Banach Spaces. Optimization 17 (1986), pp. 287-295.

THE GLOBAL RANK OF GOMORY CUTS

Joachim Piehler [1)]

The rank of a Gomory cut

(1) $$s = -\frac{\beta}{d} - \sum_{j \in N} \frac{\beta_j}{d}(-x_j)$$

in pure integer programming is the least nonnegative integer so that the diophantine equation

(2) $$\beta + \sigma d = \sum_{j \in N} \beta_j x_j$$

has a nonnegative solution x_j, $j \in N$ (cf. [1]).

N is the set of indices of nonbasic variables.

If the rank σ is known the cut (1) can be made deeper by

$$s' = -\frac{\beta}{d} - \sigma - \sum_{j \in N} \frac{\beta_j}{d}(-x_j),$$

and using deeper cuts in general gives an essential increase of effectivity in cutting methods. Some possibilities for calculating or bounding the rank are also given in [1] .

Let σ be the rank, then any nonnegative integer solution of (2) is called a rank-defining point. But such a point is not necessary a feasible integer point of the given integer programming problem because the corresponding basic variables may be noninteger or negative. If no rank-defining point corresponds to a feasible integer point then clearly the cut can be made deeper.

This leads to the following definition:

The global rank of a cut (1) is the least nonnegative integer σ_g so that

$$\beta + \sigma_g d = \sum_{j \in N} \beta_j x_j$$

has solutions in nonnegative integers x_j, $j \in N$ and at least one of

---

1) Technische Hochschule "Carl Schorlemmer" Leuna-Merseburg
Sektion Mathematik, Otto-Nuschke-Straße, DDR - 4200 Merseburg

these solutions corresponds to a feasible integer point.
The formerly regarded rank then may be called local.

By the well-known formula

$$x_B = B^{-1}b + B^{-1}N\ (-x_N)$$

for the continuous optimum in the optimal simplex tableau (here B^{-1} is the inverse of the optimal basis matrix, N the matrix of non-basic columns, b the vector of right-hand sides in the given problem and x_B, x_N are the vectors of basic resp. nonbasic variables) it is easily tested whether the nonnegative integer solutions of (2) correspond to feasible integer points. If this is the case, then $\sigma = \sigma_g$, otherwise $\sigma_g \geqq \sigma + 1$ and one may continue successively.

The essential additional expense relative to finding the (local) rank consists of the explicitely calculation of all nonnegative solutions of linear diophantine equations. Having large problems the numerical calculations, of course, may be very expensive, but on the other hand it may be possible, that this will be compensated or even overcompensated by a fewer number of simplex steps.

Therefore we now have a method for further making cuts deeper, and such a method must be used only for a small quantity of cuts (cf. [1]). At the same time we have found another very natural possibility for global considerations. It is shown in [1] that Gomory method and its corollaries are local, because only such side conditions are taken into consideration which are binding in the continuous optimum. By using generalized inverse matrices we gave a possibility of regarding all side conditions and finding deeper cuts, too.

With the aid of global rank it will be possible to get the same results, in most cases more simple and faster. Let us look to the example in [1] :

$$\begin{aligned}
& \max z = x_2 \\
\text{s.t. (a)}\quad & -x_1 + x_2 + x_3 = 1 \\
\text{(b)}\quad & 4x_1 + x_2 + x_4 = 4 \\
\text{(c)}\quad & -x_1 + x_2 - x_5 = 0 \\
\text{(d)}\quad & -4x_1 + x_2 + x_6 = 0
\end{aligned}$$

$$x_1, \ldots x_6 \geqq 0 \text{ and integer}$$

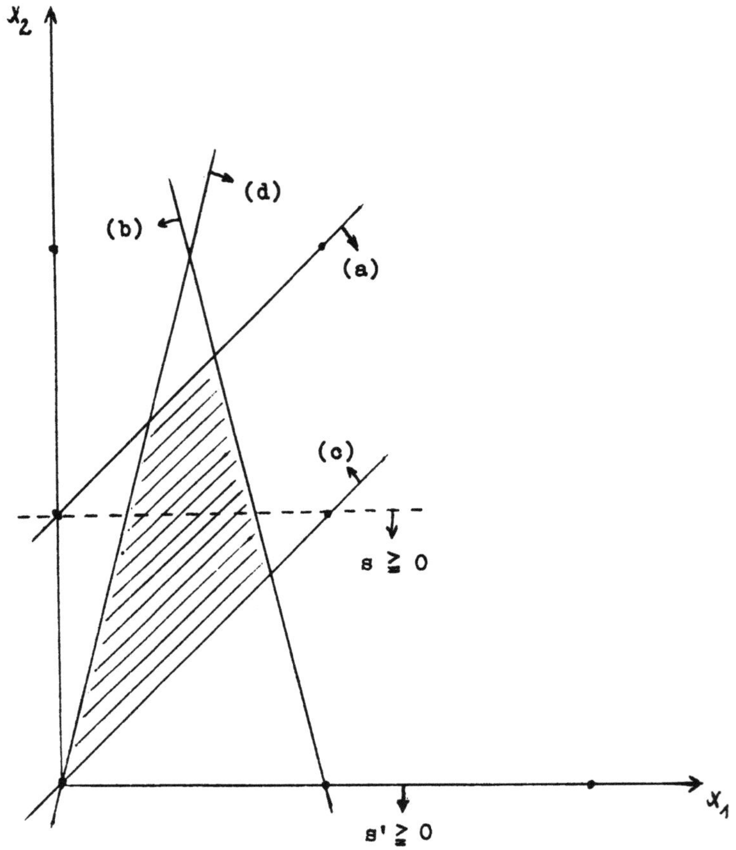

The optimal simplex tableau is

$$x_1 = \frac{3}{5} - \frac{1}{5}(-x_3) + \frac{1}{5}(-x_4)$$

$$x_2 = \frac{8}{5} + \frac{4}{5}(-x_3) + \frac{1}{5}(-x_4)$$

$$x_3 = 1 + 1\ (-x_3)$$

$$x_6 = \frac{4}{5} - \frac{8}{5}(-x_3) + \frac{3}{5}(-x_4).$$

The cut

(3) $$s = -\frac{3}{5} - \frac{4}{5}(-x_3) - \frac{1}{5}(-x_4) \geqq 0$$

or, equivalently, $x_2 \leqq 1$ cannot be made deeper by local methods.

But it is easily seen that the diophantine equation

$$3 = 4x_3 + x_4$$

has only one nonnegative integer solution

$x_3 = 0$, $x_4 = 3$ which leads to

$$x_1 = 0,\ x_2 = 1,\ x_3 = 1,\ x_4 = -1.$$

Therefore the global rank is ≥ 1 and the cut (3) can be made deeper by

$$s' = -\frac{8}{5} - \frac{4}{5}(-x_3) - \frac{1}{5}(-x_4) \geq 0 \quad \text{or} \quad x_2 \leq 0.$$

Only the optimal feasible solution $x_1 = 0$, $x_2 = 0$ then remains feasible.

References:

[1] J. Piehler: Algebraische Methoden in der ganzzahligen Optimierung, BSB B.G. Teubner Verlagsgesellschaft, Leipzig 1983 (Teubner Text 51).

NECESSARY CONDITIONS FOR AN EXTREMUM, PENALTY FUNCTIONS AND REGULARITY

Boris N. Pshenichnyj [1)]

Summary: General conditions of regularity of a point with respect to a nonlinear system of equations are formulated. It is shown that if regularity conditions are satisfied the solution of the general mathematical programming problem may be reduced to minimization of a non-smooth penalty function and the necessary conditions for an extremum may be formulated in the most general form.

The aim of this paper is to show that these three concepts are closely related when considering the most general optimization problems. Each of these concepts has been studied in numerous works where different relations among them have been established. Here we give the most general statements and we show that in the presence of regularity the necessary conditions for an extremum in the form of the Lagrange multipliers rule are always fulfilled. It should be emphasized that for the most complete description of the point of extremum in problems with non-smooth constraints the optimality conditions must be formulated not in the form of a single rule of multipliers but in the form of the whole family of such rules. Only this makes it possible to avoid a situation when necessary conditions are fulfilled at points which trivially cannot be optimal.

1. Main designations

The following explanation will be conducted in a Banach space X whose conjugate space is denoted by X^*. Here, for $x \in X$, $x^* \in X^*$ $\langle x, x^* \rangle$ is the value of a linear continuous functional x^* on the element x. Lest the details should encumber the main contents of the paper all functions considered below will be assumed to satisfy the local Lipschitz condition,

If $A, B \subseteq X$ then

$$\rho_A(x) = \inf_y \{\|y-x\|: y \in A\},$$

$$\rho(B,A) = \sup_x \{\rho_A(x): x \in B\},$$

$$\Delta(A,B) = \max \{\rho(B,A), \rho(A,B)\}.$$

[1)] V. M. Glushkov Institute of Cybernetics, Academy of Sciences of the Ukrainian SSR, 252207 Kiev 207, USSR

For the given function f we set $f^+ = f$, $f^- = -f$ and

$$Df(x,\bar{x}) = \lim_{\bar{y}\to\bar{x}} \sup_{\lambda\downarrow 0} \frac{f(x+\lambda\bar{y}) - f(x)}{\lambda},$$

$$D^0 f(x,\bar{x}) = \lim_{y\to x} \sup_{\lambda\downarrow 0} \frac{f(y+\lambda\bar{x}) - f(y)}{\lambda}.$$

The function $D^0 f$ is known to possess the following properties: it is a convex, positively homogeneous with respect to $\bar{x}$ and

$$D^0 f(x,\bar{x}) \leq L \|\bar{x}\| ,$$

where L is a local Lipschitz constant.

If $M \subseteq X$ and $x \in M$, then we set

$$T_M^0(x) = \{\bar{x}:\ D^0 \rho_M(x,\bar{x}) \leq 0\} .$$

By virtue of the property of operation D^0 $T_M^0(x)$ is a convex closed cone. The given properties of operation D^0 were studied in many works. It was firstly introduced into general use by F. H. Clarke in [1].

We call the direction $\bar{x}$ tangent to M at point x if

$$\lim_{\lambda\downarrow 0} \frac{\rho_M(x + \lambda\bar{x})}{\lambda} = 0 .$$

It is obvious that the cone $T_M^0(x)$ consists of tangent directions. We call the convex cone $T_M(x)$ the tangent cone to M at point x if it consists of tangent directions.

2. Upper Convex Approximations (u.c.a.)

Definition. Function $h(\bar{x})$ is called the upper convex approximation of f at point x if it is positively homogeneous, convex, closed and

$$h(\bar{x}) \geqslant Df(x,\bar{x}) , \quad \forall \bar{x} .$$

A set

$$\partial f(x) = \{x^* \in X^*:\ \langle\bar{x},x^*\rangle \leq h(\bar{x})\}$$

is called a subdifferential of f at point x.

From the convex analysis [2] it is well-known that under the given assumptions $\partial f(x) \neq \emptyset$ and

$$h(\bar{x}) = \max_{x^*} \{\langle\bar{x},x^*\rangle : x^* \in \partial f(x)\}.$$

It should be emphasized that u.c.a. is not defined uniquely and therefore it is desirable to construct possibly most complete families of u.c.a. for function f at a given point x. This makes it possible to

formulate the most complete necessary conditions for an extremum.
Since it is obvious that

$$D^{o}f(x,\bar{x}) \geq Df(x,\bar{x}) ,$$

then $D^{o}f(x,\bar{x})$ is always the u.c.a. Its corresponding subdifferential is denoted by $\partial^{o}f(x)$. Here it is elementary to check that

$$D^{o}f^{-}(x,\bar{x}) = D^{o}f(x,-\bar{x})$$

and therefore $\partial^{o}f^{-}(x) = -\partial^{o}f(x)$. For arbitrary subdifferentials this relation is not valid.
A class of functions which admit upper convex approximations is very broad. Methods of calculation of u.c.a. are given in considerable detail in [2]. Note also that quasidifferential functions studied in [3] also have u.c.a. which are easily calculated by their sub- and superdifferentials [3].
For illustration we give the following theorem which is easily proved.

Theorem 1. Let f be a continuous convex function. Then

$$Df(x,\bar{x}) = f'(x,\bar{x}) = \lim_{\lambda \downarrow 0} \frac{f(x+\lambda\bar{x}) - f(x)}{\lambda}$$

is the u.c.a., and the corresponding subdifferential ∂f coincides with the subdifferential $\partial^{o}f$ and with the usual subdifferential of the convex function.

At the same time

$$Df^{-}(x,\bar{x}) = -f'(x,\bar{x}) \leq \langle\bar{x},-x^{*}\rangle$$

for any element $x^{*} \in \partial f(x)$ and therefore $-x^{*}$ is the subdifferential of $f^{-} = -f$ at any choice $x^{*} \in \partial f(x)$.
This theorem shows a distinguishing feature. The Clarke subdifferential $\partial^{o}f^{-}$ is defined uniquely and equals $-\partial^{o}f(x)$. At the same time the use of the u.c.a. gives the whole family of subdifferentials.

3. Penalty Functions

Let now $f_0, f_1, \dots, f_m$ be locally Lipschitz functions, and M be an arbitrary set. Consider the problem $P(0)$: to find

$$V(0) = \inf_{x} \{f_0(x): f_i(x) \leq 0, i=1,\dots,m, x \in M\} .$$

The fact that we consider only constraints of the inequality type does not decrease the generality since the constraint

$$f(x) = 0$$

is equivalent to two inequalities

$$f(x) \leq 0 \ , \quad -f(x) \leq 0 \ ,$$

and we shall use this in future.
Let

$$V(y) = \inf_x \{f_0(x): f_i(x) \leq y_i, \ i=1,\dots,m, \ x \in M\} \ ,$$

$$F(x) = \max \{0, f_1(x), \dots, f_m(x)\} \ ,$$

$$\Phi_N(x) = f_0(x) + NF(x) \ .$$

The following chain of inequalities is evident:

$$\inf_{x \in M} \Phi_N(x) = \inf_{x,\lambda} \{f_0(x) + N\lambda: 0 \leq \lambda, \ f_1(x) \leq \lambda, \dots, f_m(x) \leq \lambda, \ x \in M\}$$

$$= \inf_{\lambda \geq 0} [V(\mathbb{1}\lambda) + N\lambda] \ , \qquad (1)$$

$$\mathbb{1} = \begin{pmatrix} 1 \\ \vdots \\ 1 \end{pmatrix} \in R^m \quad .$$

From this it follows that if the lower bound in the right-hand side of (1) is attained only at $\lambda = 0$, i. e.

$$V(\mathbb{1}\lambda) - V(0) > -N\lambda, \quad \lambda > 0, \qquad (2)$$

then

$$V(0) = \inf_x \{\Phi_N(x): x \in M\} \ . \qquad (3)$$

Thus, if (2) is satisfied then the lower bound in problem $P(0)$ coincides with the lower bound $\Phi_N(x)$ for $x \in M$.
We shall show that valid is a stronger fact that the points at which the minimum is attained in problem $P(0)$ and the points of the minimum $\Phi_N(x)$, $x \in M$ coincide.

Theorem 2. Let

$$\inf_{\lambda > 0} \frac{V(\mathbb{1}\lambda) - V(0)}{\lambda} = -L > -\infty \ , \qquad (4)$$

and $N > L$. Then the points of minimum of the problem $P(0)$ and of the problem $\inf\{\Phi_N(x), x \in M\}$ coincide.

Remark. Since $V(\mathbb{1}\lambda)$ is evidently a decreasing function of λ, then $L \geq 0$.

Proof. From (4) it follows that $V(\mathbb{1}\lambda) + N\lambda > V(0)$ for $\lambda > 0$.
Let $x_0 \in M$ and $\Phi_N(x_0) = \inf\{\Phi_N(x), x \in M\}$. Then,

$$\inf_x \{f_0(x)+NF(x):\ x\in M\} \le \inf_x \{f_0(x):\ f_i(x)\le 0,\ i=1,\dots,m,\ x\in M\},$$

i. e.,

$$V(0) \geq \inf_x \{\Phi_N(x):\ x\in M\}. \tag{5}$$

We show that $F(x_0) = 0$. Assume the opposite, i. e., that $\lambda_0 = F(x_0) > 0$. We have

$$f_0(x_0) + NF(x_0) = \min_x \{f_0(x) + NF(x):\ x\in M\} \le$$

$$\le \min_x \{f_0(x) + NF(x):\ x\in M,\ F(x)\le \lambda_0\} \le$$

$$\le \min_x \{f_0(x) + N\lambda_0:\ x\in M,\ F(x)\le \lambda_0\} = V(\mathbb{1}\lambda_0) + N\lambda_0.$$

But, on the other hand,

$$f_0(x_0) + NF(x_0) = f_0(x_0) + N\lambda_0 \geq$$

$$\geq \min_x \{f_0(x) + N\lambda_0:\ x\in M,\ F(x)\le \lambda_0\} \geq V(\mathbb{1}\lambda_0) + N\lambda_0.$$

Thus,

$$f_0(x_0) + NF(x_0) = V(\mathbb{1}\lambda_0) + N\lambda_0 .$$

From this, from (4), (5) and the fact that $N > L$ we obtain

$$V(0) \geq f_0(x_0) + NF(x_0) = V(\mathbb{1}\lambda_0) + N\lambda_0 > V(0) ,$$

since $\lambda_0 = F(x_0) > 0$. Thus, $F(x_0) = 0$, from which it follows that the point $x_0 \in M$ satisfies all constraints of the problem $P(0)$. Therefore $f_0(x_0) \geq V(0)$. But according to (5)

$$V(0) \geq f_0(x_0) + NF(x_0) = f_0(x_0) .$$

Thus, $f_0(x_0) = V(0)$. i. e., x_0 is the solution of the problem $P(0)$. Conversely, let x_0 be the solution of the problem $P(0)$. Then

$$f_0(x_0) + NF(x_0) = f_0(x_0) = V(0) \le$$

$$\le \inf_{\lambda\geq 0} [V(\mathbb{1}\lambda) + N\lambda] = \inf_x \{\Phi_N(x):\ x\in M\} ,$$

i. e., x_0 minimizes $\Phi_N(x)$ on M.

The condition (4) has a global character since at large λ it takes account of the behaviour of functions f_i at infinity. At the same time the derivation of necessary conditions for an extremum is based on a local consideration. Therefore it is necessary to introduce some additional notions.

Let further x_0 be a point of a local minimum in the problem $P(0)$, and Ω be some neighborhood of x_o.

Definition. A point x_o of a local minimum in the problem $P(0)$ is regular if there exists such its neighborhood Ω that

$$\liminf_{\lambda \downarrow 0} \frac{V_\Omega(\mathbb{1}\lambda) - V_\Omega(0)}{\lambda} > -\infty \quad , \tag{6}$$

$$V_\Omega(0) = V(0) \tag{7}$$

where

$$V_\Omega(y) = \inf_x \{f_o(x):\ f_i(x) \leq y_i,\ i=1,\dots,m,\ x \in M \cap \Omega\} .$$

It will be recalled that, by the definition of a local minimum, the neighborhood Ω satisfying (7) always exists.

Theorem 3. If the point x_o is regular, then at sufficiently large number $N > 0$ it is a point of minimum of the function $\Phi_N(x)$ on set $M \cap \Omega$.

Proof. By virtue of condition (6) there exists such number $\lambda_o > 0$ that

$$\frac{V_\Omega(\mathbb{1}\lambda) - V_\Omega(0)}{\lambda} \geq -L , \quad 0 < \lambda \leq \lambda_o .$$

Further, let

$$\alpha = \inf_x \{f_o(x):\ x \in \Omega\} .$$

α is finite since f_o satisfies the local Lipschitz condition. Then

$$\frac{V_\Omega(\mathbb{1}\lambda) - V_\Omega(0)}{\lambda} \geq \frac{\alpha - V(0)}{\lambda_o} \quad , \quad \lambda \geq \lambda_o .$$

Now take

$$N > \max \left\{L, \frac{V(0) - \alpha}{\lambda_o}\right\} .$$

Now

$$\frac{V_\Omega(\mathbb{1}\lambda) - V_\Omega(0)}{\lambda} > -N$$

for all $\lambda > 0$, and Theorem 3 follows from Theorem 2.

The importance of Theorem 3 consists in the fact that under the condition of regularity it reduces the construction of necessary conditions for a local minimum in problem $P(0)$ to the derivation of the necessary conditions for a local minimum of function $\Phi_N(x)$ on set $M \cap \Omega$.

4. Regularity

Let X, Y be Banach space, $M \subseteq X$, $x \in M$ and operator F maps some neighborhood of x into Y, here $F(x) = 0$.

Definition. A point x is called (F,M)-regular if there exists such its neighborhood Ω and the number N that

$$\rho_{D_0}(y) \leq N \|F(y)\|, \quad \forall y \in M \cap \Omega,$$

$$D_0 = \{y \in M \cap \Omega : F(y) = 0\}.$$

The notion of regularity is closely related to construction of tangent manifolds and implicit function theorems. A review of these results can be found in [4]. We present two results for illustration. The first is the Ljusternik theorem [4], the second is obtained in [5].

Theorem 4. a) Let the operator F be continuously Fréchet differentiable in the neighborhood of point x and $F'(x)X = Y$. Then the point x is (F,X)-regular.

b) Let Y be finite-dimensional, F be continuously differentiable in the neighborhood of point x and

$$F'(x)\ T^{o}_{M}(x) = Y.$$

Then points x are (F,M)-regular.

The study of (F,M)-regularity in the general case is a difficult problem important for many areas of mathematics.

Associate now the notion of regularity of solution x_0 of problem P(0) with the introduced notion of the (F,M)-regularity.

Theorem 5. If the point x_0 is (F,M)-regular where the mapping $F: X \longrightarrow R^1$ is given by the formula

$$F(x) = \max\{0, f_1(x), \ldots, f_m(x)\}$$

then x_0 is a regular solution of problem P(0).

Give the proof by contradiction. Assume that x_0 is not regular solution of problem P(0), i. e. for an arbitrarily small neighborhood Ω of point x_0

$$\liminf_{\lambda \downarrow 0} \frac{V_\Omega(\mathbb{1}\lambda) - V_\Omega(0)}{\lambda} = -\infty .$$

If $N > 0$ is fixed then there exists such small $\lambda > 0$ that

$$V_\Omega(\mathbb{1}\lambda) - V_\Omega(0) \leq -N\lambda .$$

We choose $y \in M \cap \Omega$, $F(y) \leq \lambda$ such that

$$f_o(y) - V_\Omega(1\lambda) \leq \lambda^2 .$$

Then

$$f_o(y) - \lambda^2 - V_\Omega(0) \leq - N\lambda \tag{8}$$

and

$$f_o(x) \geq V_\Omega(0) \geq f_o(y) - \lambda^2 + N\lambda$$

for any point x of the set

$$D_o = \{x \in M \cap \Omega : F(x) = 0\} .$$

Therefore

$$L_o \| y - x \| \geq f_o(x) - f_o(y) \geq \lambda N - \lambda^2 ,$$

where L_o is a Lipschitz constant for f_o in the neighborhood Ω. Hence we obtain

$$\rho_{D_o}(y) \geq \left(\frac{N}{L_o} - \frac{\lambda}{L_o}\right)\lambda \quad .$$

If λ is chosen such small that $N - \lambda \geq 0.5N$, then we finally obtain

$$\rho_{D_o}(y) \geq 0.5NL_o^{-1}\lambda \geq 0.5NL_o^{-1}F(y) \tag{9}$$

since y is chosen such that $y \in M \cap \Omega$, $F(y) \leq \lambda$. Then, by virtue of (8)

$$f_o(y) \leq V_\Omega(0) - \lambda(N - \lambda) < V_\Omega(0)$$

such that y cannot belong to D_o and therefore $F(y) > 0$. Thus, by virtue of the arbitrary smallness of neighborhood Ω and the arbitrariness of the choice N (9) contradicts the definition of the (F,M)-regularity of the point x_o. The obtained contradiction proves the theorem.

Corollary. If the point of a local minimum x_o in problem P(0) is (F,M)-regular, then there exists such neighborhood Ω that for all sufficiently large N x_o is the point of minimum of $\Phi_N(x)$ on the set $M \cap \Omega$.

5. Necessary Conditions for a Minimum

Let, as earlier, x_o be a solution of problem P(0).

Theorem 6. Let $h_i(\bar{x})$, $i=0,1,\dots,m$ be upper convex approximations for f_i at point x_o, and $T_M(x_o)$ be a convex tangent cone to M at this point. Suppose that $\operatorname{dom} h_i = X$ and the point x_o is (F,M)-regular.

Then there exist such numbers $\lambda_i \geqslant 0$, not all equal to zero, that $\lambda_0 = 1$ and

$$\sum_{i=0}^{m} \lambda_i h_i(\bar{x}) \geqslant 0 , \quad \bar{x} \in T_M(x_0) , \tag{10}$$

$$\lambda_i f_i(x_0) = 0, \quad i=1,\dots,m ,$$

$$\left(\sum_{i=0}^{m} \lambda_i \partial f_i(x_0)\right) \cap T_M^*(x_0) \neq \emptyset \tag{11}$$

where $\partial f_i(x_0)$ are subdifferentials corresponding to h_i and $T_M^*(x_0)$ is a cone conjugate to $T_M(x_0)$.

<u>Proof.</u> Relation (11) directly follows from (10) by virtue of the known results of convex analysis. Therefore we focus our attention on inequality (10). Since the point x_0 is (F,M)-regular, then there exists such its neighborhood Ω that x_0 is a point of minimum of $\Phi_N(x)$ on $M \cap \Omega$. Since x_0 is an internal point for Ω, then $T_M(x_0)$ is simultaneously a tangent cone to $M \cap \Omega$.
Denote

$$\varphi_0(x) = f_0(x)$$

$$\varphi_i(x) = f_0(x) + Nf_i(x) , \quad i=1,\dots,m.$$

Then from the fact that x_0 minimizes $\Phi_N(x)$ on $M \cap \Omega$ it follows that the point x_0, $\xi_0 = f_0(x_0)$ is a solution to the problem

$$\min_{x,\xi} \{\xi : \varphi_i(x) - \xi \leq 0, \ i=0,1,\dots,m, \ x \in M \cap \Omega\} \tag{12}$$

It is obvious that the functions $h_0(\bar{x})-\bar{\xi}$, $h_0(\bar{x}) + Nh_i(\bar{x})-\bar{\xi}$ are upper convex approximations for $\varphi_0(x)-\xi$ and $\varphi_i(\bar{x})-\xi$, $i=1,\dots,m$ at point x_0.
Now we use the known results about the necessary conditions for a minimum for problem (12) ([6]. Theorem 4.4, [7]): there exist such numbers $\lambda_i \geqslant 0$ and $\gamma \geqslant 0$, not all of them equal to zero, that

$$\gamma\bar{\xi} + \lambda_0(h_0(\bar{x})-\bar{\xi}) + \sum_{i=1}^{m} \lambda_i(h_0(\bar{x}) + Nh_i(\bar{x})-\bar{\xi}) \geqslant 0 ,$$

$$\bar{x} \in T_M(x_0) , \quad \bar{\xi} \in R^1 , \tag{13}$$

$$\lambda_i(\varphi_i(x_0) - \bar{\xi}_0) = 0, \quad i=0,1,\dots,m .$$

Since $\xi_0 = f_0(x_0)$ then from the second relation it directly follows that

$$\lambda_i N f_i(x_0) = 0 , \quad i=1,\dots,m . \tag{14}$$

We rewrite the first of these relations in the form

$$(\gamma - \sum_{i=0}^{m} \lambda_i)\bar{\xi} + (\sum_{i=0}^{m} \lambda_i) h_o(\bar{x}) + \sum_{i=1}^{m} \lambda_i N h_i(\bar{x}) \geq 0,$$

$$\bar{x} \in T_M(x_o), \quad \xi \in R^1 .$$

Since ξ is arbitrary, then the inequality is satisfied iff

$$\sum_{i=0}^{m} \lambda_i = \gamma ,$$

and here $\gamma > 0$ since if $\gamma = 0$ then all $\lambda_i = 0$ by virtue of their non-negativity. Without loss of generality, we may assume that $\gamma = 1$. Then we finally obtain

$$h_o(\bar{x}) + \sum_{i=1}^{m} \lambda_i N h_i(\bar{x}) \geq 0, \quad \bar{x} \in T_M(x_o) .$$

If we redenote $\lambda_i N$ again by λ_i, then taking account of (14) we obtain (10).

The proved theorem shows that the assumption of regularity eliminates all problems arising in the derivation of the necessary conditions for a minimum. In particular, we may, without additional efforts, take account of an arbitrary number of equality-type constraints if each of them is replaced by two inequalities as was said above. We give the corresponding result.

<u>Theorem 7.</u> Let x_o be a solution of the problem

$$\min \{f_o(x): f_i(x) \leq 0, \ i=1,\dots,m, \ f_i(x) = 0, \ i=m+1,\dots,k, \ x \in M\}.$$

Assume

$$F_o(x) = \max \{f_1(x),\dots,f_m(x), |f_{m+1}(x)|,\dots, |f_k(x)|\} .$$

If the point x_o is (F_o,M)-regular, then there exist such numbers

$$\lambda_i \geq 0, \ i=0,1,\dots,m, \ \lambda_i^+, \ \lambda_i^- \geq 0, \ i=m+1,\dots,k, \ \lambda_o = 1,$$

that

$$\sum_{i=0}^{m} \lambda_i h_i(\bar{x}) + \sum_{i=m+1}^{k} (\lambda_i^+ h_i^+(\bar{x}) + \lambda_i^- h_i^-(\bar{x})) \geq 0, \ \bar{x} \in T_M(x_o),$$

$$\lambda_i f_i(x_o) = 0 , \quad i=1,\dots,m.$$

Here h_i, h_i^+, h_i^- are u.c.a. for f_i, f_i^+, f_i^- at point x_o and it is assumed that dom h_i, dom h_i^+, dom h_i^- coincide with X.
Besides,

$$\Big[\sum_{i=0}^{m} \lambda_i^+ \partial f_i(x_o) + \sum_{i=m+1}^{k} (\lambda_i^+ \partial f_i^+(x_o) + \lambda_i^- \partial f_i^-(x_o))\Big] \cap T_M^*(x_o) \neq \emptyset.$$

Let us illustrate the use of this theorem by the following example. Let $f_o(x)$ be a smooth function and $f(x)$ be a continuous convex function. Consider the problem

$$\min \{f_o(x): f(x) = 0\}$$

and let x_o be its solution. If the point x_o is $(|f|,X)$-regular, then Theorem 7 is applicable. Further, f_o is a smooth function and therefore $\partial f_o(x_o) = \{f_o'(x_o)\}$. Since f is convex, then $\partial f^+(x_o) = \partial^o f(x_o)$ is a usual subdifferential of the convex function f. At the same time, according to Theorem 1 $-x^*$ is a subdifferential of $f^- = -f$ for any choice of x^* from $\partial^o f(x_o)$. Since the use of Theorem 7 makes it possible to choose any subdifferentials and for $M = X$ $T_M(x_o) = X$, $T_M^*(x_o) = \{0\}$ then we obtain the following result: for any $x^* \in \partial^o f(x_o)$ there exist such numbers $\lambda^+ \geqslant 0$, $\lambda^- \geqslant 0$ that

$$0 \in f_o'(x_o) + \lambda^+ \partial^o f(x_o) - \lambda^- x^* .$$

If we introduce the designation

$$\text{con } M = \{\lambda x: \lambda \geqslant 0, x \in M\} ,$$

then finally we may formulate the result in the following form: for any $x^* \in \partial^o f(x_o)$

$$(\text{con } x^*) \cap (f_o'(x_o) + \text{con } \partial^o f(x_o)) \neq \emptyset . \tag{15}$$

Compare formula (15) with what gives in this case the use of results of [1]: there exist $x^* \in \partial^o f(x_o)$ and numbers $\lambda_o \geqslant 0$, $\lambda \in R^1$ which are not at the same time equal to zero such that

$$\lambda_o f_o'(x_o) + \lambda x^* = 0 .$$

It is clear that these results are comparable only if x_o is a point of smoothness of function f. In the rest of cases the assumption (15) is considerably stronger.

6. Conclusion

The obtained results show that the regularity assumption of the problem allows for constructing the first-order necessary conditions for an extremum which are significantly more comprehensive in the case of non-smooth data entering the problem than it is possible with assumptions which do not use these conditions.
Naturally the test of regularity conditions is not a simple problem as it is shown by Theorem 4 and a further study of this problem is necessary. At the same time, the necessary conditions for a minimum obtained without the requirement for regularity conditions (especially

for non-smooth problems) are often either too weak or are fulfilled at points which are not obviously the points of minimum. Such examples can be found in abundance. Therefore, of great interest are investigations directed to the search for a reasonable trade-off of test of the regularity conditions against construction of sufficiently complete necessary conditions for an extremum without a preliminary complete test of such conditions.

References

[1] Clarke, F. H.: A new approach to Lagrange multipliers. Math. of operations research, 1976, Vol. 1, No. 2.

[2] Пшеничный, Б. Н.: Выпуклый анализ и экстремальные задачи. М. Наука, 1980, 319 стр.

[3] Dem'yanov, V. F. and A. M. Rubinov: Quasidifferential Calculus. Optimization Software, Inc. Pull. Division, N. Y. 1986, 1-289.

[4] Дмитрук, А. В., А. А. Милютин и Н. П. Осмоловский: Теорема Люстерника и теория экстремума. УМН., 1980, т.35, № 5, с. 11-46.

[5] Aubin, J.-P. and H. Frankowska: On inverse function theorems for self-valued maps. IIASA, WP-84-68, 1984, pp. 1-21.

[6] Пшеничный, Б. Н.: Необходимые условия экстремума. М. Н. 1982, с. 1-142.

[7] Neustadt, L.: Optimization: A Theory of Necessary Conditions. Princeton University Press, 1976.

ON UNIFORM STABILITY OF UNIFORMLY BOUNDED WEAK EVOLUTION OPERATORS

S. Rolewicz[1)]

Let $(X, \| \|)$ be a Banach space. By an evolution operator we shall call a family of linear continuous operators $\{U(t,s)\}$, $0 \le s \le t < +\infty$, acting in X and satisfying the following conditions:

(i) $U(t,t) = I$ (Identity operator);

(ii) $U(t,r)U(r,s) = U(t,s)$, $0 \le s \le r \le t < +\infty$;

(iii) $U(t,s)x$ is continuous with respect to t for every fixed x and s.

If, moreover,

(iv) $K_1 = \sup\limits_{0 \le s-t \le 1} \|U(t,s)\| < +\infty$

we say that the evolution operator $\{U(t,s)\}$ is uniformly bounded.

An evolution operator $\{U(t,s)\}$ is said to be uniformly exponentially equistable if there exist constants $M > 0$, $\beta > 0$ such that

$$\|U(t,s)\| \le Me^{-\beta(t-s)} \quad \text{for all } 0 \le s \le t < +\infty \tag{1}$$

In [2] there is proved the following

THEOREM 1. Let $N(u)$ be an arbitrary continuous non-decreasing real-valued function defined for $u \ge 0$. Moreover, let $N(0) = 0$ and $N(u) > 0$ for $u > 0$.

If a uniformly bounded evolution operator $\{U(t,s)\}$ is not uniformly exponentially equistable, then the set

$$X_N^0 = \{x: \sup_s \int_s^\infty N(\|U(t,s)\|)\, dt = +\infty\}$$

is dense in the whole space X.

As a consequence of Theorem 1 we obtain

THEOREM 2 ([2]). Let $N(\alpha,u)$ be a continuous real-valued function defined for $\alpha > 0$, $u \ge 0$ and nondecreasing with respect to each variable. Moreover, let $N(\alpha,0)$ for every $\alpha > 0$ and $N(\alpha,u) > 0$ for $u > 0$.

If a uniformly bounded evolution operator $\{U(t,s)\}$ has the following property:

(v) for every x there exists an $\alpha(x)$ such that

$$\sup_s \int_s^\infty N(\alpha(x), \|U(t,s)x\|)dt < +\infty$$

then the evolution operator is uniformly exponentially equistable.

1) Institute of Mathematics, Polish Academy of Sciences Sniadeckich 8, 00-950 Warszawa, Poland

In fact, how it was observed in [1], the assumption (iii) can be replaced by a weaker assumption, namely:

(iii') $\|U(t,s)\|$ is measurable with respect to t for all fixed s and x.

Operators satisfying (i), (ii), (iii') will be called <u>weak evolution operators</u>.

The proofs of Theorem 1 and 2 are going without any changes for uniformly bounded weak evolution operators.

In [1] the authors use these results for ordinary differential equations with impulses, considered of in mechanics (for instance, in the theory of gear boxes). It is easy to observe that Theorems 1 and 2 can be applied to a more general class of differential equations with impulses.

Namely, let $(X,\| \|)$ be a Banach space over reals or complexes. We shall consider a differential equation with impulses on the set $R_{t_0} = \{t: t > t_0\}$. Let Γ be a countable closed set contained in R_{t_0}. We request that on the open set $R_{t_0} \setminus \Gamma$ the following differential equation

$$\frac{dx}{dt} = A(t)x(t) \tag{2}$$

should be satisfied almost everywhere, where A(t) is a measurable and locally integrable function with values in the space of linear continuous operators acting in X and x(t) is a continuous function, almost everywhere differentiable and with values in the space X.

Assume that for all $v \in \Gamma$ there exist: the left limit

$$\lim_{t \to v-0} x(t) = x_-(v) \tag{3}$$

and the right limit

$$\lim_{t \to v+0} x(t) = x_+(v) \tag{4}$$

Suppose, moreover, that

$$x(v) = x_-(v) \tag{5}$$

and that there exists a continuous linear operator Q(v) such that

$$x_+(v) = Q(v)x_-(v) \tag{6}$$

If for a given fixed $T > 0$

$$\sup_s \int_s^{s+T} \|A(t)\| dt < +\infty \tag{7}$$

and

$$\sup_s \Big\| \prod_{\substack{s \le v \le s+T \\ v \in \Gamma}} Q(v) \Big\| < +\infty \tag{8}$$

then the evolution operator $\{U(t,s)\}$ corresponding to Equation (2) with impulses (6) and with the initial condition

$$x(s) = x$$

satisfies Conditions (i), (ii), (iii'), (iv). Thus for the stability of equations of that type we can use Theorems 1 and 2.

REFERENCES

[1] D.D. Bainov, S.I. Konstantinov: Exponential stability of the solutions of linear homogeneous differential equations with impulse effect in Banach spaces. Dokl. Bulg. Acad. Sci. 40(1987), 29-30.

[2] S. Rolewicz, On uniform N-stability. Journ. Math. Anal. Appl. 115(1986), 434-441.

On distribution sensitivity in chance constrained programming

Werner Römisch and Rüdiger Schultz*

Abstract:

Chance constrained stochastic programs are studied as parametric programming problems with respect to the underlying probability distribution. Results on continuity properties of (local) optimal values and optimal solution sets are established resp. reviewed from [15].

Let us consider the following chance constrained model

$$\min\{f(x):\ x\in\mathbb{R}^m,\ \mu(\{z\in\mathbb{R}^s:\ x\in X(z)\})\geq p_0\} \qquad (1)$$

where f is a real-valued function defined on $\mathbb{R}^m$, X is a set-valued mapping from $\mathbb{R}^s$ into $\mathbb{R}^m$, $p_0\in[0,1]$ is a prescribed probability level and μ is a probability distribution on $\mathbb{R}^s$. The multifunction X has e.g. the following shape $X(z):=\{x\in X_0,\ h_i(x,z)\geq 0,\ i\in I\}$ $(z\in\mathbb{R}^s)$, where $X_0\subset\mathbb{R}^m$ is nonempty and closed, I is a finite index set and h_i is a continuous real-valued function defined on $\mathbb{R}^m\times\mathbb{R}^s$. For theoretical results on chance constrained problems (like e.g. convexity of the constraint set) we refer to [4],[7],[11].

In the present paper we study the behaviour of (1) with respect to (small) perturbations of the probability distribution μ. Recently such studies have found considerable interest in the literature which is essentially caused by two lines of research in stochastic programming. The first one is the stability analysis for stochastic programs with incomplete information (see e.g.[5],[6]) and the second one is the use of approximation techniques ([17],[20],[21]) and of computational methods (e.g. those introduced in [18] for the calculation of probabilities etc.) for solving chance constrained programs.
Our approach to the study of perturbations of (1) relies on (qualitative and quantitative) stability results for parametric optimization problems with parameters varying in metric spaces obtained by D. Klatte in [9]. As the space of parameters we consider $\mathcal{P}(\mathbb{R}^s)$ - the space of all Borel probability measures on $\mathbb{R}^s$ equipped with a suitable metric. Of course, at first sight it seems especially desirable to cover the case of weak convergence of probability measures (see e.g.[2]) as it

---
* Humholdt-Universität Berlin, Sektion Mathematik, DDR-1086 Berlin, PSF 1297, German Democratic Republic.

has been done in the stability analysis carried out by Kall [8] and Wang [19]. In [8], the author stresses the difficulties that arise when looking for comprehensible sufficient conditions for the lower semicontinuity of the constraint-set mapping understood as a multifunction defined on $\mathcal{P}(\mathbb{R}^s)$ endowed with the topology of weak convergence. In [19] additional smoothness assumptions on the measure μ (describing the unperturbed problem) are imposed to establish the desired lower semi-continuity of the constraint set. The present paper continues the line of research begun in [15] and [16]. We present a suitable topology on $\mathcal{P}(\mathbb{R}^s)$ to gain stability results for the (locally) optimal value and the (locally) optimal solutions. To this end, let us consider the distance

$$\alpha_{\mathcal{B}_0}(\mu,\nu) := \sup_{B \in \mathcal{B}_0} |\mu(B) - \nu(B)| \qquad (\mu, \nu \in \mathcal{P}(\mathbb{R}^s))$$

where $\mathcal{B}_0$ is a proper subclass of Borel sets. In our situation $\mathcal{B}_0$ will be chosen adapted to the multifunction X, i.e. in such a way that it contains all the pre-images $X^-(x) := \{z \in \mathbb{R}^s : x \in X(z)\}$ $(x \in \mathbb{R}^m)$. Simultaneously, $\mathcal{B}_0$ is taken rich enough in the sense that $\alpha_{\mathcal{B}_0}$ forms a metric. With $\alpha_{\mathcal{B}_0}$ selected according to these objectives we state a fairly general qualitative stability result (Proposition 1) and a quantitative continuity property of the optimal value (Theorem 7). In Remark 2 we reveal additional hypotheses to identify those measures μ at which stability with respect to the topology of weak convergence holds. We present an example (Example 3) to show that under the hypotheses of Proposition 1 stability with respect to weak convergence may collapse.

Next we introduce some basic concepts and notations which are used throughout. For $\nu \in \mathcal{P}(\mathbb{R}^s)$ and $p \in [0,1]$ the set $C_p(\nu)$ is given as $C_p(\nu) := \{x \in \mathbb{R}^m : \nu(X^-(x)) \geq p\}$, hence problem (1) becomes $\min\{f(x) : x \in C_{p_0}(\mu)\}$. Given $V \subseteq \mathbb{R}^m$ and $\nu \in \mathcal{P}(\mathbb{R}^s)$ we denote

$\varphi_V(\nu) := \inf\{f(x) : x \in C_{p_0}(\nu) \cap \operatorname{cl} V\}$ and

$\psi_V(\nu) := \{x \in C_{p_0}(\nu) \cap \operatorname{cl} V : f(x) = \varphi_V(\nu)\}$.

By cl V and bd V we understand the closure resp. the boundary of V. Following [13],[9] we call a nonempty subset M of $\mathbb{R}^m$ a complete local minimizing set (CLM set) for f on $C_{p_0}(\mu)$ if there is an open set $Q \supset M$ such that $M = \psi_Q(\mu)$. Examples for CLM sets are the set of global minimizers or strict local minimizing points.

We recall that a multifunction Γ from a metric space T to $\mathbb{R}^m$ is said to be closed at $t_0 \in T$ if $t_k \to t_0$, $x_k \to x_0$, $x_k \in \Gamma(t_k)$ $(k \in \mathbb{N})$ imply $x_0 \in \Gamma(t_0)$, Γ is said to be upper semicontinuous (usc) at t_0 if for any open set $\Omega \supset \Gamma(t_0)$ there exists a neighbourhood $U(t_0)$ such that $\Gamma(t) \subset \Omega$ whenever $t \in U(t_0)$, Γ is said to be locally lower semicontinuous (locally lsc) at $(x_0,t_0) \in \Gamma(t_0) \times T$ if for any open set $\Omega \supset \{x_0\}$

there exists a neighbourhood $U(t_o)$ such that $\Omega \cap \Gamma(t) \neq \emptyset$ whenever $t \in U(t_o)$ and finally Γ is said to be pseudo-Lipschitzian at $(x_o, t_o) \in \Gamma(t_o) \times T$ (cf. [14]) if there are neighbourhoods $U = U(t_o)$, $V = V(x_o)$ and a positive constant L such that

$$\Gamma(t') \cap V \subset \Gamma(t'') + L\, d(t',t'')B_m \quad \text{whenever } t',t'' \in U,$$

here d denotes the metric in T and B_m is the closed unit ball in $\mathbb{R}^m$. Note that the above property especially implies that Γ is locally lsc at (x_o, t_o).

Our first proposition is an adaptation of a general parametric programming result ([9], Prop. 1, [13] Prop. 3.2 and Prop. 3.3) to the situation of distribution sensitivity in stochastic programming.

Proposition 1:

Let $\mu \in \mathcal{P}(\mathbb{R}^s)$, $p_o \in (0,1)$ and $\{X^-(x): x \in \mathbb{R}^m\} \subseteq \mathcal{B}_o$. Let X be a closed multifunction and f be continuous. Assume that there exists a bounded open set $V \subset \mathbb{R}^m$ such that $\psi_V(\mu)$ is a CLM set for f on $C_{p_o}(\mu)$. Let the multifunction $p \mapsto C_p(\mu)$ be locally lsc at (x_o, p_o) for some $x_o \in \psi_V(\mu)$.

Then (i) φ_V is continuous at μ and ψ_V is usc at μ with respect to the metric $\alpha_{\mathcal{B}_o}$ on $\mathcal{P}(\mathbb{R}^s)$, and

(ii) there exists $\delta > 0$ such that $\psi_V(\nu)$ is a CLM set for f on $C_{p_o}(\nu)$ whenever $\nu \in \mathcal{P}(\mathbb{R}^s)$ and $\alpha_{\mathcal{B}_o}(\mu,\nu) < \delta$.

Proof:

Let us consider the metric space $(\mathcal{P}(\mathbb{R}^s), \alpha_{\mathcal{B}_o})$. It is known from Berge's classical stability theory for parametric optimization problems (applied to $\min\{f(x): x \in C_{p_o}(\nu) \cap \mathrm{cl}\, V\}$, $\nu \in \mathcal{P}(\mathbb{R}^s)$) that assertion (i) holds if the multifunction (from $\mathcal{P}(\mathbb{R}^s)$ into $\mathbb{R}^m$) $\nu \mapsto C_{p_o}(\nu) \cap \mathrm{cl}\, V$ is closed at μ and locally lsc at (x_o, μ). (cf. e.g. [1] , Theorem 4.2.2). In the proof of Theorem 5.4 in [15] it was shown that $\nu \mapsto C_{p_o}(\nu)$ is closed at μ since X is closed. Since $x_o \in V$, it remains to show that $\nu \mapsto C_{p_o}(\nu)$ is locally lsc at (x_o, μ). Let $\varepsilon > 0$ be arbitrary, but fixed. Since $p \mapsto C_p(\mu)$ is locally lsc at (x_o, p_o), there exists $\delta_o > 0$ such that $C_p(\mu) \cap B(x_o, \varepsilon) \neq \emptyset$ for every $p \in \mathbb{R}$ with $|p - p_o| < \delta_o$. Let $\nu \in \mathcal{P}(\mathbb{R}^s)$ be such that $\tilde{\delta} := \alpha_{\mathcal{B}_o}(\mu,\nu) < \delta_o$. Put $p := p_o + \tilde{\delta}$. Then we have for every $y \in C_p(\mu)$ that

$$\nu(X^-(y)) \geq \mu(X^-(y)) - |\mu(X^-(y)) - \nu(X^-(y))| \geq p - \alpha_{\mathcal{B}_o}(\mu,\nu) = p_o.$$

Hence we obtain $y \in C_{p_o}(\nu)$ and $C_{p_o}(\nu) \cap B(x_o, \varepsilon) \supseteq C_p(\mu) \cap B(x_o, \varepsilon) \neq \emptyset$.

Thus the multifunction $\nu \mapsto C_{p_o}(\nu)$ is locally lsc at (x_o, μ). Since ψ_V is usc at μ (with respect to $\alpha_{\mathcal{B}_o}$) and $\psi_V(\mu)$ is contained in the open set V, there exists $\delta>0$ such that $\psi_V(\nu)\subset V$ for every $\nu\in\mathcal{P}(\mathbb{R}^s)$ such that $\alpha_{\mathfrak{B}_o}(\mu,\nu)<\delta$. This completes the proof. □

<u>Remark 2:</u>
Proposition 1 may also be viewed as a qualitative distribution stability result with respect to perturbations of μ in the space $(\mathcal{P}(\mathbb{R}^s),\tau_w)$ (τ_w denoting the topology of weak convergence, cf. [2]) if $\mathfrak{B}_o$ is a μ-uniformity class of Borel subsets of $\mathbb{R}^s$. Recall that $\mathfrak{B}_o$ is a μ-uniformity class if $\alpha_{\mathfrak{B}_o}(\mu_n,\mu)\rightarrow 0$ holds for every sequence (μ_n) converging weakly to μ ([3]). Let us consider two particular classes $\mathfrak{B}_o$ of Borel sets:

(i) $\mathfrak{B}_R := \{\emptyset, (-\infty,z] : z\in\mathbb{R}^s\}$ is a μ-uniformity class if the distribution function of $\mu\in\mathcal{P}(\mathbb{R}^s)$ is continuous.

(ii) $\mathfrak{B}_C := \{B\subseteq\mathbb{R}^s$: B is convex and Borel$\}$ is a μ-uniformity class if $\mu\in\mathcal{P}(\mathbb{R}^s)$ is absolutely continuous w.r. to Lebesgue measure on $\mathbb{R}^s$.

Hence, at least, if $\mu\in\mathcal{P}(\mathbb{R}^s)$ has a density and all pre-images $X^-(x)$ ($x\in\mathbb{R}^m$) are convex, Prop. 5.1 appears to be a useful stability result w.r. to perturbations of μ in $(\mathcal{P}(\mathbb{R}^s),\tau_w)$.
Furthermore, the above proof shows that it is just the shape of the metric $\alpha_{\mathfrak{B}_o}$ which allows to carry over the local lower semicontinuity of the mapping $p\mapsto C_p(\mu)$ at some (x_o,p_o) to the mapping $\nu\mapsto C_{p_o}(\nu)$ at (x_o,μ). In the general situation, local lower semicontinuity of $p\mapsto C_p(\mu)$ at (x_o,p_o) implies $x_o\in \mathrm{cl}(\{x\in\mathbb{R}^m: \mu(X^-(x))>p_o\})$; if now a metric d on $\mathcal{P}(\mathbb{R}^s)$ is chosen such that the functions $t_{\mu_n}(x):=\mu_n(X^-(x))$ converge pointwise to $t_\mu(x):=\mu(X^-(x))$ as $d(\mu_n,\mu)\rightarrow 0$ one obtains that $\nu\mapsto C_{p_o}(\nu)$ is locally lsc at (x_o,μ) with respect to $(\mathcal{P}(\mathbb{R}^s),d)$ (cf. [17], Corr. 3.2.1). For the topology of weak convergence, however, the desired pointwise convergence of t_{μ_n} as $\mu_n \xrightarrow{\tau_w} \mu$ holds, if $\mu(\mathrm{bd}\,X^-(x))=0$ for all $x\in\mathbb{R}^m$, i.e. all sets $X^-(x)$ are μ-continuity sets ([2], Portmonteau Theorem). We refer to a corresponding discussion in [8], p. 404.

By the following example we illustrate what has been said in the above remark. We present a multifunction $\nu\mapsto C_{p_o}(\nu)$ (from $(\mathcal{P}(\mathbb{R}^s),\tau_w)$ to $\mathbb{R}^m$) not locally lsc at some (x_o,μ) although the multifunction $p\mapsto C_p(\mu)$ is locally lsc at (x_o,p_o).

<u>Example 3:</u>
Let $X(z):=\{x\in(-\infty,0] : x\geq z\}$, $\mu_o\in\mathcal{P}(\mathbb{R})$ be the uniform distribution on $[-1,0]$, $\delta_x\in\mathcal{P}(\mathbb{R})$ the distribution with unit mass at $x\in\mathbb{R}$.

Define $\mu := \frac{1}{2}\delta_0 + \frac{1}{2}\mu_0$

$$\mu_n := (\tfrac{1}{2} + \tfrac{1}{n})\delta_{\frac{1}{n}} + (\tfrac{1}{2} - \tfrac{1}{n})\mu_0 \quad (n \in \mathbb{N}).$$

One confirms $\mu_n \xrightarrow{\tau_w} \mu$ as $n \to \infty$, but $\alpha_{\mathcal{B}_R}(\mu_n, \mu) \not\to 0$. Furthermore, $p \mapsto C_p(\mu)$ is locally lsc at $(x_o, p_o) := (0, \frac{1}{2})$ since $C_p(\mu) = \{0\}$ for $p \in [\frac{1}{2}, 1]$ and $C_p(\mu) = [2p-1, 0]$ for $p \in (0, \frac{1}{2})$. On the other hand, $C_{p_o}(\mu_n) = \emptyset$ for all $n \in \mathbb{N}$. Thus, the mapping $\nu \mapsto C_{p_o}(\nu)$ is not locally lsc at (x_o, μ).

To see that assertion (i) in Proposition 1 does no longer hold when the multifunction $p \mapsto C_p(\mu)$ is not locally lsc at (x_o, p_o) at any $x_o \in \Psi_V(\mu)$ consider the next example.

<u>Example 4:</u>

Let $\mu \in \mathcal{P}(\mathbb{R})$ be given by the distribution function

$$F_\mu(x) = \begin{cases} 0 & x < -1 \\ x+1 & -1 \le x < -\frac{1}{2} \\ \frac{1}{2} & -\frac{1}{2} \le x < \frac{1}{2} \\ x & \frac{1}{2} \le x < 1 \\ 1 & 1 \le x \end{cases}$$

Let further $p_o := \frac{1}{2}$, $X(z) := \{x \in \mathbb{R} : x \ge z\}$ and $f(x) := (x + \frac{1}{8})^2$. then with $V := (-\frac{1}{4}, \frac{1}{4})$ the set $\Psi_V(\mu) = \{-\frac{1}{8}\}$ is a CLM set for f on $C_{p_o}(\mu)$. However, the mapping $p \mapsto C_p(\mu)$ is not locally lsc at (x_o, p_o) with $x_o = -\frac{1}{8}$.
Let

$$F_{\mu_n}(x) = \begin{cases} 0 & x < -1 \\ 1 - \frac{1}{n} + (1 - \frac{1}{n})x & -1 \le x < -\frac{1}{2} \\ \frac{1}{2} + \frac{1}{n}x & -\frac{1}{2} \le x < \frac{1}{2} \\ \frac{1}{n} + (1 - \frac{1}{n})x & \frac{1}{2} \le x < 1 \\ 1 & 1 \le x \end{cases} \qquad (n \in \mathbb{N})$$

With $\mathcal{B}_0 = \{(-\infty, x] : x \in \mathbb{R}\}$ we have $\{X^-(x) : x \in \mathbb{R}\} \subseteq \mathcal{B}_0$ and $\alpha_{\mathcal{B}_0}(\mu, \mu_n) \to 0$ $(n \to \infty)$. On the other hand $\Psi_V(\mu_n) = \{0\}$ and $\varphi_V(\mu_n) = \frac{1}{64}$ for all n. Hence Ψ_V and φ_V are not usc resp. continuous at μ.

In view of the consequences for the stability of local optimal values resp. locally optimal solutions it is interesting to ask for conditions

under which multifunctions (especially the mapping $p \mapsto C_p(\mu)$) behave locally lsc. In the literature this question is related to certain constraint qualifications ([1],[12],[14]).
The next Lemma, which is due to Rockafellar [14], reflects this approach and gives also quantitative information about the continuity of certain constraint-set-mappings.

Lemma 5:

Let $g: \mathbb{R}^m \times \mathbb{R}^d \to \mathbb{R}^r$ be continuously differentiable and $(x_o,p_o) \in \mathbb{R}^m \times \mathbb{R}^d$ with $g(x_o,p_o) \geq 0$. Assume that the "Mangasarian-Fromowitz constraint qualification" holds, i.e.
there exists a $y \in \mathbb{R}^m$ such that

$$y^T \nabla_x g_i(x_o,p_o) > 0 \quad \text{for all } i \in \{1,\dots,r\} \text{ with } g_i(x_o,p_o) = 0.$$

Then the multifunction $p \mapsto \{x \in \mathbb{R}^m: g(x,p) \geq 0\}$ (from $\mathbb{R}^d$ to $\mathbb{R}^m$) is pseudo-Lipschitzian at (x_o,p_o).

Proof: follows immediately from Corollary 3.5. and Remark 3.6. in [14].

As we will see lateron it is possible to derive a local Lipschitz property for φ_V at μ if the mapping $p \mapsto C_p(\mu)$ is even pseudo-Lipschitzian at suitable (x_o,p_o). In this context let us recall that without Mangasarian-Fromowitz condition the constraint-set-mapping may fail to be pseudo-Lipschitzian even when it is already locally lsc. This will be illustrated by the next example which is a mapping of the type $p \mapsto C_p(\mu)$.

Example 6:

Let $\mu \in \mathcal{P}(\mathbb{R})$ be given by a distribution function $F_\mu(\cdot)$ which is continuously differentiable and for which $F_\mu(x) = x^3 + \frac{1}{2}$ on some neighbourhood around 0. Let further $p_o = \frac{1}{2}$, $X(z) := \{x \in \mathbb{R}: x \geq z\}$. Then $C_p(\mu) = \{x \in \mathbb{R}: F_\mu(x) \geq p\}$. Consider $x_o = 0$. Then $p \mapsto C_p(\mu)$ is locally lsc at (x_o,p_o) but not pseudo-Lipschitzian at (x_o,p_o).

Theorem 7:

Assume the hypotheses of Proposition 1 and let additionally f be locally Lipschitzian and the multifunction $p \mapsto C_p(\mu)$ be pseudo-Lipschitzian at each (x_o,p_o) belonging to $\psi_V(\mu) \times \{p_o\}$. Then there exist constants $L > 0$ and $\delta > 0$ such that

$$|\varphi_V(\mu) - \varphi_V(\nu)| \leq L \alpha_{\mathfrak{B}_o}(\mu,\nu)$$

whenever $\alpha_{\mathfrak{B}_o}(\mu,\nu) < \delta$.

The above theorem has been proven in [15] (Theorem 5.4) relying on quantitative stability results obtained by D. Klatte in [9] (Theorem 1).

Now, let us consider a particular chance constrained model with random right-hand side. The multifunction X (from $\mathbb{R}^s$ into $\mathbb{R}^m$) is given as

$$X(z) := \{x \in \mathbb{R}^m : g_i(x) \geq 0,\ i=1,\dots,r,\ Ax \geq z\}$$

where $A \in L(\mathbb{R}^m, \mathbb{R}^s)$ and $g_i : \mathbb{R}^m \to \mathbb{R}$ continuously differentiable ($i=1,\dots,r$). Note that $X^-(x) \in \mathcal{B}_R$ for each $x \in \mathbb{R}^m$ and

$$C_p(\mu) = \{x \in \mathbb{R}^m : g_i(x) \geq 0,\ i=1,\dots,r,\ F_\mu(Ax) \geq p\}$$

where F_μ denotes the distribution function of $\mu \in \mathcal{P}(\mathbb{R}^s)$ and $p \in (0,1)$.

Corollary 8:

Let $\mu \in \mathcal{P}(\mathbb{R}^s)$ with continuously differentiable F_μ, $p_o \in (0,1)$ and f be locally Lipschitzian. Assume that there exists a bounded open set $V \subset \mathbb{R}^m$ such that $\Psi_V(\mu)$ is a CLM set for f on $C_{p_o}(\mu)$. Suppose further that for each $x_o \in \Psi_V(\mu)$ there exists $y \in \mathbb{R}^m$ such that

$$y^T \nabla g_i(x_o) > 0, \quad \text{for all } i \in \{1,\dots,r\} \text{ with } g_i(x_o) = 0, \text{ and}$$

$$y^T A^T \nabla F_\mu(Ax_o) > 0 \quad \text{if} \quad F_\mu(Ax_o) = p_o.$$

Then Ψ_V is usc at μ and there exist constants $L > 0$ and $\delta > 0$ such that $\Psi_V(\nu)$ is a CLM set for f on $C_{p_o}(\nu)$ and

$$|\varphi_V(\mu) - \varphi_V(\nu)| \leq L \sup_{z \in \mathbb{R}^s} |F_\mu(z) - F_\nu(z)|$$

whenever $\sup_{z \in \mathbb{R}^s} |F_\mu(z) - F_\nu(z)| < \delta$.

Proof:

According to Lemma 5 the mapping $p \mapsto C_p(\mu)$ is pseudo-Lipschitzian at any $(x_o, p_o) \in \Psi_V(\mu) \times \{p_o\}$. The assertion now follows from Proposition 1 and Theorem 7 with $\mathcal{B}_o := \mathcal{B}_R$ (noting that $\alpha_{\mathcal{B}_R}(\mu,\nu) = \sup_{z \in \mathbb{R}^s} |F_\mu(z) - F_\nu(z)|$). □

For another chance constrained model which was introduced in [10] a quantitative stability result relying on Theorem 7 (with $\mathcal{B}_o := \mathcal{B}_C$) has been developed in [15].

Acknowledgement:

The authors wish to thank Bernd Kummer (Humboldt-Universität Berlin) for stimulating discussions.

References:

[1] B. Bank J. Guddat, D. Klatte, B. Kummer and K. Tammer, Non-Linear Parametric Optimization, Akademie-Verlag, Berlin, 1982.

[2] P. Billingsley, Convergence of Probability Measures, Wiley, New York, 1968.

[3] P. Billingsley and F. Topsøe, Uniformity in weak convergence, Z. Wahrscheinlichkeitsth. 7(1967),1-16.

[4] A. Charnes and W.W. Cooper, Chance-constrained programming, Management Sci. 6(1959), 73-89.

[5] J. Dupačová, Stability in stochastic programming-probabilistic constraints, in: Stochastic Optimization, Lecture Notes in Control and Inform. Sci. 81, Springer-Verlag, Berlin, 1986, 314-325.

[6] J. Dupačová, Stochastic programming with incomplete information: A survey of results on postoptimization and sensitivity analysis, Optimization 18(1987), 507-532.

[7] P. Kall, Stochastic Linear Programming, Springer-Verlag, Berlin,1976.

[8] P. Kall, On approximations and stability in stochastic programming, in: Parametric Optimization and Related Topics (Eds. J. Guddat et al.), Akademie-Verlag, Berlin, 1987, 387-407.

[9] D. Klatte, A note on quantitative stability results in nonlinear optimization, in: Proceedings of the 19. Jahrestagung "Mathematische Optimierung" (Ed. K. Lommatzsch), Humboldt-Universität Berlin, Sekt. Mathematik, Seminarbericht 1987 (to appear).

[10] C. van de Panne and W. Popp, Minimum cost cattle feed under probabilistic protein constraints, Management Sci. 9(1963), 405-430.

[11] A. Prékopa, Logarithmic concave measures with applications to stochastic programming, Acta Sci. Math. 32(1971), 301-316.

[12] S.M. Robinson, Stability theory for systems of inequalities, Part II: Differentiable nonlinear systems, SIAM J. Numer. Anal. 13(1976), 497-513.

[13] S.M. Robinson, Local epi-continuity and local optimization, Math. Programming 37(1987), 208-223.

[14] R.T. Rockafellar, Lipschitzian properties of multifunctions, Nonlinear Analysis 9(1985), 867-885.

[15] W. Römisch and R. Schultz, Distribution sensitivity in stochastic programming, Humboldt-Universität Berlin, Sekt. Mathematik, Preprint Nr. 160, 1987.

[16] W. Römisch and A. Wakolbinger, Obtaining convergence rates of approximations in stochastic programming, in: Parametric Optimization and Related Topics (Eds. J. Guddat et al.), Akademie-Verlag, Berlin, 1987, 327-343

[17] G. Salinetti, Approximations for chance-constrained programming problems, Stochastics 10(1983), 157-179.

[18] T. Szántai, Calculation of the multivariate probability distribution function values and their gradient vectors, IIASA-Working Paper, WP-87-82, 1987.

[19] J. Wang, Continuity of feasible solution sets of probabilistic constrained programs, manuscript, University of Nanjing, P.R. China, 1986.

[20] J. Wang, A successive approximation method for solving probabilistic constrained programs, manuscript, Univ. of Nanjing, P.R. China, 1987.

[21] R. Wets, Stochastic programming: solution techniques and approximation schemes, in: Mathematical Programming: The State-of-the-Art 1982, Springer-Verlag, Berlin, 1983, 566-603.

THE CANONICAL GROUP OF INCREASING FUNCTIONS OF BOUNDED CONVEX VARIATION AND SOME OF ITS APPLICATIONS

G.S. Rubinstein [1)]

Abstract: In the group of order automorphisms of the unit segment $T = [0,1]$ which is ordered in the natural way, we distinguish and study several right-ordered subgroups which are closely related to the known Finetti-Fenchel problem of characterizing functional preferences (on convex sets), which may be given by convex functions.

1. Used Concepts and Notation from the Theory of Ordered Sets

1.1. We recall that a nonempty set M is said to be ordered if there is defined an ordering on M, i.e., if one has some reflexive, antisymmetric and transitive binary relation $N \subset M \times M$.
If $(x,y) \in N$, we say that x majorizes y, but y minorizes x, and we use the symbols $x \geqslant y$ and $y \leqslant x$ (or, similarly, $\succcurlyeq$ and $\preccurlyeq$). Given an ordered set $M = (M, \geqslant)$, one also considers the so-called strict order

$$x > y \ (y < x) \iff x \geqslant y, \ x \neq y.$$

Further, we assign to each element x of an ordered set $M = (M, \geqslant)$ two sets

$$\uparrow x = \{y \in M \mid y \geqslant x\} \quad , \quad \downarrow x = \{z \in M \mid z \leqslant x\} ,$$

whose elements are called majorants and minorants of the element x, respectively. Moreover, we assign to each nonempty set $X \subset M$ the sets

$$\uparrow X = \bigcap_{x \in X} \uparrow x, \qquad \downarrow X = \bigcap_{x \in X} \downarrow x,$$

whose elements are called upper and lower bounds of the set X, respectively. A nonempty set $X \subset M$ having at least one upper bound (lower bound) is said to be bounded above (or bounded below, respectively).
For $X = \emptyset$ we put

$$\uparrow \emptyset = \downarrow \emptyset = M.$$

Finally, if a nonempty set $X \subset M$ is bounded above (bounded below), an upper bound $y \in \uparrow X$ (lower bound $z \in \downarrow X$) is called the least upper bound or the supremum (greatest lower bound or infimum), if $y \in \downarrow(\uparrow X)$ (if $z \in \uparrow(\downarrow X)$).

---

1) Mathematical Institute of the Sibirian Branch of the Academy of Sciences of the USSR, USSR-630090 Novosibirsk, Universitetsky Prospect 4

Obviously, a nonempty set X which is bounded above (bounded below) need not have a supremum (infimum), in general. However, if there exists a supremum (infimum), then it is uniquely determined and will be symbolized by $\bigvee X$ (or $\bigwedge X$, respectively). In the case of a finite set $X = \{x_s\}_{s=\overline{1,r}}$, we use the symbols

$$\bigvee X = \bigvee_{s=1}^{r} x_s = x_1 \vee x_2 \vee \ldots \vee x_r,$$

$$\bigvee X = \bigvee_{s=1}^{r} x_s = x_1 \wedge x_2 \wedge \ldots \wedge x_r.$$

For any ordered set $M = (M, \geqq)$, we have that, by the above definitions, each subset consisting of only one element possesses both the supremum and the infimum, which coincide with this element. On the other hand, even in the case of a subset which consists of two elements, this subset may be unbounded and need not have supremum and infimum.

An ordered set is called a lattice, if each subset consisting of two elements (and so, each nonempty finite subset) possesses its supremum and infimum. A lattice is said to be complete if all nonempty subsets are bounded and have their suprema and infima. A lattice which contains unbounded subsets is said to be conditionally complete if any nonempty, bounded above subset possesses its supremum, but any nonempty, bounded from below subset possesses its infimum.

Examples of complete lattices are the following families ordered by inclusion: the family of all convex sets of an arbitrary affine space, and its subfamily of affinely closed convex sets. However, the subfamily of bounded convex sets and the subfamily of affinely closed bounded convex sets are conditionally complete lattices, without being complete ones.

1.2. We recall now some notions from the theory of ordered linear spaces. Let $E = (E,+,.)$ be any real linear space, and let K be a pointed and affinely closed cone in E, this means

$$x \in K,\ y \in K,\ \lambda \in \mathbb{R}_+ \implies x + y \in K,\ \lambda x \in K,$$

$$x \in K,\ -x \in K \implies x = \mathbb{0},$$

$$E = \{z = x - y \mid x \in K,\ y \in K\},$$

and the set

$$\Lambda(x,y,K) = \{\lambda \in \mathbb{R} \mid (1-\lambda)x + \lambda y \in K\}$$

is closed for any $x, y \in K$.

The cone K defines an order relation on E by

$$x \geqq y \iff x \in E,\ y \in E,\ x - y \in K,$$

which is compatible with the linear operations on E:

$$x \geqq y,\ z \geqq u,\ \lambda \in \mathbb{R}_+ \Longrightarrow x + z \geqq y + u,\ \lambda x + \lambda y,$$

where

$$K = \{x \in E \mid x \geqq \mathbb{0}\}.$$

Further, the affine closedness of K implies that the ordering induced by K is compatible with the affine topology on E, i.e., for any finitely dimensional subspace E' of E and for any two convergent sequences $\{x_i\}_{i \in \mathbb{N}}, \{y_i\}_{i \in \mathbb{N}} \subset E'$, we have

$$x_i \geqq y_i,\ i \in \mathbb{N} \Longrightarrow \lim x_i \geqq \lim y_i .$$

Moreover, if the one-dimensional space E' is generated by some element $x \neq \mathbb{0}$ of E, then

$$\uparrow E' = \downarrow E' = \emptyset .$$

Therefore, a non-trivial complete linear lattice does not exist. Conditionally complete linear lattices, do exist, indeed, and play a crucial role in functional analysis and in many of its applications. The theory of such lattices has been developed by the Academic L.V. Kantorovich already in the middle of the thirties (cf. [1]-[4]). One of the conditionally complete linear lattices, which have been studied by L.V. Kantorovich, are of interest now.

This lattice is constructed by the linear space $V = V(T)$ of functions x which are defined on the unit segment $T = [0,1]$, which have bounded variation, and which fulfil the property

$$x(t) = \frac{1}{2}\,[x(t-0) + x(t+0)]\ ,\ t \in \,]0,1[\qquad x(0) = x(+0),\ x(1) = x(1-0), \tag{1.1}$$

where $x(t-0)$ and $x(t+0)$ denote the left and right limit at $t \in T$, respectively.

The cone $K = V_+$ generated by all increasing (more precise: by all non-decreasing) functions $x \in V(T)$ plays the role of the cone of non-negative elements of V. One easily varifies that, in this case, the supremum (infimum) of a nonempty family $X \subset V(T)$, which is bounded above (bounded below), coincides with the point-wise supremum (or the point-wise infimum, respectively) of the functions $x \in X$. For the conditionally complete linear lattice generated in this way, we use the symbol $V = V(T,\geqq)$, where

$$x \geqq y \Longleftrightarrow x \in V(T),\ y \in V(T),\ x-y \in V_+ .$$

2. Some Right-Ordered Group and its Subgroup of Piecewise-Affine Functions

2.1. Any order automorphism of the unit segment $T = [0,1]$ which is ordered in the natural way may obviously be identified with a continuous, strictly increasing function $f \in T^T$ which satisfies the condition

$$f(0) = 0,\ f(1) = 1. \qquad (2.1)$$

Together with the superposition operation

$$(f \circ g)(t) = f(g(t)),\ t \in T,$$

these functions form a non-commutative group F, where

$$\mathbb{1}(t) = t,\ T \in T,$$

defines the unit element of F.

We are interested in studying a subgroup of F, namely, the canonical group of increasing functions of bounded convex variations (for shortness: canonical bcv-functions). This subgroup, denoted by G=G(T), is generated (i) by the semigroup G_+ which consists of all convex functions $\alpha \in F$ satisfying

$$0 < \alpha'(+0) \leq \alpha'(1-0) < +\infty,$$

and (ii) by the semigroup $G_- = G_+^{-1}$ which consists of all concave functions $\beta \in F$ satisfying

$$+\infty > \beta'(+0) \geq \beta'(1-0) > 0.$$

This means that a function is a canonical bcv-function if and only if it can be represented as a finite superposition of convex and concave functions belonging to $G_+ \cup G_-$.

By means of the generating semigroup G_+ we define the following ordering in the non-commutative group G:

$$g \succcurlyeq h \Longleftrightarrow g \in G,\ h \in G,\ g \circ h^{-1} \in G_+.$$

The ordered group obtained in this way is denoted by $G = G(T, \succcurlyeq)$. We shall say that G is right-ordered; this is justified in virtue of

$$g \succcurlyeq h \Longrightarrow g \circ f \succcurlyeq h \circ f,\ f \in G.$$

On the other hand, this ordering is not compatible with the multiplication from the left, because there exist elements g, h and f of G such that

$$g \circ h^{-1} \in G_+,\ (f \circ g) \circ (f \circ h)^{-1} \notin G_+.$$

2.2. Although the following facts with respect to the simplest subgroup of G will not be used in the next sections, we present these facts, because they illuminate the nature of the integral construction used below and may serve as a basis for the constructive study of bcv-function.

Let A denote the subgroup of piecewise-affine functions f F, which are defined by

$$f(t) = f(t_i) + k_i(t-t_i),\ t \in [t_i, t_{i+1}],\ i=\overline{0,r-1},$$

where $r \in \mathbb{N}$ and

$$0 = t_0 < t_1 < \ldots < t_r = 1,\ k_i > 0,\ i=\overline{0,r-1}$$

are given numbers satisfying

$$\sum_{i=1}^{r} k_i(t_{i+1}-t_i) = 1.$$

A piecewise-affine function defined in this way is convex (concave) if the coefficients k_i do not decrease (not increase).

By A_+ and A_- we denote the semigroups of piecewise-affine convex and concave functions, respectively. Obviously, A_+ and A_- are contained in the semigroups G_+ and G_- introduced above and generate the whole group A. The ordering in A, which is induced by A_+, coincides, of course, with the ordering '$\preccurlyeq$' in G defined above, and we shall use the same symbol.

Finally, we state a proposition which can easily be proved and which can be taken as a basis for the constructive investigation of the group G.

Proposition 1. The functions g, $h \in G = G(T, \succcurlyeq)$ satisfy the condition $g \succcurlyeq h$ if and only if for any $r \in \mathbb{N}$ and for any decomposition

$$0 = t_o < t_1 < \ldots < t_r = 1$$

of the unit segment, the piecewise-affine functions

$$f_g(t)=g(t_i)+k_i(g)(t-t_i),\ t \in [t_i,t_{i+1}]\ ,\ i = \overline{0,r-1},$$

$$f_h(t)=h(t_i)+k_i(h)(t-t_i),\ t \in [t_i,t_{i+1}]\ ,\ i = \overline{0,r-1},$$

satisfy the condition $f_g \succcurlyeq f_h$, where

$$k_i(g) = \frac{g(t_{i+1})-g(t_i)}{t_{i+1}-t_i}\ ,\ k_i(h) = \frac{h(t_{i+1})-h(t_i)}{t_{i+1}-t_i}\ ,\ i = \overline{0,r-1}.$$

3. Main Theorem and its Consequences

The study of the right-ordered canonical group of bcv-functions, which was introduced in § 2.1, is based on determining its relations to the conditionally complete linear lattice $V = V(T,\geq)$ considered in § 1.2. Recall that V is formed by all functions $x \in T^{\mathbb{R}}$ which have bounded variation on T and which, in addition, satisfy (1.1).

Given $g \in G$, we assign to g a function of bounded variation

$$z_g(t)= \begin{cases} 0{,}5\ \ln[g'(t-0)g'(t+0)] & \text{if } t \in]0,1[\ , \\ \ln g'(+0) & \text{if } t = 0, \\ \ln g'(1-0) & \text{if } t = 1, \end{cases} \qquad (3.1)$$

and the function $x_g \in V$ defined by

$$x_g(t) = z_g(t) - z_g(0),\ t \in T. \qquad (3.2)$$

In the following we analyze the mapping

$$g \in G \longmapsto x_g \in V, \tag{3.3}$$

which plays a principal role in the subsequent investigations.

Lemma 3.1. Suppose that g,h and α belong to G. Then we have:

$$x_{g \circ h}(t) = x_G(h(t)) + x_h(t), \; t \in T, \tag{3.4}$$

$$\alpha \in G_+ \Longleftrightarrow x_\alpha \in V_+, \tag{3.5}$$

$$g \succcurlyeq h \text{ in } G \Longleftrightarrow x_g \geqslant x_h \text{ in } V. \tag{3.6}$$

Proof. (3.4) immediately follows from the definitions (3.1), (3.2), and from the chain rule in the differential calculus. (3.5) and (3.6) are simple consequences of (3.1), (3.2) and standard arguments on convex functions.

Lemma 3.2. The mapping (3.3) is bijective.

Proof. Obviously, $g \in G$ has the integral representations

$$g(t) = \int_0^t e^{z_g(\tau)} d\tau$$

$$= \int_0^t e^{x_g(\tau)+z_g(\tau)} d\tau = g'(+0) \int_0^t e^{x_g(\tau)} d\tau, \; t \in T,$$

and, vice versa, it is easy to verify that for any $x \in V(T)$ (with $c_x = -\ln \int_0^1 e^{x(\tau)} d\tau$) the function

$$g(t) = \int_0^t e^{x(\tau)+c_x} d\tau \qquad , \; t \in T,$$

is the unique function $g \in G(T)$ which satisfies $x_g = x$. This completes the proof.

Now the subsequent theorem immediately follows from Lemma 3.1 and Lemma 3.2.

Theorem 3.1. The mapping (3.3) which is defined via (3.1) and (3.2) determines an order isomorphism between the right-ordered group $G = G(T, \succcurlyeq)$ of canonical bcv-functions (introduced in § 2.1) and the conditionally complete linear lattice $V = V(T, \geqslant)$ (introduced in § 1.2).

Corollary. The right-ordered canonical group $G = G(T, \succcurlyeq)$ is also a conditionally complete lattice.

Theorem 3.2. Suppose that $g \in G$. The sets

$$M(g) = \{(\alpha, \beta) \in G_+ \times G_- \mid g = \alpha \circ \beta\}$$

$$N(g) = \{(\alpha, \beta) \in G_+ \times G_- \mid g = \beta \circ \alpha\}$$

are nonempty and have the parametric representation

$$M(g) = \{(\alpha_g^* \circ \gamma, \gamma^{-1} \circ \beta_g^*) \mid \gamma \in G_+\},$$

$$N(g) = \{(\gamma \circ \alpha_g^{**}, \beta_g^{**} \circ \gamma^{-1}) \mid \gamma \in G_+\},$$

where $(\alpha_g^*, \beta_g^*) \in M(g)$, $(\alpha_g^{**}, \beta_g^{**}) \in N(g)$ are uniquely determined elements.

Proof. One easily verifies that the extremal convex and concave functions, which we are interested in, may be represented by the formulae

$$\beta_g^* = g \wedge \mathbf{1}, \quad \alpha_g^{**} = g \vee \mathbf{1},$$

$$\alpha_g^* = g \circ (\beta_g^*)^{-1}, \quad \beta_g^{**} = g \circ (\alpha_g^{**})^{-1}.$$

Remark. The given proof only uses the fact that the right-ordered group $G = G(T, \succcurlyeq)$ is a lattice.

4. Canonical Functions of Locally Bounded Convex Variation

Let $\tilde{G}_+$ $(\tilde{G}_-)$ be the semigroup arising from $G_+(G_-)$ by supplementing all convex (concave) functions which belong to $F \setminus (G_+ \cup G_-)$. We consider the group $\tilde{G}$ generated by the semigroups $\tilde{G}_+$ and $\tilde{G}_-$. By

$$g \succcurlyeq h \iff g \in \tilde{G},\ h \in \tilde{G},\ g \circ h^{-1} \in \tilde{G}_+$$

we obtain a right ordering in $\tilde{G}$, whose restriction to G coincides with the right ordering introduced in § 2.1. The right-ordered group $\tilde{G} = \tilde{G}(T, \succcurlyeq)$ defined in this way will be called the canonical group of increasing functions of locally bounded convex variation (for shortness: lbcv-functions).

The investigation of $\tilde{G}$ follows the scheme known from the analysis of its subgroup $G = G(T, \succcurlyeq)$. Therefore, we only point out the existing differences in some details. The main difference consists in the fact that the right-ordered group $\tilde{G}(T, \succcurlyeq)$ is not related to the conditionally complete linear lattice $V = V(T, \geq)$, but to some natural extension of V.

To construct this extension, we consider the linear set $\tilde{V} = \tilde{V}(T)$ of all functions x which are defined on the open interval $T_0 =]0,1[$, and which satisfy the following conditions:

(i) For any $\varepsilon \in]0, \frac{1}{2}[$, the function x is of bounded variation on the segment $[\varepsilon, 1-\varepsilon]$, where $x(t) = \frac{1}{2}[x(t-0)+x(t+0)]$, $t \in T_0$, $x(0,5) = 0$;

(ii) it corresponds to x a finite improper integral

$$\int_0^1 e^{x(t)} dt = \lim_{\varepsilon \to +0} \int_\varepsilon^{1-\varepsilon} e^{x(t)} dt.$$

By means of the cone $K = \tilde{V}_+$ of non-decreasing functions $u \in \tilde{V}$ we induce an order:

$$x \geq y \iff x \in \tilde{V},\ y \in \tilde{V},\ x-y \in \tilde{V}_+.$$

It is easy to see that the ordered linear space $\tilde{V} = \tilde{V}(T_o, \geq)$ defined in this way is a conditionally complete linear lattice.
Concerning the proof of this fact we only note that the supremum (infimum) of any nonempty set $X \subset \tilde{V}$, which is bounded above (bounded below), coincides with the point-wise supremum (point-wise infimum) of the corresponding functions $x \in X$.

<u>Theorem 4.1.</u> The right-ordered group $\tilde{G} = \tilde{G}(T, \succsim)$ and its subgroup $G = G(T, \succsim)$ are conditionally complete lattice.

<u>Proof.</u> We assign to each lbcv-function $g \in \tilde{G}$ a function belonging to the conditionally complete linear lattice $\tilde{V} = \tilde{V}(T_o, \geq)$ by

$$x_g(t) = z_g(t) - z_g(0,5), \quad t \in T_o, \tag{4.1}$$

where

$$z_g(t) = 0,5 \cdot \ln(g'(t-0) \cdot g'(t+0)), \quad t \in T_o. \tag{4.2}$$

Moreover, for any function $x \in \tilde{V}$, there exists a uniquely determined lbcv-function $g \in \tilde{G}$ satisfying $x_g = x$, namely

$$g(t) = \int_0^t e^{x(\tau)+c_x} d\tau \quad , t \in T,$$

where $c_x = -\ln \int_0^1 e^{x(\tau)} d\tau$. Hence, the mapping from $\tilde{G} = \tilde{G}(t_o, \succsim)$ to $\tilde{V} = \tilde{V}(T_o, \geq)$, which is defined by (4,1), is a bijective mapping onto $\tilde{V}$. To complete the proof, it remains to remark that for any g,h and α from $\tilde{G}$, one has

$$\tilde{x}_{g \circ h}(t) = x_g(h(t)) + x_h(t), \ t \in T_o,$$

$$\alpha \in \tilde{G}_+ \Longleftrightarrow \tilde{x}_\alpha \in V_+,$$

$$g \succsim h \text{ in } \tilde{G} \Longleftrightarrow x_g \geq x_h \text{ in } \tilde{V},$$

we refer to the proof of Lemma 3.1.

<u>Corollary.</u> With respect to $\tilde{G} = \tilde{G}(T, \succsim)$ a proposition analogous to Theorem 3.2 holds.

5. Relation to the Finetti-Fenchel Problem

The investigation of conditionally complete group lattices is closely related to the known Finetti-Fenchel problem of characterizing functional preferences (on convex sets), which may be given by convex functions, cf. [5] - [16] .

First we study some variant of the announced problem in the non-commutative group $F = F(T, \circ)$ of continuous, strict increasing functions $f \in T^T$ which satisfy (2.1). A nonempty family $\Phi \subset F$ is called a Finetti-Fenchel family if the set

$$U(\Phi) = \{u \in F \mid u \circ \varphi \in G_+, \ \varphi \in \Phi\} \tag{5.1}$$

is nonempty. This definition, in particular, means that the nonempty family $\Phi \subset F$ is a Finetti-Fenchel family if and only if $f \circ F$ has this property for all $f \in F$. Thus, in studying these families, we may restrict our considerations to families which contain the unit element $\mathbb{1}$ of the group F.

The results of the preceding paragraph immediately entail the subsequent criterion for the Finetti-Fenchel property.

Theorem 5.1. Let $\Phi \subset F$ be a nonempty family containing $\mathbb{1}$. Φ is a Finetti-Fenchel family if and only if it satisfies the following two conditions:

(i) Φ is contained in the conditionally complete group lattice $\tilde{G} = \tilde{G}(T, \succeq)$,

(ii) the set $\Psi_\Phi \subset \tilde{G}$ formed by the functions $\psi_\varphi = \varphi^{-1}$, $\varphi \in \Phi$, is bounded above.

Moreover, the nonempty set $U(\Phi)$ given by (5.1) has a parametric representation

$$U(\Phi) = \{u = \alpha \circ u^* \,|\, \alpha \in \tilde{G}_+\},$$

where the convex function $u^* \in \tilde{G}_+$ coincides with the supremum of the set Ψ_Φ.

Now we switch to the investigation of the classical Finetti-Fenchel problem. We consider a real linear space E of finite or infinite dimension and an affinely open convex set $X \subset E$. By $C(X)$ we denote the space of all affinely continuous functions $f: X \to R$. Two functions $f, \varphi \in C(X)$ are called order equivalent (in symbols: $f \sim \varphi$) if

$$f(x) > f(y) \iff \varphi(x) > \varphi(y).$$

It is easy to see that the functions $f, \varphi \in C(X)$ are order equivalent if and only if for some continuous, strict increasing function $u: f(X) \to \mathbb{R}$, the equations

$$(u \circ f)(x) = u(f(x)) = \varphi(x), \; x \in X,$$

hold.

It is worth noting that the cone $C_+(X)$ of all convex functions $f \in C(X)$ is not closed with respect to order equivalence, i.e., $f \in C_+(X)$ and $f \sim \varphi$ do not in general imply that $\varphi \in C_+(X)$. However, the larger cone $\tilde{C}_+$ formed by all quasiconvex functions $f \in C(X)$ is closed with respect to order equivalence,

We recall that $f \in C(X)$ is said to be quasiconvex, if for any $x \in X$, $y \in X$ and $t \in T_0 =]0,1[$ the inequality

$$f((1-t)x+ty) \leq \max\{f(x), f(y)\}$$

holds, where for $f(x) \neq f(y)$ this inequality is strictly fulfilled.

The classical Finetti-Fenchel problem consists in finding conditions, under which a given quasiconvex function $f \in \tilde{C}_+(X)$ is order equivalent with some convex function $\varphi \in C_+(X)$. Our proposal of investigating this problem is essentially based on the Finetti-Fenchel families studied above, in particular, the criterion given in Theorem 5.1 is used.

We now consider an arbitrary quasiconvex function

$$f \in \tilde{C}_+(X) \setminus C_+(X). \tag{5.2}$$

By $f(X)$ we denote the set $\{f(x) \mid x \in X\}$. Further, we assign to each pair $c,d \in f(X)$ with $c<d$ the set $Z_f(c,d)$ of those pairs

$$(x,y) \in f^{-1}(c) \times f^{-1}(d),$$

for which the functions

$$g_{x,y}(t) = \frac{f((1-t)x+ty)-c}{d-c}, \quad t \in T, \tag{5.3}$$

(which satisfy $g_{x,y}(0) = 0$ and $g_{x,y}(1) = 1$) are strictly increasing and, therefore, belong to the group F considered above.

<u>Theorem 5.2.</u> A quasiconvex function (5.2) is order equivalent with some convex function $\varphi \in C_+(X)$ if and only if for any pair $c,d \in f(X)$ with $c<d$ the set of strictly increasing functions (5.3) (which belong to F) is a Finetti-Fenchel family. Moreover, the set $U(f)$ of corresponding strictly increasing functions $u: f(X) \to \mathbb{R}$ which fulfil the condition $u \circ f \in C_+(T)$ has the following parametric representation

$$U(f) = \{\alpha \circ u^* \mid \alpha \in A_{u^*}\},$$

where the continuous, strictly increasing function u^* is determined with an accuracy up to a left affine multiplier, and where A_{u^*} denotes the family of all strictly increasing affine functions defined on $u^*(T)$.

References

[1] Л.В. Канторович. О полуупорядоченных линейных пространствах и их применениях в теории линейных операций. Доклады АН СССР, 4 /1935/, 11 - 14.

[2] Л.В. Канторович. Линейные полуупорядоченные пространства. Математический сборник, 2 /44/, /1937/, 121- 168.

[3] Л.В. Канторович, Б.З. Вулих и А.Г.Пинскер. Функциональный анализ в полуупорядоченных пространствах. Москва - Ленинград /1950/, 1 - 548.

[4] Б.З.Вулих. Введение в теорию полуупорядоченных пространств. Москва /1961/, 1- 407.

[5] B. de Finetti. Sulle stratificazion convesse. Annali di Matematica Pura ed Applicata, 30, Serie 4 (1949), 173-183.

[6] W. Fenchel. Convex cones, sets and functions. Mimeographed lecture notes (1953), Princeton, N.Y.

[7] W. Fenchel. Über konvexe Funktionen mit vorgeschriebenen Niveaumannigfaltigkeiten. Mathematische Zeitschrift 63 (1956), 496-506.

[8] Б. А. Вертгейм и Г.Ш.Рубинштейн. К определению квазивыпуклых функций. Математическое программирование. Москва, 1966, с. 121-134.

[9] Г.Ш.Рубинштейн. Характеристика насыщения класса выпуклых функций. Оптимизация 9/26/, 165-180. Новосибирск, 1973.

[10] Г.Ш.Рубинштейн и Ю.Н. Владимиров. О возрастающих функциях ограниченной выпуклой вариации. Доклады АН СССР, том 236, № 1/1977/, с. 27-30.

[11] Г.Ш. Рубинштейн. Некоторые вопросы теории выпуклых множеств, выпуклых функций и квазивыпуклых предпорядков. Сообщение на Международном симпозиуме по математическому программированию и его приложениям. Бонн, 1982.

[12] Y. Kannai. Approximation of convex preferences. Journal of Mathematical Economics 1(1974), 101-106.

[13] G. Debreu. Least concave utility functions. Journal of Mathematical Economics 3(1976), 121-129.

[14] Y. Kannai. Construction d'une fonction concave à surfaces de niveau données. C.R. Acad. Sc. Paris, 282(1976), Série A, 1355-1358.

[15] J.P. Crouzeix. Contributions à l'étude des Fonctions Quasiconvexes. Thèse, Université de Clermont-Ferrand, 1977.

[16] Y. Kannai. Concave utility functions - existence, constructions and cardinality. Generalized concavity in Optimization and Economics (edited by S. Schaible and W.T. Ziemba), 544-611. Academic Press, 1981.

THEORETICAL FOUNDATIONS OF TWO-LEVEL METHODS IN NONCONVEX OPTIMIZATION

Jan Rückmann [1), Klaus Tammer [1)

Abstract

In the present paper we give a highly uniform representation and foundation of generalizations of the primal and the dual decomposition method. Both approaches are two-level methods to generate a stationary point of a nonlinear optimization problem. They have comparable numerical advantages as the original version of the corresponding method but do not require convexity assumptions.

1. Introduction

Both the primal and the dual decomposition methods may be efficient solution approaches to nonlinear optimization problems of a special structure. Both methods are characterized by the fact that the original is replaced by a suitable two-level problem (cf. /6/,/13/,/16/). In the following these methods will be briefly described in a highly uniform way. Only for brevity's sake equality constraints have been omitted throughout the paper.

The primal method has been developed to solve optimization problems of the following type:

$$\min \left\{ f_o(x,y) \ / \ f_i(x,y) \leq 0, \ i=1,\dots,p, \ g_j(y) \leq 0, \ j=1,\dots,q \right\} \qquad (1.1)$$

using the parametric problem of the lower level:

$$P(y): \min_x \left\{ f_o(x,y) \ / \ f_i(x,y) \leq 0, \ i=1,\dots,p \right\} \qquad (1.2)$$

and the problem of the upper level:

$$\min \left\{ \varphi(y) \ / \ g_j(y) \leq 0, \ j=1,\dots,q \right\} \qquad (1.3)$$

where $\varphi(y) = \inf_x \left\{ f_o(x,y) \ / \ f_i(x,y) \leq 0, \ i=1,\dots,p \right\}$.

Without any additional assumption it can be easily shown that the original problem (1.1) and the two-level problem (1.2),(1.3) are equivalent.

The dual method has been developed to solve optimization problems of the type:

$$\min \left\{ h_o(x) \ / \ f_i(x) \leq 0, \ i=1,\dots,p, \ g_k(x) \leq 0, \ k=1,\dots,m \right\}. \qquad (1.4)$$

Using the same notation as for the primal method, the corresponding parametric problem of the lower level has the form:

1) TH Leipzig, Sektion Mathematik und Informatik, PF 66, 7030 Leipzig DDR

$$P(y): \min_{x} \{ f_o(x,y) \ / \ f_i(x) \leq 0, \ i=1,\ldots,p \} \tag{1.5}$$

with $f_o(x,y) = h_o(x) + \sum_{k=1}^{m} y_k \, g_k(x)$

where the related problem of the upper level is now (represented again as a minimum problem):

$$\min \{ -\varphi(y) \ / \ y \geq 0 \} . \tag{1.6}$$

Note that (1.5) is of type (1.2) having the special properties that f_o is linearly affine in y and f_i ($i=1,\ldots,p$) depends on x only.
For this approach, equivalence of the original problem (1.4) and the two-level problem (1.5),(1.6) holds <u>iff</u> f_o possesses a saddle point, which can be generally ensured under convexity and regularity assumptions on (1.4) only (cf. /6/,/13/).

Both decomposition methods may be efficient if the lower level problem $P(y)$ can be solved much more easily than the original problem, even for a sequence of parameter values. In this connection we are especially interested in problems of such a type that the vector x of the original problem can be decomposed into subvectors x^r ($r=1,\ldots,N$) such that (1.1) and (1.4) have the form

$$\min\{\sum_{r=1}^{N} f_o^{\,r}(x^r,y) \ / \ F^r(x^r,y) \leq 0, \ r=1,\ldots,N, \ g_j(y) \leq 0, \forall j \} \tag{1.1'}$$

and

$$\min\{\sum_{r=1}^{N} h_o^{\,r}(x^r) \ / \ F^r(x^r) \leq 0, \ r=1,\ldots,N, \sum_{r=1}^{N} G^r(x^r) \leq 0 \}, \tag{1.4'}$$

where F^r, G^r are suitable vector functions. In this case for both approaches the corresponding lower level problem splits into N independent subproblems with respect to x^r which can be solved separately.

However, as already mentioned the dual method is theoretically founded for the convex case only. Moreover, and this holds for both methods, because of the required global minimization at the lower and upper levels, an efficient numerical realization is possible again for convex problems only.
Usually, apart from expansive approaches to global optimization, the realistic aim for high-dimensional nonconvex problems is the computation of a stationary point or of a locally minimal point. In this sense several previous papers dealt with two-level approaches for the computation of stationary (locally minimal) points of (1.1) and (1.4), resp. We mention the results in /4/,/5/,/6/ for a generalization of the dual method using augmented Lagrangians for which the separability of the lower level can be ensured by introducing additional variables and a third level.
In our paper we refer to the propositions for a generalization of the primal method given in /8/,/19/ and those for a generalization of the dual method given in /15/,/17/,/18/. These approaches have the same nu-

merical advantage of splitting the lower level in the cases (1.1') and (1.4') as the original versions of the two methods and do not require neither additional variables nor a third level.
In Sections 2 and 3 of our paper the mathematical foundations of these generalizations are represented in a uniform way on the common basic assumption of strong stability at the lower level following Kojima /11/. Section 2 deals with the lower level and Section 3 with the upper one and with the corresponding equivalence statements. In Section 4 the relation between the type of the stationary solution of the original problem (local minimum, saddle point, local maximum) and that of the corresponding stationary solutions of the lower-level and upper-level problems are considered. In this connection we apply the concept of the stationary index introduced by Kojima /11/ and extend it to the upper-level problem for which the objective function belongs to the class $C^{1,1}$ and generally not to C^2.
As in /9/,/14/ we use the symbols $In(K) = (p(K), \eta(K), o(K))$ and $In(K/T)$ for the inertia of a real symmetric matrix K and the inertia ofK with respect to a linear subspace, resp. We shall apply the following two inertia relations for a partitioned matrix $K = \begin{pmatrix} U & V \\ V' & W \end{pmatrix}$:

1. The regularity of U implies $In(K) = In(U) + In(S)$ (1.7)
   where $S = W - V'U^{-1}V$ is the Schur complement of U in K. (cf. /14/)
2. If W=O and $T = \{x \,/\, V'x = 0\}$ then $In(K) = In(U/T) + (r,r,s-r)$ (1.8)
   where r is the rank and s is the number of columns of V (cf. /9/).

The symbols ∇f and $\nabla^2 f$ will be used for the gradient and the Hessian of f, resp., where a subscript like that one in $\nabla_x f$ denotes the subvector of ∇f of the corresponding partial derivations of f. The symbol C^k denotes the space of all functions which are k-fold continuously differentiable where $C^{1,1}$ is the subset of C^1 for which ∇f is locally Lipschitzian and $\partial^2 f$ denotes the generalized Hessian of $f \in C^{1,1}$ according to Clarke /2/.
Now let us consider a nonlinear optimization problem:

$$\text{(P)} \quad \min \{ \varphi_0(x) \,/\, \varphi_i(x) \le 0,\ i=1,\dots,l \} \quad \text{with } \varphi_i \in C^1(R^n).$$

A pair $(\hat{x},\hat{u}) \in R^n \times R^l$ satisfying the Kuhn-Tucker conditions for (P) is called a stationary point of (P) with $\hat{x}$ being a stationary solution of (P) and $\hat{u} = (\hat{u}_1,\dots,\hat{u}_l)'$ a corresponding multiplier. For a given stationary point $(\hat{x},\hat{u})$ of (P) the following symbols are used:

$I_0(\hat{x}) = \{ i \,/\, \varphi_i(\hat{x}) = 0,\ 1 \le i \le l \}$, $I_+(\hat{u}) = \{ i \,/\, \hat{u}_i > 0,\ 1 \le i \le l \}$ and

$T(I) = \{ w \,/\, \nabla \varphi_i(x)'w = 0,\ i \in I \}$ where $I_+(\hat{u}) \subset I \subset I_0(\hat{x})$.

Now given that in (P) even $\varphi_i \in C^2(R^n)$, and let $H = \nabla_x^2 \phi(\hat{x},\hat{u})$ be the Hessian of the Lagrangian $\phi(x,u) = \varphi_0(x) + \sum_{i=1}^{l} u_i \varphi_i(x)$ of (P).

The following conclusions for a given stationary point $(\hat{x},\hat{u})$ of (P)

are well-known:

(C1) If $\hat{x}$ is a locally minimal point of (P), then $\eta(H/T(I_0(\hat{x})))= 0$.

(C2) If $\eta(H/T(I_+(\hat{u})))= o(H/T(I_+(\hat{u})))= 0$, then $\hat{x}$ is a strictly locally minimal point of (P).

Generally, there is a difference between the necessary optimality condition in (C1) and the strong sufficient optimality condition in (C2). In /11/ Kojima has defined the class of strongly stable stationary solutions of (P) which is of great interest from the theoretical as well as from the numerical point of view. On the assumption that the Linear Independence Constraint Qualification (LICQ) or the Mangasarian-Fromovitz Constraint Qualification (MFCQ) is satisfied, the strong stability can be expressed as follows (cf. /11/, Sections 4,7):

(S1) A stationary solution $\hat{x}$ of (P) satisfying LICQ is strongly stable <u>iff</u> $\eta(H/T(I))=$ const, $o(H/T(I))= 0$ for all I with $I_+(\hat{u}) \subset I \subset I_0(\hat{x})$ where $\hat{u}$ is the corresponding unique multiplier.

(S2) A stationary solution $\hat{x}$ of (P) satisfying MFCQ – and not LICQ – is strongly stable <u>iff</u> $\eta(H/T(I))= o(H/T(I))= 0$ for all I with $I_+(\hat{u}) \subset I \subset I_0(\hat{x})$ and each corresponding multiplier $\hat{u}$.

Using these relations, Kojima /11/ introduced the stationary index

$$\text{s.index } (\hat{x}/P) := \eta(H/T(I_+(\hat{u}))) \tag{1.9}$$

for a strongly stable stationary solution $\hat{x}$ of (P) satisfying at least MFCQ. If there is no obscurity concerning the underlying problem we write s.index $(\hat{x})$. With regard to (C1) and (C2) the type of a strongly stable stationary solution $\hat{x}$ of (P) satisfying MFCQ can be uniquely characterized by its stationary index (cf. /12/, Theorem 3.1):

$$\begin{array}{lll} \hat{x} \text{ is a locally minimal point of (P)} & \underline{\text{iff}} & \text{s.index } (\hat{x}) = 0 \\ \hat{x} \text{ is a saddle point of (P)} & \underline{\text{iff}} & 0 < \text{s.index } (\hat{x}) < n \\ \hat{x} \text{ is a locally maximal point of (P)} & \underline{\text{iff}} & \text{s.index } (\hat{x}) = n\ . \end{array} \tag{1.10}$$

2. Strongly Stable Stationary Solutions of the Lower-level Problem

In the following we assume that the functions in (1.1) and (1.4) are of class C^2 with respect to all appearing variables. Of course, a weakening of this assumption for some statements is possible. However, convexity assumptions shall <u>not</u> be made. As in this case the global minimization is in general not realistic, our aim is to compute a stationary point $(\hat{x},\hat{y},\hat{u},\hat{v})$ of (1.1) and $(\hat{x},\hat{y},\hat{u})$ of (1.4), resp.
For (1.1) the Kuhn-Tucker conditions are:

$$\nabla_x f_0(x,y) + \sum_{i=1}^{p} u_i \nabla_x f_i(x,y) = 0,\ f_i(x,y) \le 0,\ u_i \ge 0,\ u_i f_i(x,y) = 0 \quad i=1,\dots,p \tag{2.1a}$$

$$\nabla_y f_0(x,y) + \sum_{i=1}^{p} u_i \nabla_y f_i(x,y) + \sum_{j=1}^{q} v_j \nabla g_j(y) = 0,\ g_j(y) \le 0,\ v_j \ge 0,\ v_j g_j(y) = 0\ \forall j\ . \tag{2.1b}$$

For (1.4) they are:

$$\nabla_x f_o(x,y) + \sum_{i=1}^{r} u_i \nabla f_i(x) = 0, \ f_i(x) \leq 0, \ u_i \geq 0, \ u_i f_i(x) = 0 \ \forall i \quad (2.2a)$$

$$g_k(x) \leq 0, \ y_k \geq 0, \ y_k g_k(x) = 0, \ k=1,\dots,m \ . \quad (2.2b)$$

As (2.1a) and (2.2a), resp., are the Kuhn-Tucker conditions for the corresponding problem P(y) of the lower level it can be seen immediately that for both cases the subvector $(\hat{x},\hat{u})$ of a stationary point of the original problem is always a stationary point of $P(\hat{y})$.
Let us denote the set of all stationary points of P(y) by $\Sigma(y)$ and $\Sigma = \{(x,u,y) \ / \ (x,u) \in \Sigma(y), \ y \in R^m\}$. The common idea of our generalized primal and dual decomposition methods can be expressed now more precisely in generating such an element $(\hat{x},\hat{u},\hat{y}) \in \Sigma$ which satisfies the remaining conditions (2.1b) of (2.1) and (2.2b) of (2.2), resp. As shown in Section 3 this can be realized by finding a stationary point of a suitable redefined upper-level problem. However, to define the objective function of the upper level at least locally, in a unique way, suitable assumptions on the lower level must be made (cf. /15/,/20/). The following special case of the rather general but complicated assumptions V1),V2) in /20/, which is of particular interest, will be used here.
We say that Condition A1) is fulfilled at a point (x,u,y) if (x,u) is a strongly stable stationary point of P(y) and x satisfies LICQ. Throughout this paper it will be assumed now that for a given initial point (x^o,u^o,y^o) A1) is fulfilled. According to /3/,/11/, this basic assumption has the following consequences.

<u>Lemma 2.1:</u> Condition A1) for (x^o,u^o,y^o) implies:
1. There exist a region Y with $y^o \in Y$ and exactly one pair of vector functions $\bar{x}(.),\bar{u}(.)$ which are defined and continuous on Y with $\bar{x}(y^o)=x^o$, $\bar{u}(y^o)=u^o$ such that for each $y \in Y$ $(\bar{x}(y),\bar{u}(y)) \in \Sigma(y)$.
2. The vector functions $\bar{x}(.),\bar{u}(.)$ are even locally Lipschitzian on Y.
3. For each $y \in Y$ Condition A1) is fulfilled at $(\bar{x}(y),\bar{u}(y),y)$.
4. For each $y \in Y$ $s.index(\bar{x}(y)) = s.index(x^o)$.

Instead of the global minimization problems (1.2) and (1.5) the lower-level problems of the generalized primal and dual methods are now formulated as follows:

$\bar{P}(y)$: Find the stationary point $(\bar{x}(y),\bar{u}(y))$ of P(y) for given $y \in Y$!

Obviously, similar to P(y), problem $\bar{P}(y)$ splits into N independent subproblems assuming the special form (1.1') and (1.4') of the original problem.

3. The Generalized Upper Level and Equivalence Statements

In order to formulate the generalized upper level for both the primal approach and the dual one the function

$$\bar{\varphi}(y) = f_o(\bar{x}(y),y) \qquad (3.1)$$

is used instead of the global optimal value function φ in (1.3) and (1.6) where $\bar{x}(y)$ is given by Lemma 2.1. At the upper level our aim is to find a stationary point of the problem

$$\min\left\{ \bar{\varphi}(y) \;/\; g_j(y)\leq 0,\ j=1,\dots,q,\ y\in Y\right\} \qquad (3.2)$$

for the generalized primal method and a stationary point of

$$\min\left\{-\bar{\varphi}(y) \;/\; y\geq 0,\ y\in Y\right\} \qquad (3.3)$$

for the generalized dual method.
For further considerations the following notations will be used:
L and $\bar{L}$ are the Lagrangians for the lower-level problem $P(y)$ and for the upper-level problem (3.2) or (3.3), resp.,
$I_+(y)=\{i \;/\; \bar{u}_i(y)>0,\ 1\leq i\leq p\}$, $I_o(y)=\{i \;/\; f_i(\bar{x}(y),y)=0,\ 1\leq i\leq p\}$.
Further, the following matrices are used:

$$A=\nabla_x^2 L,\quad B(I)=(\nabla_x f_i,\ i\in I),\quad C=\nabla_{xy} L,\quad D(I)=(\nabla_y f_i,\ i\in I),$$

$$E=\nabla_y^2 L,\quad M_o(I)=\begin{pmatrix} A & B(I)\\ B'(I) & 0\end{pmatrix} \text{ and } M(I)=\begin{pmatrix} A & B(I) & C\\ B'(I) & 0 & D'(I)\\ C' & D(I) & E\end{pmatrix}$$

where all derivations are taken for $x=\bar{x}(y)$, $u=\bar{u}(y)$, $y\in Y$ and I satisfies $I_+(y)\subset I\subset I_o(y)$.
For the special case of $P(y)$ in (1.5) we have $D(I)=0$, $E=0$ and $C=(\nabla g_k(x)',\ k=1,\dots,m)$.
Because of Lemma 2.1 (3.) the matrices $M_o(I)$ are regular (cf. /11/) such that the Schur complement $S(I)$ of $M_o(I)$ in $M(I)$ can be formed.
The following lemma characterizes some properties of $\bar{\varphi}$.

<u>Lemma 3.1:</u> The function $\bar{\varphi}$ in (3.1) belongs to the class $C^{1,1}(Y)$ where

$$\nabla\bar{\varphi}(y) = \nabla_y L(x,y,u)\Big|_{x=\bar{x}(y),u=\bar{u}(y)} \qquad (3.4)$$

and

$$\partial^2\bar{\varphi}(y) = \operatorname{conv}\left\{S(I),\ I\in V(y)\right\} \qquad (3.5)$$

where $V(y)$ is a suitable nonempty subset of $\{I \;/\; I_+(y)\subset I\subset I_o(y)\}$.

<u>Proof:</u> The relations $\bar{\varphi}\in C^{1,1}(Y)$ and (3.4) have been proved in /8/. Relation (3.5) follows from the fact that $\bar{\varphi}$ is a continuous selection of a finite number of C^2-functions (cf. /3/). According to /10/, this implies $\partial^2\bar{\varphi}(y)=\operatorname{conv}\{\nabla^2\varphi_I(y) \;/\; I\in V(y)\}$ with a suitable set $V(y)$ and - hence - (3.5) if the definition of $S(I)$ is compared with the representation of $\nabla^2\varphi_I(y)$ given in /8, formula (3.13)/.

Using formula (3.4), we are able now to prove the announced equivalence.

Theorem 3.1:

1. $(\hat{x},\hat{y},\hat{u},\hat{v})$ is a stationary point of (1.1) such that $(\hat{x},\hat{u},\hat{y})$ satisfies A1) iff $(\hat{x},\hat{u})$ solves $P(\hat{y})$ and $(\hat{y},\hat{v})$ is a stationary point of (3.2).
2. $(\hat{x},\hat{y},\hat{u})$ is a stationary point of (1.4) such that $(\hat{x},\hat{u},\hat{y})$ satisfies A1) iff $(\hat{x},\hat{u})$ solves $P(\hat{y})$ and $\hat{y}$ is a stationary point of (3.3).

Proof: Using Formula (3.4), this equivalence follows immediately from the fact that (2.1b) and (2.2b), resp., together with $y \in Y$ are just equivalent to the Kuhn-Tucker conditions for (3.2) and (3.3), resp.

4. Inertia Statements

In this section it is assumed that $(\hat{x},\hat{y},\hat{u},\hat{v})$ or $(\hat{x},\hat{y},\hat{u})$ is a stationary point of (1.1) and (1.4), resp., such that $(\hat{x},\hat{u},\hat{y})$ satisfies A1). The matrices introduced in Section 3 are used and taken now specially for $y=\hat{y}$ with $I_+=I_+(\hat{y})$, $I_o=I_o(\hat{y})$. Moreover, let be

$$J_+=\begin{cases}\{j \,/\, \hat{v}_j>0\}\\ \{k \,/\, \hat{y}_k>0\}\end{cases},\quad J_o=\begin{cases}\{j/\ g_j(\hat{y})=0\}\\ \{k/\ g_k(\hat{x})=0\}\end{cases},\quad F(J)=\begin{cases}(\nabla g_j(\hat{y}),j\in J)\\ (e_j,\ j\in J)\end{cases},$$

where e_j is the j-th unit vector in R^m and $J_+\subset J\subset J_o$. Here the upper definition concerns the primal approach and the lower the dual one.

In order to formulate the inertia statements for the generalized primal approach, the following linear subspaces are used additionally:

$$T(I,J)=\{(x,y) \,/\, B'(I)\ x + D'(I)\ y= 0,\ F(J)'y= 0\},$$

$$T_x(I)=\{x \,/\, B'(I)\ x= 0\},\quad T_y(J)=\{y \,/\, F'(J)= 0\}.$$

Moreover we introduce the matrices $G = \begin{pmatrix} A & C\\ C' & \overline{E}\end{pmatrix}$ with $\overline{E}= E+\sum_{j=1}^{q} \hat{v}_j\nabla^2 g_j(\hat{y})$

and $\overline{S}(I)$ defined as the Schur complement of $M_o(I)$ in $\overline{M}(I)=\begin{pmatrix} A & B(I) & C\\ B'(I) & 0 & D'(I)\\ C' & D(I) & \overline{E}\end{pmatrix}$.

Theorem 4.1 (Primal Approach):

1. For each pair I,J with $I_+\subset I\subset I_o$, $J_+\subset J\subset J_o$, we have
$In(G/\ T(I,J)) = In(A/\ T_x(I)) + In(S(I)/\ T_y(J))$.
2. If $(\hat{x},\hat{y},\hat{u},\hat{v})$ is even a strongly stable stationary point of (1.1) and $(\hat{x},\hat{y})$ satisfies MFCQ, we have,
$\eta(S/\ T_y(J))=$ s.index$(\hat{x},\hat{y})$ − s.index$(\hat{x})$ and $o(S/\ T_y(J))= 0$
for each $S\in \partial_y^2\ \overline{L}(\hat{y},\hat{v})$ and each J with $J_+\subset J\subset J_o$.

Proof: 1. According to /11/, assumption A1) on $(\hat{x},\hat{u},\hat{y})$ implies that $o(A/\ T_x(I))=0$ and B(I) has the full rank. Now the proof can follow the line of the proof of Lemma 3.3 in /8/ taking relation (1.7) into account.
2. From (S1), (S2) and (1.9) follows $\eta(A/\ T_x(I))=$ s.index$(\hat{x})$, $o(A/\ T_x(I))=0$, and on the assumption of 2. $\eta(G/\ T(I,J))=$ s.index$(\hat{x},\hat{y})$ and $o(G/\ T(I,J))=0$ for $I_+\subset I\subset I_o$, $J_+\subset J\subset J_o$. Hence Statement 1. implies

$$\eta(\overline{S}(I)/\ T_y(J))= \text{s.index}(\hat{x},\hat{y}) - \text{s.index}(\hat{x}),\quad o(\overline{S}(I)/\ T_y(J))=0. \qquad (4.1)$$

For each pair $I_1, I_2 \in \{I / I_+ \subset I \subset I_0\}$ with $I_1 = I_2 \setminus \{i\}$ $\bar{S}(I_2) - \bar{S}(I_1) =$
$= \bar{E} - (C', D(I_2))\, M_0(I_2)^{-1} \binom{C}{D'(I_2)} - (\bar{E} - (C', D(I_1))\, M_0(I_1)^{-1} \binom{C}{D'(I_1)}) =$

$= \alpha\, zz'$ with $\alpha = 1/ ((\nabla_x f_i(\hat{x}, \hat{y}))'\, M_0(I_1)^{-1} (\nabla_x f_i(\hat{x}, \hat{y})))$ and

$z = (C', D(I_1))\, M_0(I_1)^{-1} (\nabla_x f_i(\hat{x}, \hat{y})) - \nabla_y f_i(\hat{x}, \hat{y})$ if $M_0(I_2)^{-1}$ is expressed by $M_0(I_1)^{-1}$ and $\nabla_x f_i(\hat{x}, \hat{y})$ using Theorem 2.7 in /14/. If we write $\tilde{M}_0(I) = \begin{pmatrix} A & B(I) \\ -B'(I) & 0 \end{pmatrix}$, assumption A1) implies (cf. /11, Section 4/) $\operatorname{sgn} |\tilde{M}_0(I_1)| = \operatorname{sgn} |\tilde{M}_0(I_2)| \neq 0$. Moreover, we have (cf. Theorem 2.1 in /14/) $\alpha = 1/ ((\nabla_x f_i(\hat{x}, \hat{y}))'\, M_0(I_1)^{-1} (\nabla_x f_i(\hat{x}, \hat{y}))) = \frac{|\tilde{M}_0(I_1)|}{|\tilde{M}_0(I_2)|} > 0$ such that $\bar{S}(I_2) - \bar{S}(I_1)$ is positive semidefinite. Because of this fact and because of (3.5) we see now that each $S \in \partial_y^2 \bar{L}(\hat{y}, \hat{v})$ can be written

$$S = \bar{S}(I_+) + Q_1 = \bar{S}(I_0) - Q_2 \qquad (4.2)$$

with Q_1, Q_2 positive semidefinite. This implies, using relation (1.A.45) in /1, p.34/:
$p(\bar{S}(I_+)/T_y(J)) \leq p(S/T_y(J)) \leq p(\bar{S}(I_0)/T_y(J))$ and $\eta(\bar{S}(I_0)/T_y(J)) \leq \eta(S/T_y(J)) \leq \eta(\bar{S}(I_+)/T_y(J))$ for an arbitrary index set J, $J_+ \subset J \subset J_0$, and each $S \in \partial_y^2 \bar{L}(\hat{y}, \hat{v})$ such that 2. follows taking relation (4.1) into account.

In order to formulate the inertia statements for the generalized dual approach, we use additionally the matrices
$C(J) = (\nabla g_k(\hat{x}), k \in J)$, $E(J) = (\nabla_{y_i y_j} L(\hat{x}, \hat{y}, \hat{u}),\ i, j \notin J)$, $D(I, J) =$
$= (\nabla_{y_j} f_i(\hat{x}, \hat{y}),\ j \notin J, i \in I)$, $M(I, J) = \begin{pmatrix} A & B(I) & C(J) \\ B'(I) & 0 & D'(I, J) \\ C'(J) & D(I, J) & E(J) \end{pmatrix}$ and $S(I, J)$ as the Schur complemente of $M_0(I)$ in $M(I, J)$, where the latter is formed just by the rows and columns $j \in \{1, \ldots, m\} \setminus J$ of $S(I)$. Moreover, the linear subspaces $T_x(I)$, $T_y(J)$ are used as done above, and
$T_0(I, J) = \{x / B'(I)\, x = 0,\ C(J)' x = 0\}$, $r(I, J) = \operatorname{rank}(B(I), C(J))$.

<u>Theorem 4.2</u> (Dual Approach):
1. For each pair I, J with $I_+ \subset I \subset I_0$, $J_+ \subset J \subset J_0$, we have
   $\operatorname{In}(A/T_0(I, J)) = \operatorname{In}(A/T_x(I)) + \operatorname{In}(S(I)/T_y(J)) + (|I| - r(I, J), |I| - r(I, J), r(I, J) - |I| - |J|)$.
2. If $(\hat{x}, \hat{y}, \hat{u})$ is even a strongly stable stationary point of (1.4) and $(\hat{x}, \hat{y})$ satisfies MFCQ, then for each $S \in \partial^2 \bar{L}(\hat{y})$ and each J with $J_+ \subset J \subset J_0$ we have $\eta(S/T_y(J)) = \text{s.index}(\hat{x}/P(\hat{y})) - \text{s.index}(\hat{x}/1.4)$ and, furthermore, $o(S/T_y(J)) = 0$ if $(\hat{x}, \hat{y})$ satisfies even LICQ.

<u>Proof:</u> 1. Because of $T_y(J) = \{y / y_j = 0,\ j \in J\}$ in this case, we have

$In(S(I)/T_y(J)) = In(S(I,J))$. Relation (1.8) yields
$In(M_0(I)) = In(A/T_x(I)) + (|I|,|I|,0)$ and $In(M(I,J)) = In(A/T_0(I,J)) + (r(I,J), r(I,J), |I|+|J|-r(I,J))$, and relation (1.7) implies $In(M(I,J)) = In(M_0(I)) + In(S(I,J))$ for $I_+ \subset I \subset I_0$, $J_+ \subset J \subset J_0$. From these results statement 1. follows immediately.
2. Here (S1),(S2) and (1.9) are used. A1) implies $o(A/T_x(I)) = 0$ and $\eta(A/T_x(I)) = s.index(\hat{x}/P(\hat{y}))$ for $I_+ \subset I \subset I_0$. Moreover, we have $\eta(S(I,J)) = p(-S(I,J))$, $p(S(I,J)) = \eta(-S(I,J))$, $o(S(I,J)) = o(-S(I,J))$. If $(\hat{x},\hat{y})$ satisfies MFCQ, then $\eta(A/T_0(I,J)) = s.index(\hat{x}/1.4)$ and $o(A/T_0(I,J)) = 0$ for $I_+ \subset I \subset I_0$, $J_+ \subset J \subset J_0$, and if $(\hat{x},\hat{y})$ satisfies LICQ, then even $r(I,J) = |I|+|J|$ holds such that in view of statement 1. it follows $\eta(-S(I)/T_y(J)) = s.index(\hat{x}/P(\hat{y})) - s.index(\hat{x}/1.4)$ and, using LICQ, $o(-S(I)/T_y(J)) = 0$ for $I_+ \subset I \subset I_0$, $J_+ \subset J \subset J_0$. Now relation (3.5) is used and the proof can be continued as done in Theorem 4.1, taking into consideration $\partial^2 \bar{L}(\hat{y}) = -\partial^2 \bar{\varphi}(\hat{y})$.

Since on the assumptions of Statement 2. in Theorem 4.1 and Theorem 4.2 we have the same property as in (S1), we are able now to generalize the concept of the stationary index for the stationary solution $\hat{y}$ of (3.2):

$$s.index(\hat{y}) := \eta(\bar{S}(I_+)/\ T_y(J_+)) \tag{4.3a}$$

and for the stationary solution $\hat{y}$ of (3.3):

$$s.index(\hat{y}) = \eta(-S(I_+)/\ T_y(J_+)). \tag{4.3b}$$

In generalization of (1.10), for both cases this stationary index characterizes the type of the stationary solution of the upper-level problem as follows.

<u>Theorem 4.3:</u> On the assumptions of Statement 2. in Theorem 4.1 and Theorem 4.2, resp., the following holds:

$\hat{y}$ is a locally minimal point <u>iff</u> $s.index(\hat{y}) = 0$
$\hat{y}$ is a saddle point <u>iff</u> $0 < s.index(\hat{y}) < m$ (4.4)
$\hat{y}$ is a locally maximal point <u>iff</u> $s.index(\hat{y}) = m$.

<u>Proof:</u> Without loss of generality we carry out the proof using the notation of the primal approach.
If $s.index(\hat{y}) = 0$, $\hat{y}$ satisfies the generalized second-order sufficient optimality condition in /10,Theorem 1/ and hence, $\hat{y}$ is a locally minimal point of (3.2). On the other hand, given that $s.index(\hat{y}) > 0$, then there is an $h \in T_y(J_+)$ with $h'\bar{S}(I_0)h < 0$ and - because of relation (4.2)- even $h'Sh < 0$ for each $S \in \partial_y{}^2\, \bar{L}(\hat{y},\hat{v})$. The assumptions in 2. of Theorem 4.1 imply that $\hat{y}$ satisfies MFCQ and so Corollary 3.1 in /7/ can be applied and it can be deduced that $\hat{y}$ is not a locally minimal point of (3.2).
In a similar way it can be shown that $\hat{y}$ is a locally maximal point if and only if $s.index(\hat{y}) = m$. Hence Theorem 4.3 is proved.

5. Concluding Remarks

Remark 1: Using Definition (4.3), Statement 2. of Theorem 4.1 and 4.2 can be formulated as follows:
for the primal approach

$$\boxed{\text{s.index}(\hat{x},\hat{y}) \;=\; \text{s.index}(\hat{x}) \;+\; \text{s.index}(\hat{y})} \tag{5.1}$$

original problem — lower level — upper level

for the dual approach

$$\boxed{\text{s.index}(\hat{x}/1.4) \;=\; \text{s.index}(\hat{x}/P(\hat{y})) \;-\; \text{s.index}(\hat{y})}\,. \tag{5.2}$$

original problem — lower level — upper level

Remark 2: If (x^o,y^o,u^o) is a given initial point satisfying A1), then from Lemma 2.1 (4.) it necessarily follows s.index$(\hat{x}/P(\hat{y}))$= s.index(x^o). If now our aim is to generate a stationary solution of the original problem with a prescribed stationary index s, then in our two-level approach a stationary solution $\hat{y}$ of the upper level with s.index$(\hat{y})$= s− s.index$(x^o/P(y^o))$ for the primal approach and with s.index$(\hat{y})$= s.index$(x^o/P(y^o))$− s for the dual approach must be computed. Especially, if we want to generate a locally minimal point of the original problem (s=0), the following relations can be used:

a) $(\hat{x},\hat{y})$ is a locally minimal point of (1.1) iff $\hat{x}$ is a locally minimal point of $P(\hat{y})$ and $\hat{y}$ is a locally minimal point of (3.2).
b) $\hat{x}$ is a locally minimal point of (1.4) with corresponding multipliers $\hat{u},\hat{y}$ iff s.index$(\hat{x}/P(\hat{y}))$= s.index$(\hat{y})$.

Relation a) is already proved in /8/ on somewhat weaker assumptions. Relation b) implies that an initial point (x^o,y^o,u^o) with s.index(x^o)> m cannot lead to a locally minimal point $\hat{x}$ of (1.4).

References

/1/ Bandemer,H.(Herausg.): Theorie und Anwendung der optimalen Versuchsplanung I, Akademie Verlag, Berlin, 1977.
/2/ Clarke,F.H.: Optimization and Nonsmooth Analysis, Wiley,New York,1983
/3/ Dontchev,A.L.,and H.Th.Jongen: On the regularity of the Kuhn-Tucker curve, SIAM J.Control and Optim. 24(1986), 169-176.
/4/ Engelmann,B.,and W.Schäfer: Zur Erweiterung der Theorie der dualen Funktion bei der Mehrebenenoptimierung, Wiss.Z. TH Leipzig 4(1980), 93-100.
/5/ Engelmann,B.: Local statements on dual methods in problems of multilevel optimization, MOS, ser.optimization 15(1984), 115-134.
/6/ Findeisen,W. et al: Control and Coordination in Hierarchical Systems, Wiley, London, 1980.
/7/ Hiriatt-Urruty,J.B.,J.J.Strodiot,and V.Hien Nguyen: Generalized Hessian matrix and second-order optimality conditions for problems with $C^{1,1}$-data, Appl.Math.Optimization 11(1984), 43-56.
/8/ Jongen,H.Th.,T.Möbert,and K.Tammer: On iterated minimization in nonconvex optimization, Math.Oper.Res. 11(1986), 679-691.
/9/ Jongen,H.Th.,T.Möbert,J.Rückmann,and K.Tammer: On inertia and Schur complement in optimization, to appear in Linear Algebra Appl.
/10/ Klatte,D.,and K.Tammer: On second-order sufficient optimality conditions for $C^{1,1}$-optimization problems, to appear in optimization.

/11/ Kojima,M.: Strongly stable stationary solutions in nonlinear programs, in: S.M.Robinson(Ed.), Analysis and Computation of Fixed Points, Academic Press, New York, 1980, 93-138.
/12/ Kojima,M.,and R.Hirabayashi: Continuous deformation of nonlinear programs, Math.Progr.Study 21(1984), 150-198.
/13/ Lasdon,L.S.: Optimization Theory for Large Systems, The Macmillan Co., New York, 1970.
/14/ Ouellette,D.V.: Schur complements and statistics, Linear Algebra Appl. 36(1981), 187-295.
/15/ Rinke,B.,K.Tammer,and S.Wolf: Definition und Eigenschaften einer lokal dualen Funktion als Grundlage zur Verallgemeinerung der dualen Dekompositionsmethode für den nichtkonvexen Fall, Seminarbericht Nr.50 (1983) d.Sektion Math. d.Humboldt-Universität Berlin, Teil 2, 248-258.
/16/ Szymanowski,J.,and A.Ruszcynski: Convergence analysis for two-level algorithms of mathematical programming, Math.Progr.Study 10 (1979), 158-171.
/17/ Tammer,K.: A local dual method for structured nonconvex optimization problems, in: A.Straszak(Ed.), Proceedings of the 3rd IFAC/IFORS Symposium, Warsaw, July 1983, IFAC Proceedings Series, No. 10, Pergamon Press, Oxford, New York, 1984.
/18/ Tammer,K.: Zur Erzeugung streng regulärer stationärer Punkte mit Hilfe einer Verallgemeinerung der dualen Methode, Proceedings of the Conference "Modellierung und Optimierung von Systemen", Lichte, März 1984, 66-71.
/19/ Tammer,K.: The application of parametric optimization and imbedding to the foundation and realization of a generalized primal decomposition approach, in: J.Guddat et al(Eds.), Parametric Optimization and Related Topics, Akademie-Verlag, Berlin, 1987, 376-386.
/20/ Tammer,K.: Zweiebenenprobleme in der nichtkonvexen Optimierung - Problemformulierung und theoretische Grundlagen, to appear in Seminarberichte d.Sektion Math. d.Humboldt-Universität Berlin.

AN INVESTIGATION TO THE FROBENIUS PROBLEM

Manfred Schoch [1)]

For diophantine equations

$$\sum_{j=0}^{n} a_j x_j = b, \ x_j \geqq 0, \text{ integer}, \ j = 0,1,\dots,n, \tag{A}$$

with positive integers $a_0, a_1, \dots, a_n, b$ and $\gcd(a_0, a_1, \dots, a_n) = 1$ the following statement is well-known:

There exists an integer $b_0 = b_0(a_0, a_1, \dots, a_n)$ such that (A) is solvable for all integers $b > b_0$.

Frobenius posed the question about the smallest integer b_0 with this property. This problem is known as the Frobenius problem. Until now for the general case the solution of this problem is unknown. However the value of $b_0(a_0, a_1, \dots, a_n)$ is known for some special cases. For instance in [1] there is given a formula for $b_0(a_0, a_1, \dots, a_n)$ in the case

$$a_i = a + id, \quad i=0,1,\dots,n$$

with positive integers a and b. With $d=1$ from this formula for the special diophantine equation

$$\sum_{j=0}^{m} (a+j)x_j = b, \quad x_j \geqq 0, \text{ integer}, \quad j=0,1,\dots,m, \tag{B}$$

there results the Frobenius solution

$$b_0(a, a+1, a+2, \dots, a+m) = a([\tfrac{a-2}{m}] + 1) - 1 \tag{1}$$

where $[\frac{c}{d}]$ means the integer g with

$$g \leqq \frac{c}{d} < g + 1 .$$

In the present paper a simple proof for (1) will be given and by a detailed study of (B) for diverse b the Frobenius solution for a further special case of (A) will be obtained. First we prove a lemma, that is proved in [2] under an additional condition only. Let $\langle \frac{e}{f} \rangle$ be the integer h satisfying

$$h - 1 < \frac{e}{f} \leqq h .$$

<u>Lemma:</u> Let $a > 0$, $m > 0$, $b > 0$ be given integers. The diophantine equation (B) is solvable if and only if either

$$\frac{b}{a} = \left\langle \frac{b}{a} \right\rangle \quad \text{or} \quad \left\langle \frac{b}{a} \right\rangle > \left\langle \frac{b}{a+m} \right\rangle .$$

---

1) Bergakademie Freiberg, Sektion Mathematik, Bernhard-von-Cotta-Str. 2
DDR 9200 Freiberg

Proof.

1) In the case $\frac{b}{a} = \left\langle\frac{b}{a}\right\rangle$ obviously

$$x_0 = \frac{b}{a}, \quad x_j = 0, \quad j=1,2,\dots,m,$$

is a solution of (B).

2) In the case $\left\langle\frac{b}{a}\right\rangle > \left\langle\frac{b}{a+m}\right\rangle$ there exists an integer p with $0 \leqq p \leqq m-1$ such that

$$\left\langle\frac{b}{a+p}\right\rangle > \left\langle\frac{b}{a+m}\right\rangle \quad \text{and} \quad \left\langle\frac{b}{a+p+1}\right\rangle = \left\langle\frac{b}{a+m}\right\rangle .$$

We put

$$\Delta = b - (a+p)\left\langle\frac{b}{a+m}\right\rangle .$$

There exists an $\varepsilon > 0$ such that the relation

$$\left\langle\frac{b}{a+m}\right\rangle = \frac{b}{a+p} - \varepsilon$$

holds. Therefore we have

$$\Delta = \varepsilon(a+p) > 0,$$

and since $\frac{b}{a+p+1} \leqq \left\langle\frac{b}{a+m}\right\rangle$ also the inequality

$$\Delta \leqq \left\langle\frac{b}{a+m}\right\rangle$$

is true. Therefore we have $0 < \Delta \leqq \left\langle\frac{b}{a+m}\right\rangle$, Δ integer, and

$$x_p = \left\langle\frac{b}{a+m}\right\rangle - \Delta, \; x_{p+1} = \Delta, \; x_j = 0, \; j=0,1,\dots,p-1,p+2,\dots,m,$$

is a solution of (B).

3) Since

$$(a+m) \sum_{j=0}^{m} x_j \geqq b ,$$

for every solution of (B) we have

$$\sum_{j=0}^{m} x_j \geqq \left\langle\frac{b}{a+m}\right\rangle .$$

In the case $\left\langle\frac{b}{a}\right\rangle = \left\langle\frac{b}{a+m}\right\rangle$, $\frac{b}{a} < \left\langle\frac{b}{a}\right\rangle$ there exists no solution of (B) with $x_j = 0$, $j=1,2,\dots,m$. For any vector $\tilde{x}$ with nonnegative integer components $\tilde{x}_j$, $j=0,1,\dots,m$, and

$$\sum_{j=0}^{m} \tilde{x}_j \geqq \left\langle\frac{b}{a+m}\right\rangle, \quad \sum_{j=1}^{m} \tilde{x}_j \geqq 1$$

we obtain

$$\sum_{j=0}^{m} (a+j)\tilde{x}_j \geqq a \sum_{j=0}^{m} \tilde{x}_j + \sum_{j=1}^{m} \tilde{x}_j \geqq a\left\langle\frac{b}{a+m}\right\rangle + 1 .$$

Because of $\frac{b}{a} < \left\langle\frac{b}{a+m}\right\rangle$ we get

$$\sum_{j=0}^{m} (a+j)\tilde{x}_j > b+1 .$$

Therefore in the case $\left\langle\frac{b}{a}\right\rangle = \left\langle\frac{b}{a+m}\right\rangle$, $\frac{b}{a} < \left\langle\frac{b}{a}\right\rangle$ there is no solution of (B), and the lemma is proved.

In virtue of this lemma the solution $b_a(a,a+1,\dots,a+m)$ of the special Frobenius problem (B) results as the largest positive integer b satisfying $\left\langle\frac{b}{a}\right\rangle = \left\langle\frac{b}{a+m}\right\rangle$ and $\frac{b}{a} < \left\langle\frac{b}{a}\right\rangle$. That means, $b_o(a,a+1,\dots,a+m)$ is the solution of the following discrete programme

$$b_o(a,a+1,\dots,a+m) = \max_{b\in N} \left\{ b \;\middle|\; \left\langle\frac{b}{a}\right\rangle = \left\langle\frac{b}{a+m}\right\rangle \wedge \frac{b}{a} < \left\langle\frac{b}{a}\right\rangle \right\}. \qquad (2)$$

Here N denotes the set of natural numbers. We will solve this programme now.

If $\left\langle\frac{b}{a}\right\rangle = \left\langle\frac{b}{a+m}\right\rangle$, there exists an integer z with

$$z-1 < \frac{b}{a} \leqq z \quad \text{and} \quad z-1 < \frac{b}{a+m} \leqq z .$$

Therefore we have

$$(z-1)(a+m) < b \leqq za \qquad (3)$$

and

$$z < \frac{a+m}{m} \quad \text{respectively} \quad z \leqq \left\langle\frac{a+m}{m}\right\rangle - 1 .$$

Since

$$\left\langle\frac{a+m}{m}\right\rangle - 1 = \left[\frac{a+m-1}{m}\right]$$

we have

$$z \leqq \left[\frac{a+m-1}{m}\right] .$$

Further, due to $\frac{b}{a} < \left\langle\frac{b}{a}\right\rangle = z$ there must be $b < za$ respectively

$$b \leqq za - 1 . \qquad (4)$$

Obviously, b takes its maximal value if the same is true for z. Hence we can put

$$z = \left[\frac{a+m-1}{m}\right] - \nu ,$$

where ν is a nonnegative integer. Because of (3) and (4) this nonnegative number can be determined as the smallest integer satisfying

$$\left(\left[\frac{a+m-1}{m}\right] - \nu - 1\right)(a+m) < \left(\left[\frac{a+m-1}{m}\right] - \nu\right) a - 1$$

There results that $\nu = 1$ if $\frac{a-1}{m}$ is an integer and $\nu = 0$ otherwise.

Accordingly, for integer $\frac{a-1}{m}$ we obtain

$$z = \left[\frac{a+m-1}{m}\right] - 1 = \frac{a-1}{m} = \left[\frac{a-2}{m}\right] + 1$$

and for noninteger $\frac{a-1}{m}$ there follows

$$z = \left[\frac{a+m-1}{m}\right] = \left[\frac{a-1}{m}\right] + 1 = \left[\frac{a-2}{m}\right] + 1 .$$

Therefore, in fact, the relation

$$b_o(a,a+1,\dots,a+m) = a\left(\left[\frac{a-2}{m}\right] + 1\right) - 1$$

holds true.

Let the assumption $a_o < a_1 < \dots < a_n$ be fulfilled. If the general diophantine equation (A) for a given b is solvable, then also the corresponding "augmented" equation

$$\sum_{j=0}^{a_n-a_o} (a_o+j)x_j = b, \quad x_j \geqq 0, \text{ integer}, \ j=0,1,\dots,a_n-a_o,$$

for this b is solvable. Consequently, $b_o(a_o, a_o+1, a_o+2, \dots, a_n)$ is a lower bound for $b_o(a_o, a_1, a_2, \dots, a_n)$:

$$a_o\left(\left[\frac{a_o-2}{a_n-a_o}\right]+1\right)-1 \leqq b_o(a_o,a_1,a_2,\dots,a_n).$$

Now we examine the equation (B) for all $b \geqq a([\frac{a-2}{m}]+1)$ and derive a special solution for each such b. So the Frobenius solution for another special case of (A) will be obtained.

For

$$b = a([\tfrac{a-2}{m}]+1) + ta, \quad t=0,1,2,\dots,$$

obviously the vector x with

$$x_o = [\tfrac{a-2}{m}]+1+t, \quad x_j = 0, \ j=1,2,\dots,m,$$

is a solution of (B).

Now we consider the cases

$$b = a([\tfrac{a-2}{m}]+1)+ta+r, \ t=0,1,2,\dots, \ r=1,2,\dots,a-1,$$

and show that a solution of (B) is given by

$$x_{l-1} = k, \ x_l = [\tfrac{a-2}{m}]+1+t-k, \ x_j=0, \ j=0,1,\dots,l-2,l+1,\dots,m,$$

where

$$l = \left\langle \frac{r}{[\frac{a-2}{m}]+1+t} \right\rangle, \quad k=l([\tfrac{a-2}{m}]+1+t) - r.$$

Obviously, $l \geqq 1$ and we have to show that alsways $l \leqq m$. Evidently

$$\tilde{l} = \left\langle \frac{a-1}{[\frac{a-2}{m}]+1} \right\rangle$$

is the greatest possible value of l. Furthermore, there holds

$$[\tfrac{a-2}{m}] = \frac{a-2}{m} - \frac{\lambda}{m} \quad \text{with} \quad 0 \leqq \lambda \leqq m-1,$$

and therefore $\tilde{l} \leqq m$. We have

$$k \geqq \frac{r}{[\frac{a-2}{m}]+1+t}([\tfrac{a-2}{m}]+1+t) - r = 0,$$

so that $k \geqq 0$. Moreover with some ε, $0 \leqq \varepsilon < 1$, we obtain

$$x_l = [\tfrac{a-2}{m}]+1+t - \left(\frac{r}{[\frac{a-2}{m}]+1+t}+\varepsilon\right)([\tfrac{a-2}{m}]+1+t)+r$$

$$= (1-\varepsilon)([\tfrac{a-2}{m}]+1+t)$$

and therefore also $x_l \geqq 0$. Finally, we have

$$\sum_{j=0}^{m} (a+j)x_j = (a+l-1)k + (a+l)([\frac{a-2}{m}] + 1 + t - k)$$
$$= (a+l)([\frac{a-2}{m}] + 1 + t) - l([\frac{a-2}{m}] + 1 + t) + r$$
$$= a([\frac{a-2}{m}] + 1) + ta + r = b.$$

With the stated special solutions of (B) for all $b \geqq a([\frac{a-2}{m}] + 1)$ we proved the following statement:

Let the coefficients in (A) be ordered, $a_0 < a_1 < \ldots < a_n$. If the special coefficients

$$a_i = a_o + i, \quad i=0,1,\ldots,\tilde{l} = \left\langle \frac{a_o - 1}{\left[\frac{a_o-2}{a_n-a_o}\right] + 1} \right\rangle \leqq n$$

occur in (A), then the Frobenius solution is given by

$$b_o(a_o,a_1,\ldots,a_n) = a_o(\left[\frac{a_o-2}{a_n-a_o}\right] + 1) - 1 .$$

On account of

$$k \leq \frac{a_o-2}{a_n-a_o} < k + 1$$

this at once implies the further result.

If the inequality

$$\frac{ka_n+2}{k+1} \leq a_o < \frac{(k-1)a_n+2}{k+2}$$

holds, there is

$$b_o = (k+1)a_o - 1 .$$

References

[1] Roberts, J. B.: Note on linear forms. Proc. Amer. Math. Soc. 7 (1956), 465-469.

[2] Schoch, M.: Zur Lösung einer speziellen linearen diskreten Optimierungsaufgabe mit Zusatzforderungen. Math. Operationsforsch. Statist. Series Optimization Vol. 13 (1982) No 3, 373-392.

A COMBINATION OF THE BUNDLE APPROACH AND THE TRUST REGION CONCEPT

H. Schramm, J. Zowe*

1 SUBJECT AND AIM OF THE PAPER

Let f be a convex (not necessarily differentiable) function from $\mathbb{R}^n$ into $\mathbb{R}$ and let us make the often encountered assumption that

$$\textit{we know } f(x) \textit{ and one (arbitrary) subgradient } g(x) \in \partial f(x) \textit{ at every } x. \tag{1.1}$$

Two closely related methods have been proposed during recent years how to minimize such f: an *ε–descent–like method* (L), due to LEMARECHAL [5], and a *modified cutting plane idea* (M) which also goes back to Lemaréchal and has been studied extensively by MIFFLIN [7] and KIWIEL [3]. Both ideas follow the *Bundle concept*; at iteration point x_k they use the information contained in the *bundle* of the current subgradient g_k and the previously computed subgradients $g_{k-1},\dots,g_1$ to build up a model of f. This model is used to determine a direction d_k of descent for f. A line search along d_k either provides a new iterate with a sufficient decrease in f, or leads to a *Null Step* (the second characteristic ingredient of all bundle methods): we stay at x_k but we improve the model by adding another subgradient to the bundle.

Method (L) uses the information contained in the bundle of subgradients $g_k,\dots,g_1$ to approximate the ε–subdifferential of f at x_k by a polytope ($\epsilon_k > 0$ is a fixed parameter; the α_i^k measure how "close" g_i is to $\partial f(x_k)$, cf. (2.3) and (3.2))

$$G(\epsilon_k;g_k,\dots,g_1;x_k) := \left\{\Sigma_{i=1}^k \lambda_i g_i \,|\, \lambda_i \geq 0,\ \Sigma\lambda_i=1,\ \Sigma\lambda_i\alpha_i^k \leq \epsilon_k\right\}. \tag{1.2}$$

The role of ϵ_k is to control the contribution of the single g_i's to the approximation G; for $\epsilon_k \downarrow 0$ the set G reduces to $\{g_k\}$ and for $\epsilon_k \uparrow \infty$ one has $G = \text{conv}\,\{g_k,\dots,g_1\}$. The second method (M) works with a *piecewise linear approximation* of f at x_k plus a *penalty term* (with fixed parameter $t_k > 0$)

$$m(t_k;g_k,\dots,g_1;x_k;d) := \max_{1\leq i\leq k} \{f(x_i)+g_i^T(x_k+d-x_i)\} + \frac{1}{2t_k}\|d\|^2 \textit{ for } d \in \mathbb{R}^n. \tag{1.3}$$

For small t_k the minimizer d_k of (1.3) will be small in norm. This ensures that we stay in a region where we can still trust the model m. Both ideas have been implemented and successfully tested during recent years.

There are, however, two major (closely connected) objections against these two approaches. First, no (theoretically justified) selection rules for the parameters ϵ_k and t_k in (1.2) and (1.3) are known. Yet, a proper choice of these quantities is vital for the quality of the models and thus for the success of the methods. Second, why is a line search necessary at all? If e.g. (1.3) would model the problem sufficiently well (i.e. if the bundle $g_k,\dots,g_1$ would contain enough information and if in addition t_k would be properly chosen), then a line search would automatically yield the steplength 1 and thus become superfluous. In the following we will bypass these difficulties via a *Trust Region–like approach* (this terminology will be justified in Section 3). Instead of choosing t_k before minimizing (1.3), we consider t as variable in m. The concept of a Null Step is kept (to improve the information contained in the bundle $g_k,\dots,g_1$, if necessary), but now also t is tuned in a systematic way till the model will be

---

Mathematisches Institut, Universität Bayreuth, Postfach 10 12 51, D–8580 Bayreuth
This research has been partly supported by a grant from the DFG under contract Zo 27/5–1

efficient enough to provide some d_k with a substantial decrease for f. This, as a by-product, makes a line search superflous. The same idea applies of course to (L), i.e. to ϵ and the model G. All this can be done in such a way that the convergence results, known for (L) and (M), remain true. First numerical results look very promising. Similar good testing was reported by KIWIEL [4], who also tries to combine the Bundle concept with the Trust Region idea.

In Section 2 and 3 we shortly summarize a *conceptual ϵ-descent method* and show that (L) is indeed the child of this concept. This helps to realize why it is so difficult to establish sound rules for a proper ϵ_k-choice in (1.2). A switch from (L) to (M) does not help, since, as we will see, (L) and (M) are brothers: they are dual to each other. Hence a good t_k-choice in (1.3) is as difficult as a good ϵ_k-choice in (1.2).

Section 4 makes a first step in the direction we have in mind. We follow the path of the minimizer d(t) of model (1.3) when t varies on $(0,\infty)$. A similar result will hold for $d(\epsilon)$ coming from the "equivalent" model (1.2). For lack of space we restrict ourselves to the variant (M).

Some bibliographical remarks follow. The conceptual ϵ-descent method and its interpretation as spiritual father of (L) can be found e.g. in LEMARECHAL [6] and ZOWE [9]. Methods (L) and (M) are discussed in detail in LEMARECHAL [5], [6] and MIFFLIN [7]; see also the extensive monography by KIWIEL [3]. The "equivalence" of (L) and (M) was already mentioned in [5]. As far as we know the results of Section 4 have been put down here for the first time. The algorithmic idea of Section 5 was developped in cooperation with Lemaréchal and the convergence proofs in Section 5 use techniques introduced in [3].

Finally we mention that the nonconvex situation and the case with constraints are just under consideration and that their treatment does not imply conceptual difficulties.

2 A CONCEPTUAL ϵ-DESCENT METHOD

The ideas of this section are discussed in detail in [6] and [9]. Suppose we want to copy the *steepest descent idea*. This makes sense since the gradient of the convex f exists almost everywhere and, where $\nabla f(x)$ does not exist, we have as substitute the *subdifferential*

$$\partial f(x) := \{g \in \mathbb{R}^n \mid g^T(z-x) \leq f(z)-f(x) \ \forall z \in \mathbb{R}^n\}.$$

To unburden the notation we will write henceforth gz instead of g^Tz etc. It is well-known that $\partial f(x)$ is a nonempty convex compact set which shrinks to $\{\nabla f(x)\}$ whenever the gradient exists. Now let proj $(0|\partial f(x_k))$ denote the unique element in $\partial f(x)$, which is nearest to the origin in the Euclidean norm $\|\cdot\|$. A closer analysis shows that d:= $-$proj$(0|\partial f(x))$ is the (first order) steepest descent direction of f at x. Hence we are led to the *iteration scheme* (we assume that we know the whole $\partial f(x_k)$):

$$\begin{aligned} &(1)\ \textit{Compute}\ d_k := -\operatorname{proj}(0|\partial f(x_k)). \\ &(2)\ \textit{Compute}\ s_k\ \textit{with}\ f(x_k+s_kd_k)=\min_{s\geq 0} f(x_k+sd_k)\ \textit{and put}\ x_{k+1}:=x_k+s_kd_k. \end{aligned} \qquad (2.1)$$

If $f \in C^1$, then (2.1) is just the *steepest descent method* with exact line search and convergence to optimal points is ensured under standard assumptions. However, for $f \notin C^1$, iteration (2.1) can provide a sequence $\{x_k\}$ which converges to a nonstationary (hence nonoptimal) point $\bar{x}$; see [5] and [9] for illustrative examples. This is caused by the loss of information in the first order sets $\partial f(x_k)$, when the x_k approach some $\bar{x}$ where f is not differentiable. The cure consists in replacing the subdifferential in

(2.1) (1) by the larger *ε-subdifferential*

$$\partial_\epsilon f(x) := \{g \in \mathbb{R}^n \mid g(z-x) \le f(z) - f(x) + \epsilon \ \ \forall z \in \mathbb{R}^n\} \text{ for } \epsilon \ge 0. \tag{2.2}$$

This set collects the (sub-)gradient information from a whole neighbourhood of x, more precisely:

$$\forall x \, \forall \epsilon > 0 \, \exists \rho > 0: \ \partial_\epsilon f(x) \supset \bigcup_{\|y-x\| \le \rho} \partial f(y) \tag{2.3}$$

Hence, when x_k is close to $\bar{x}$, then the set $\partial_\epsilon f(x_k)$ will detect the nondifferentiabilty at $\bar{x}$ and one can avoid convergence to a nonoptimal $\bar{x}$. The proof of the basic result (2.3) is easy; only the Lipschitz continuity of the convex f and the subgradient inequality are used. (2.3) motivates the following *conceptual ε-descent method* (it is assumed that we know $\partial_{\epsilon_k} f(x_k)$):

(0) *Choose some* $\epsilon_k > 0$.
(1) *Compute* $d_k := -\operatorname{proj}(0 \mid \partial_{\epsilon_k} f(x_k))$. (2.4)
(2) *Compute* s_k *with* $f(x_k + s_k d_k) = \min_{s \ge 0} f(x_k + s d_k)$ *and put* $x_{k+1} := x_k + s_k d_k$.

The iteration can be stopped as soon as $d_k = 0$ (i.e. $0 \in \partial_{\epsilon_k} f(x_k)$). This is motivated by (see (2.2))

$$0 \in \partial_{\epsilon_k} f(x_k) \implies f(x_k) \le f(x) + \epsilon \text{ for all } x. \tag{2.5}$$

A more detailed analysis yields as "complement" of (2.5):

$$0 \notin \partial_{\epsilon_k} f(x_k) \implies \begin{array}{l} d_k \ne 0 \ \ in \ (2.4)(1) \\ and \ f(x_k + s d_k) \le f(x_k) - \epsilon_k \ \text{for suitable } s \ge 0. \end{array} \tag{2.6}$$

(2.5) and (2.6) show that now convergence to a nonoptimal $\bar{x}$ is excluded. Indeed, it follows from (2.5) and (2.6) that, if $f^* := \inf f(z) > -\infty$ and $0 < \underline{\epsilon} \le \epsilon_k \le \bar{\epsilon} < \infty$ for all k, then:

$$\exists \, m \in \mathbb{N}: f(x_m) \le f^* + \bar{\epsilon}. \tag{2.7}$$

We close this section with a remark concerning the ϵ_k-choice in the conceptual algorithm (2.4).

Remark. Suppose $f^* := \inf f(z) > -\infty$ and we can make a good guess for f^*. Then, obviously, ϵ_k slightly smaller than $f(x_k) - f^*$ will be a good choice in (2.4). For such ϵ_k one has $0 \notin \partial_{\epsilon_k} f(x_k)$ (follows from (2.5)), hence $d_k \ne 0$ and by (2.6) the line search will yield an $x_{k+1} = x_k + s_k d_k$ with $f(x_{k+1}) \le f(x_k) - \epsilon_k$ close to f^* by choice of ϵ_k.

3 AN IMPLEMENTATION OF THE ε-DESCENT METHOD: ALGORITHMS (L) AND (M)

The algorithm (L) tries to realize as much as possible from the conceptual idea (2.4) under the restrictive assumption (1.1). At x_k we have at hand the sequence of iterates $x_1,\dots,x_k$ and an auxiliary sequence $y_1,\dots,y_k$ (the role of the y_i's will become clearer in (3.3) and (3.7)). Further let $g_i \in \partial f(y_i)$, $i = 1,\dots,k$, be the collection of previously computed subgradients. Then put

$$\alpha_i^k := f(x_k) - f(y_i) - g_i(x_k - y_i) \text{ for } i = 1,\dots,k. \tag{3.1}$$

The α_i^k measure how well the linearizations $f(y_i) + g_i(x-y_i)$ of f at y_i approximate f at x_k. Now fix some $\epsilon_k > 0$ and define the polytope (we omit the summation index in the following)

$$G(\epsilon_k;g_k,...,g_1;x_k) := \{\Sigma\lambda_i g_i \mid \lambda_i \geq 0, \Sigma\lambda_i = 1, \Sigma\lambda_i\alpha_i^k \leq \epsilon_k\}.$$

It is a straightforward exercise to show that

$$g_i \in \partial_{\alpha_i^k} f(x_k) \quad and \quad G(\epsilon_k;g_k,...,g_1;x_k) \subset \partial_{\epsilon_k} f(x_k). \tag{3.2}$$

This suggests to replace $\partial_{\epsilon_k} f(x_k)$ in (2.4) (1) by its inner approximation $G := G(\epsilon_k;g_k,...,g_1;x_k)$. If, however, this G badly approximates $\partial_{\epsilon_k} f(x_k)$ then, different from the implication (2.6), a line search along $d_k := -\operatorname{proj}(0|G)$ may not lead to an ϵ_k-decrease in f. This defect is taken care of by a *Null Step*: we stay at x_k, compute a further subgradient g_{k+1} at $x_k + sd_k$ for some small $s > 0$ and repeat (2.4) (1) and (2) with the "improved" model $G(\epsilon_k;g_{k+1},g_k,...,g_1;x_k)$ for $\partial_{\epsilon_k} f(x_k)$; note that, by (2.3), g_{k+1} is almost a subgradient at x_k. We obtain the *method* (L):

(0) *Choose some* $\epsilon_k > 0$.
(1) *Compute* $d_k := -\operatorname{proj}(0|G(\epsilon_k;g_k,...,g_1;x_k))$. (3.3)
(2) *Determine* $y_{k+1} := x_k + s_k d_k$ *via a "line search" and compute* $g_{k+1} \in \partial f(y_{k+1})$;
  (a) *if* $f(y_{k+1})$ *is "sufficiently smaller" than* $f(x_k)$, *then put* $x_{k+1} := y_{k+1}$;
  (b) *if* $f(y_{k+1})$ *is "not sufficiently smaller" than* $f(x_k)$, *then put* $x_{k+1} := x_k$, add g_{k+1} *to the bundle* $g_k,...,g_1$ *and go back to* (1).

For all (of course very important) details concerning the computation of d_k, the line search, the interpretation of "sufficiently smaller" and the convergence results see [5]. Our concern here goes to the role played by ϵ_k. In the conceptual algorithm (2.4) an ϵ_k slightly smaller than $f(x_k) - f^*$ was a good choice; see the Remark following (2.7). However, this is no longer true for (L). For such ϵ_k the origin will be close to $\partial_{\epsilon_k} f(x_k)$, hence the projection problem is badly conditioned and replacing $\partial_{\epsilon_k} f(x_k)$ by $G(\epsilon_k;g_k,...,g_1;x_k)$ may lead to a disastrous direction which has nothing to do with the direction in (2.4) (1). On the other hand, for $\epsilon_k > 0$ near to 0, the d_k in (3.3) (1) will be close to $-g_k$ ($= -\nabla f(x_k)$ almost everywhere) and we are again close to the steepest descent method, which we do not like because of the reasons discussed in Section 2. Before we consider how to bypass this dilemma, we shortly discuss an alternative to (3.3).

Obviously, we can also use $g_k,...,g_1$ to define a *piecewise linear* function

$$\check{f}(x_k;d) := \max_{1 \leq i \leq k} \{f(y_i) + g_i(x_k+d-y_i)\} - f(x_k) \; for \; d \in \mathbb{R}^n,$$

which approximates $f(x_k+d) - f(x_k)$ from below (follows from the subgradient inequality). With the α_i^k from (3.1) this $\check{f}$ can be written in a more convenient way (we omit the maximization index in the following):

$$\check{f}(x_k;d) = \max \{g_i d - \alpha_i^k\} \; for \; d \in \mathbb{R}^n. \tag{3.4}$$

Minimization of $\check{f}(x;\cdot)$ is the *Cutting Plane algorithm*, which may not have a solution at all or which may lead to some d_k so large in norm that $\check{f}(x_k;\cdot)$ cannot be considered any more as an approximation of $f(x_k+\cdot) - f(x_k)$ at d_k. To prevent these two defects one adds in (3.4) a term $\frac{1}{2t_k}\|d\|^2$ with some fixed $t_k > 0$. This leads to the *model*

$$m(t_k;g_k,\dots,g_1;x_k;d) := \hat{f}(x_k;d) + \frac{1}{2t_k}\|d\|^2 \text{ for } d \in \mathbb{R}^n. \tag{3.5}$$

Minimization of m yields a direction d_k along which one makes a line search. This d_k (together with some v_k) is most easily computed via the *quadratic programming problem* in $(v,d) \in \mathbb{R}^{1+n}$:

$$\min v + \frac{1}{2t_k}\|d\|^2 \text{ subject to } v \geq g_i d - \alpha_i^k \text{ for } i = 1,\dots,k. \tag{3.6}$$

The same arguments as above (concerning Null Steps to improve the model m etc.) apply, and we obtain the *method* (M); see e.g. [3] and [7] for details:

(0) *Choose some* $t_k > 0$.
(1) *Compute* d_k (*and* v_k) *from* (3.6). (3.7)
(2) *Determine* $y_{k+1} := x_k + s_k d_k$ *via a "line search" and compute* $g_{k+1} \in \partial f(y_{k+1})$;
  (a) *if* $f(y_{k+1})$ *is "sufficiently smaller" than* $f(x_k)$, *then put* $x_{k+1} := y_{k+1}$;
  (b) *if* $f(y_{k+1})$ *is "not sufficiently smaller" than* $f(x_k)$, *then put* $x_{k+1} := x_k$, *add* g_{k+1} *in* (3.5) *and* (3.6) *respectively, and go back to* (1).

The methods (L) and (M) are closely related. First note that the projection $-d_k$ in (3.3)(1) is of the form $\Sigma\lambda_i g_i$ where $\lambda \in \mathbb{R}^k$ solves the quadratic programming problem (it is convenient to add here the factor 1/2 and to use $\|\cdot\|^2$ instead of $\|\cdot\|$):

$$\min \frac{1}{2}\|\Sigma\lambda_i g_i\|^2 \text{ subject to } \lambda_i \geq 0, \Sigma\lambda_i = 1, \Sigma\lambda_i\alpha_i^k \leq \epsilon_k. \tag{3.8}$$

For suitable choice of t_k and ϵ_k program (3.8) is equivalent to (follows by comparing the Kuhn-Tucker conditions of (3.8) and (3.9))

$$\min \frac{1}{2}\|\Sigma\lambda_i g_i\|^2 + \frac{1}{t_k}\Sigma\lambda_i\alpha_i^k \text{ subject to } \lambda_i \geq 0, \Sigma\lambda_i = 1. \tag{3.9}$$

A further look shows that (3.9) is basically the dual of (3.6). Hence for, corresponding t_k- and ϵ_k-values, (3.3)(1) and (3.7)(1) provide (up to the factor t_k) the same d_k and thus (L) and (M) can be considered as equivalent realization of the conceptual idea of Section 2; for details see [5]. In the following we restrict our attention to the variant (M), since here the dependence of the minimizer d(t) of $m(t;g_k,\dots,g_1;x_k;t)$ on t can be fully analysed. Similar results can be expected for the $d(\epsilon)$ coming from the model $G(\epsilon;g_k,\dots,g_1;x_k)$.

A final word should be made concerning the word *trust region* in the title of the paper. A method which solves

$$\min \hat{f}(x_k;d) \text{ subject to } \frac{1}{2}\|d\|^2 \leq \rho \tag{3.10}$$

for various ρ's, till one has reached a sufficient decrease in f, is usually called a trust region method. Now a tuning of ρ in (3.10) is equivalent to a tuning of t in m or in (3.6), respectively (follows by comparing the Kuhn-Tucker conditions for (3.6) and (3.10)). It is only because of numerical reasons that we prefer to have the t (via a penalty term) in the objective function since this leads to the quadratic programming problem (3.6), for which one has a lot of reliable software.

4 THE TRAJECTORY d(t)

Throughout this Section let k, x_k, $g_k,\dots,g_1$ etc. be fixed. We put $I := \{1,\dots,k\}$ and write $m(t;d)$ for $m(t;g_k,\dots,g_1;x_k;d)$, i.e.

$$m(t;d) = \max\{g_i d - \alpha_i^k\} + \frac{1}{2t}\|d\|^2 \text{ for } d \in \mathbb{R}^n \text{ and } t \in (0,\infty).$$

It is an easy exercise to show that the strictly convex $m(t;\cdot)$ has a unique minimizer for each $t > 0$, which will be denoted by $d(t)$ in the following. For a clever tuning of t in the model m it is important to learn more about $d(t)$ for varying t. This precisely is the aim of this Section. The results which follow will be used in Section 5.

To minimize $m(t;\cdot)$ is equivalent to solve (3.6): if (v,d) is a solution of (3.6) then, obviously, d minimizes $m(t;\cdot)$. Conversely, if d is a minimizer of $m(t;\cdot)$ then (v,d) with $v := \max\{g_i d - \alpha_i^k\}$ solves (3.6). Therefore $d(t)$ can be characterized via the Kuhn-Tucker conditions for (3.6). We use the notation

$$I(t) := \{j \in I \mid g_j d(t) - \alpha_j^k = \max_{i \in I}\{g_i d(t) - \alpha_i^k\}\}$$

and

$$\Lambda := \{\lambda \in \mathbb{R}^k \mid \lambda_i \ge 0 \text{ for } i \in I \text{ and } \Sigma\, \lambda_i = 1\}.$$

Then for each fixed $t > 0$:

LEMMA 4.1. *The vector* d *minimizes* $m(t;\cdot)$, i.e. $d = d(t)$, *if and only if there exists* $\lambda(t) \in \mathbb{R}^k$ *such that*

$$d = -t \sum_{i \in I} \lambda_i(t) g_i, \quad \lambda(t) \in \Lambda \text{ and } \lambda_i(t) = 0 \text{ for } i \in I \setminus I(t). \tag{4.1}$$

PROOF. The Kuhn-Tucker conditions for (3.6) hold at (v,d), if there exists $\lambda \in \mathbb{R}^k$ such that

$$\begin{pmatrix} 1 \\ t^{-1} d \end{pmatrix} + \Sigma\, \lambda_i \begin{pmatrix} -1 \\ g_i \end{pmatrix} = 0_{\mathbb{R}^{1+n}}, \quad \lambda_i \ge 0 \text{ and } \lambda_i(-v + g_i d - \alpha_i^k) = 0 \text{ for } i \in I. \tag{4.2}$$

Since (3.6) is a convex problem with affin-linear constraints, these conditions are necessary and sufficient for the optimality of some feasible (v,d) and thus for minimality of d in $m(t;\cdot)$. A second look shows that (4.1), supplemented with $v := \max\{g_i d - \alpha_i^k\}$, is but a reformulation of (4.2). ∎

The next result says that the minimizer $d(t)$ of $m(t;\cdot)$ depends continuously on t.

THEOREM 4.2. *The function* $d(\cdot) : (0,\infty) \to \mathbb{R}^n$ *is continuous.*

PROOF. Let t converge to some $\bar{t} > 0$. From (4.1) we conclude that $\|d(t)\|$ is bounded, when t varies in a neighbourhood of $\bar{t}$, and thus we can assume $d(t) \xrightarrow[t \to \bar{t}]{} d^\star$ for some $d^\star$. It remains to show $d^\star = d(\bar{t})$. By optimality of $d(t)$, $m(t;d(t)) \le m(t;d(\bar{t}))$ for all $t > 0$. Since m is continuous in (t,d) we get for $t \to \bar{t}$ the inequality $m(\bar{t};d^\star) \le m(\bar{t};d(\bar{t}))$, and thus $d^\star = d(\bar{t})$ because of uniqueness of the minimizer. ∎

Since $|I| < \infty$ there exist only finitely many different subsets of I, say $I^1, I^2,\dots,I^m$. If we put

$$T^j := \{t \in (0,\infty) \mid I(t) = I^j\} \text{ for } j = 1,\dots,m,$$

then

$$(0,\infty) = T^1 \cup T^2 \cup \dots \cup T^m. \tag{4.3}$$

The next result says more about the decomposition (4.3) and the structure of $d(\cdot)$.

PROPOSITION 4.3. *Suppose* $0 < t^1 < t^2$ *and let* t^1, t^2 *belong to the same* T^j. *Then* $[t^1,t^2] \subset T^j$ *and* $d(\mu t^1 + (1-\mu)t^2) = \mu d(t^1) + (1-\mu)d(t^2)$ *for all* $0 \le \mu \le 1$.

PROOF. Put $d^i := d(t^i)$ and $\lambda^i := \lambda(t^i)$, $i = 1,2$, with $\lambda(t)$ from (4.1). Then for $i = 1,2$

$$g_\beta d^i - \alpha_\beta^k = g_\gamma d^i - \alpha_\gamma^k > g_\delta d^i - \alpha_\delta^k \text{ for all } \beta,\gamma \in I^j \text{ and } \delta \notin I^j. \tag{4.4}$$

Now let $t_\mu := \mu t^1 + (1-\mu)t^2$ for some $\mu \in [0,1]$ and put $d_\mu := \mu d^1 + (1-\mu)d^2$. Then (4.4) remains true with d^i replaced by d_μ. Further, it is straightforward to check that (4.1) holds for d_μ, t_μ and $\lambda_\mu := \frac{1}{t_\mu}\,[\mu t^1\lambda^1 + (1-\mu)t^2\lambda^2]$. This proves the assertion. ∎

We conclude that the T^j in (4.3) must be intervals. Further we can assume, of course that $T^j \neq \emptyset$ for all j and $s < t$ for $s \in T^j$ and $t \in T^{j+1}$. Thus we obtain, in addition to the continuity statement of Theorem 4.2:

THEOREM 4.4. *There are finitely many intervals* T^j *of increasing order such that* $d(t)$ *is affin-linear on each* T^j, $I(t) = I^j$ *for* $t \in T^j$ *and* $(0,\infty) = T^1 \cup T^2 \cup \ldots \cup T^m$. ∎

PROPOSITION 4.5. *The function* $\|d(\cdot)\| : (0,\infty) \to \mathbb{R}$ *is monotonically increasing.*

PROOF. Let $0 < t^1 < t^2 < \infty$ and put $d^1 := d(t^1)$, $d^2 := d(t^2)$. Optimality implies

$$m(t^1,d^2) \ge m(t^1,d^1) \text{ and } m(t^2,d^1) \ge m(t^2,d^2).$$

If we add these inequalities, then the max–terms cancel and we obtain after reordering

$$\left(\frac{1}{2t^1} - \frac{1}{2t^2}\right)\left(\|d^2\|^2 - \|d^1\|^2\right) \ge 0,$$

which proves $\|d^2\| \ge \|d^1\|$. ∎

It is easy to extend this analysis and to show that $\|d(s)\| < \|d(t)\|$ for $s \in T^j$ and $t \in T^{j+1}$. The same is true for $s < t$ belonging to the same T^j, if the g_i for $i \in I^j$ are linearly independent.
Of special interest is of course the limit behaviour for $t \downarrow 0$ and $t \uparrow \infty$.

CASE $t \downarrow 0$: By Theorem 4.4 there are $a,b \in \mathbb{R}^n$ such that $d(t) = a+tb$ for $t \in T^1$. It follows from (4.1) that $d(t) \to 0$ for $t \downarrow 0$ and thus $d(t) = tb$ on T^1. Now let us assume that $I^0 := \{i \in I \mid \alpha_i^k = 0\}$ is nonempty; if $x_k = y_k$ then e.g. $\alpha_k^k \in I^0$. Then, as is easily seen, $I(t) \subset I^0$ for small $t > 0$ and thus $I^1 \subset I^0$. Now fix some $t \in T^1$ and choose $\lambda := \lambda(t)$ as in (4.1). Then $b = -\sum_I \lambda_i g_i$ and with $v = v(t) := \max_I \{g_i tb - \alpha_i^k\}$ (we use $\alpha_i^k = 0$ for $i \in I^1 \subset I^0$):

$$v = tg_i b \;\text{ for } i \in I^1 \text{ and } v > tg_i b - \alpha_i^k \text{ for } i \notin I^1.$$

This implies for λ from (4.1) and for arbitrary $\mu \in \Lambda$ with $\mu_i = 0$ for $i \notin I^0$

$$t\sum_I \lambda_i g_i b = \sum_I \lambda_i v = v \ge t\sum_I \mu_i g_i b,$$

and thus,

$$-bb \geq gb, \text{ i.e. } (-b)(g-(-b)) \geq 0 \text{ for all } g \in \text{conv } \{g_i \mid i \in I^0\}.$$

Hence $-b$ is the projection of the origin onto conv $\{g_i \mid i \in I^0\}$ and Theorem 4.4 can be supplemented:

$$d(t) = -t \text{ proj } (0 \mid \text{conv } \{g_i \mid i \in I^0\}) \text{ for all } t \in T^1. \tag{4.5}$$

CASE $t \uparrow \infty$: Theorem 4.4 says that

$$\textit{there are } a,b \in \mathbb{R}^n \textit{ such that } d(t) = a + tb \textit{ for all } t \in T^m. \tag{4.6}$$

Now consider the Cutting Plane problem

$$\min_d \max_I \{g_i d - \alpha_i^k\}. \tag{4.7}$$

We know that (4.7) has a solution d_{cp} in $\mathbb{R}^n$ if and only if $0 \in \text{conv } \{g_i \mid i \in I\}$. In terms of $d(\cdot)$ we can characterize the existence of such a d_{cp} as follows.

PROPOSITION 4.6. *(4.7) has a solution* d_{cp} *if and only if* $b = 0$ *in (4.6). In this case also the* a *from (4.6) solves (4.7).*

PROOF. Let d_{cp} solve (4.7). Then $\check{f}(x_k;d_{cp}) \leq \check{f}(x_k;d(t))$ for all $t > 0$ (recall definition (3.4)). This together with $m(t;d(t)) \leq m(t;d_{cp})$ yields

$$\check{f}(x_k;d_{cp}) + \tfrac{1}{2t}\|d(t)\|^2 \leq \check{f}(x_k;d(t)) + \tfrac{1}{2t}\|d(t)\|^2 \leq \check{f}(x_k;d_{cp}) + \tfrac{1}{2t}\|d_{cp}\|^2 \tag{4.8}$$

and thus $\|d(t)\| \leq \|d_{cp}\|$ for all $t > 0$. Hence $b = 0$ in (4.5), i.e. $d(t) = a$ for $t \in T^m$. Now put $d(t) = a$ in (4.8) and let $t \uparrow \infty$ to see that a solves (4.7).

Conversely, put $b = 0$ in (4.6). Then $\check{f}(x_k;a) + \frac{1}{2t}\|a\|^2 \leq \check{f}(x_k;d) + \frac{1}{2t}\|d\|^2$ for all $t \in T^m$ and all d. This can hold only if $\check{f}(x_k;a) \leq \check{f}(x_k;d)$ for all d, i.e. a solves (4.7). ∎

In case $b \neq 0$ in (4.6) then there is no Cutting Plane solution.

EXAMPLE (1) $n = 1$, $k = 2$, $x_2 = 0$, $g_2 = -1$, $g_1 = 1$, $\alpha_2 = 0$, $\alpha_1 = 2$. Then

$$m(t;d) = \max \{d-2, -d\} + \tfrac{1}{2t}|d|^2.$$

$I^0 = \{2\}$ and thus, by (4.5), $d(t) = t$ for small t. Further (4.6) implies $d(t) = 1$ for t large enough. A closer looks shows that

$$d(t) = t \textit{ for } 0 < t \leq 1 \textit{ and } d(t) = 1 \textit{ for } t \geq 1,$$

$$(0,\infty) = T^1 \cup T^2 = (0,1) \cup [1,\infty) \textit{ and } I^1 = \{2\},\ I^2 = \{1,2\}.$$

(2) Now replace g_1, α_1 in (1) by $g_1 = -1/2$ and $\alpha_1 = 1/2$. Then

$$m(t;d) = \max\{-\tfrac{1}{2}d - \tfrac{1}{2}, -d\} + \tfrac{1}{2t}|d|^2.$$

It is easy to see that this time $d(t) = t$ for $0 < t \leq 1$, $d(t) = 1/2\ t$ for $2 \leq t < \infty$ and $I(1) = I(2) = \{1,2\}$. Hence, by Proposition 4.3 and Theorem 4.4, $d(t) = 1$ on $[1,2]$ and

$$(0,\infty) = T^1 \cup T^2 \cup T^3 = (0,1) \cup [1,2] \cup (2,\infty) \textit{ and } I^1 = \{2\},\ I^2 = \{1,2\},\ I^3 = \{1\}.$$

5 THE ALGORITHM BT

We will now present our algorithm which combines the decisive features of the Bundle- and the Trust Region concept. Suppose x_k is the current iterate, $g_k,\dots,g_1$ the bundle of previously computed subgradients and $\alpha_k^k,\dots,\alpha_1^k$ the linearization errors defined in (3.1). Let $\underline{t},\bar{t}$ and m_1,m_2 be constants with $0 < \underline{t} < \bar{t}$ and $0 < m_1 < m_2 < 1$; further let μ and β be small positive numbers. In Sections 5.1 and 5.2 we will describe one step of the algorithm that leads to the next x_{k+1}. Section 5.3 summarizes the overall algorithm and 5.4 sketches the convergence analysis.

5.1 STEP k→k+1: SERIOUS STEP AND NULL STEP

Our aim is to tune at x_k the parameter t in the model $m(t;g_k,\dots,g_1;x_k;d)$ (or in (3.6), respectively) till either the minimizer d_k of m provides a "sufficient" decrease for f or, till we have reached a Null Step situation (i.e. we can improve the model m by adding a further subgradient to the bundle; cf. (3.3) and (3.7)). This *inner iteration*, which replaces the line search in (L) and (M), is realized as follows. Here k is fixed and j is the running index; the criteria (5.2) and (5.3) will be explained below and the choice of t_k^1 in (0) and t_k^{j+1} in (2)(c) will be discussed in 5.2.

(0) *Choose* $t_k^1 > 0$.

(1) *Compute the minimizer* $d_k^j := d(t_k^j)$ of $m(t_k^j;g_k,\dots,g_1;x_k;d)$ (*respectively the solution* d_k^j,v_k^j *of* (3.6)), *put* $y_{k+1}^j := x_k + d_k^j$ *and compute* $g_{k+1}^j \in \partial f(y_{k+1}^j)$.

(2)(a) *If* (5.2) *holds, then make a* ***Serious Step***: *put* $x_{k+1} := y_{k+1} := y_{k+1}^j, g_{k+1} := g_{k+1}^j$ *and stop*; (5.1)

(b) *if* (5.3) *holds, then make a* ***Null Step***: *put* $x_{k+1} := x_k,\ y_{k+1} := y_{k+1}^j,\ g_{k+1} := g_{k+1}^j$ *and stop*;

(c) *if neither* (5.2) *nor* (5.3) *holds, then choose a "better"* t_k^{j+1}, *put* $j := j+1$ *and go to* (1).

The criteria (5.2) and (5.3) are:

$$\text{(a): } f(y_{k+1}^j)-f(x_k) < m_1 v_k^j \qquad \text{(b): } g_{k+1}^j d_k^j \ge m_2 v_k^j \ \textit{ or } \ \bar{t}-t_k^j \le \mu \tag{5.2}$$

$$\text{(a): } f(y_{k+1}^j)-f(x_k) \ge m_1 v_k^j \qquad \text{(b): } f(x_k)-f(y_{k+1}^j)+g_{k+1}^j d_k^j \le \beta \ \textit{ or } \ t_k^j-\underline{t} \le \mu. \tag{5.3}$$

In case of a Serious Step, the decrease $f(x_k) - f(x_{k+1})$ in f is at least m_1-times the decrease $-v_k^j = -\hat{f}(x_k;d_k^j)$ in the model function $\hat{f}(x_k;\cdot)$; this follows from (5.2 a) and this is the main motivation for making a Serious Step. Condition (5.2 b) plays only a minor role in (2)(a); namely if (5.2 b) holds then (we use $\alpha_{k+1}^{k+1} = 0$, $0 < m_2 < 1$, $v_k^j \ge g_i d_k^j - \alpha_i^k$ and $0 > v_k^j$ as long as x_k is not yet optimal; see (5.6) below)

$$g_{k+1}^j d_k^j - \alpha_{k+1}^{k+1} > g_i d_k^j - \alpha_i^k \ \textit{ for } i = 1,\dots,k. \tag{5.4}$$

This implies that in the next step k+1, (3.6) will provide some d which differs from the present d_k^j. The purpose of the second part of (5.2 b) is to prevent the t_k^j from becoming too large (recall that d_k^j coincides with the Cutting Plane solution for large t_k^j, which we do not like; see Section 4).

(5.3 a) describes the situation where, due to a bad model, we cannot reach a decrease as in step (2)(a).

If, in addition, (5.3 b) holds then we make a Null Step. This is motivated by the following: (5.3 a) implies again (5.4); this follows from $x_{k+1} = x_k$, $y_{k+1} = x_k + d_k^j$ and from (3.1). Hence step k+1 will provide some d different from the unsuccessful present one. Further, because of (5.3 b), at least one of the reals t_k^j and α_{k+1}^{k+1} will be small (use that μ, β and $\underline{t}$ are small) and thus g_{k+1} will be close to $\partial f(x_k)$ (see (2.3) and (3.2)). Together this gives us hope that $m(t_k^j; g_{k+1},\ldots,g_1; x_{k+1}; d)$ (respectively (3.6) with the additional constraint $v \geq g_{k+1}d - \alpha_{k+1}^{k+1}$) will be a substantially better model at $x_{k+1} (= x_k)$. If neither a Serious Step nor a Null Step is possible then we improve the model by a better t_k^{j+1}. Before discussing how to choose this new t such that (5.1) becomes a finite subroutine we mention a simple *stopping criterion*. Let t_k^j, d_k^j and v_k^j be as in (5.1)(1). Then

LEMMA 5.1. *If* $|v_k^j| \leq \epsilon$ *for some* $\epsilon \geq 0$ *then* $f(y) \geq f(x_k) - (\epsilon/t_k^j)^{1/2} \|y - x_k\| - \epsilon$ *for all* $y \in \mathbb{R}^n$.

PROOF. The Kuhn-Tucker conditions (4.2) imply

$$v_k^j = -t_k^j \|\sum_I \lambda_i g_i\|^2 - \sum_I \lambda_i \alpha_i^k. \tag{5.5}$$

By the definition of α_i^k one has $g_i \in \partial_{\alpha_i^k} f(x_k)$ for $i = 1,\ldots,k$. Thus $\sum_I \lambda_i g_i \in \partial_\sigma f(x_k)$ for $\sigma := \sum_I \lambda_i \alpha_i^k$. Simple manipulation, using (5.5) and the ϵ–subgradient inequality, leads from this to the assertion. ∎

As immediate consequence we note (put $\epsilon = 0$ in Lemma 5.1) $v_k^j = 0 \implies x_k$ *is optimal.* (5.6)

5.2 STEP k→k+1: CHOICE OF t_k^j IN (5.1)

Since k is fixed in (5.1) we omit the subscript k and write + for k+1. We start with $t^1 \in [\underline{t}, \bar{t}]$ and define $t^2, t^3, \ldots$ via bisection. We mention that the bisection may be replaced by a more efficient strategy. The flow chart in Figure 1 summarizes (5.1) with this t-choice and the stopping criterion from Lemma 5.1.

THEOREM 5.2. *The iteration in* Figure 1 *leads in finitely many steps either to a Serious Step or to a Null Step or to a stop with the information that* x_k *is* ϵ*-optimal in the sense of* Lemma 5.1.

PROOF. We suppose the algorithm does not stop and consider the three cases: (i): $l^j = \underline{t}$ for all j; (ii): $u^j = \bar{t}$ for all j; (iii): neither (i) nor (ii) holds.

Ad (i): We are all the time on the right branch of the flow chart and thus $t^{j+1} = B(\underline{t}, t^j)$ for all j. Therefore $t^j \downarrow \underline{t}$ and $f(y_+^j) - f(x) \geq m_1 v^j$ for $j = 1,2,\ldots$. Hence for sufficiently large j, (5.3 a,b) will be satisfied and the algorithm will stop with a Null Step.

Ad (ii): Now we are all the time on the left branch of the flow chart and thus $t^{j+1} := B(t^j, \bar{t})$ for all j. Therefore $t^j \uparrow \bar{t}$ and $f(y_+^j) - f(x) < m_1 v^j$ for $j = 1,2,\ldots$. Hence for large j, (5.2 a,b) will be satisfied and the algorithm will stop with a Serious Step.

Ad (iii): A closer inspection of Figure 1 shows that in this case $\underline{t} < l^j < u^j < \bar{t}$ for all j large enough. It follows $l^j \uparrow t^\star$ and $u^j \downarrow t^\star$ for suitable $t^\star \in (\underline{t},\bar{t})$. Continuity of $d(\cdot)$ and $v(\cdot)$ yields for $d^\star := d(t^\star)$ and $v^\star := v(t^\star)$

$$f(x+d^\star) - f(x) = m_1 v^\star. \tag{5.7}$$

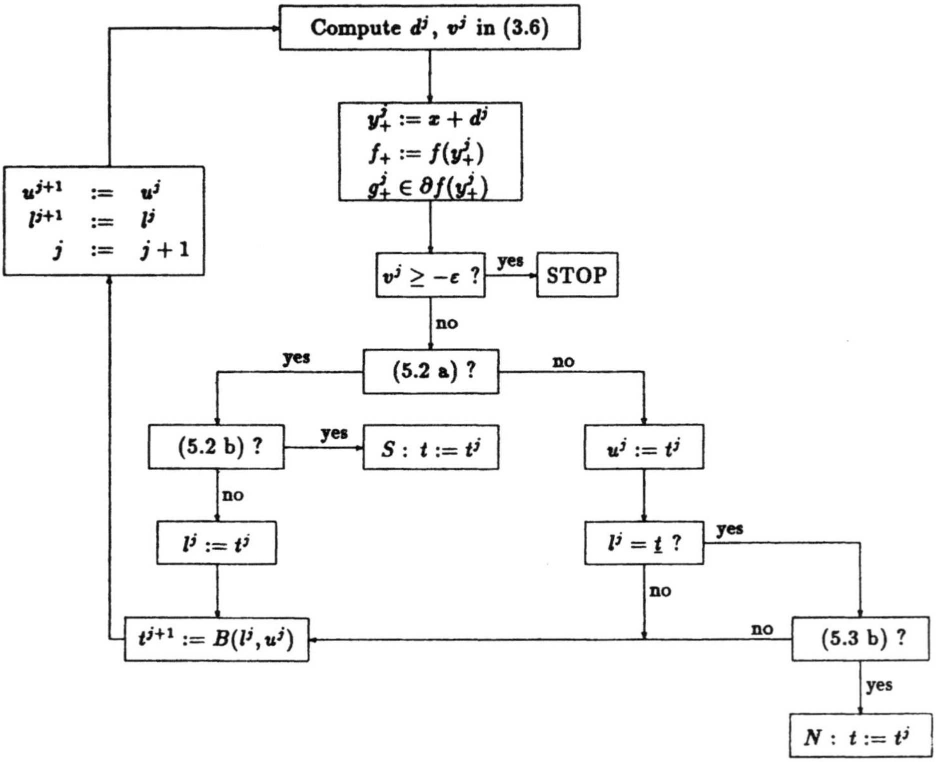

Put $u^1 := \bar{t}$, $l^1 := \underline{t}$, $j := 1$ and $\epsilon \ge 0$. $B(\cdot,\cdot)$ is defined by $B(l^j,u^j) := l^j + \frac{u^j - l^j}{2}$. S respectively N denotes a Serious respectively Null Step.

Figure 1

Now let $\{g^{j(i)}\}_{i=1,2,\ldots}$ (respectively $\{d^{j(i)}\}_{i=1,2,\ldots}$ and $\{v^{j(i)}\}_{i=1,2,\ldots}$) be the possibly finite sequence for which $t^{j(i)} = l^{j(i)}$. Since $l^j \uparrow t^\star$, the $g^{j(i)}$ have an accumulation point $g^\star$ (the proof uses the Lipschitz-continuity of f). The upper semicontinuity of the multivalued map $x \to \partial f(x)$ implies $g^\star \in \partial f(x+d^\star)$. From this we get

$$f(x) \ge f(x+d^\star) - g^\star d^\star$$

and thus because of (5.7)

$$g^\star d^\star \ge m_1 v^\star. \tag{5.8}$$

Since by assumption no stop occurs, we have $v(l^1) < 0$ and, since $v(\cdot)$ is monotonically decreasing (proof similar to Proposition 4.5), we conclude $v^\star < 0$. This together with (5.8) and $0 < m_1 < m_2 < 1$ implies

$$g^\star d^\star > m_2 v^\star.$$

A continuity argument shows that $g^{j(i)}d^{j(i)} \geq m_2 v^{j(i)}$ for sufficiently large i. Hence the algorithm will stop with a Serious Step. We conclude that in all three situations we have a finite stop in contradiction to the assumption. ∎

We close 5.2 with two remarks. (1) The computation of $d(t^j)$ can be simplified by taking advantage of the piecewise linearity of $d(\cdot)$; cp. Proposition 4.3. (2) If (5.2 a) is satisfied for some t_k^j, then the above algorithm will always end with a Serious Step; follows from the proof of Theorem 5.2.

5.3 THE ALGORITHM

We can now shortly summarize the whole algorithm BT:

Initialization: *Choose* $x_1 \in \mathbb{R}^n$, $\bar{t} > \underline{t} > 0$, $0 < m_1 < m_2 < 1$, $\mu > 0$, $\beta > 0$ *and* $\epsilon \geq 0$.

(0) *Put* $y_1 := x_1$; *compute* $f(x_1)$, $g_1 \in \partial f(x_1)$ *and* α_1^1; *put* $G_1 := \{g_1\}$ *and* $k := 1$.

(1) *Compute* x_{k+1} *or stop as described in Section* 5.2. (5.9)

(2) *Compute* α_i^{k+1} *for* $i = 1,...,k+1$ *and put* $G_{k+1} := G_k \cup \{g_{k+1}\}$.

(3) *Put* $k := k+1$ *and go to* (1).

REMARK.

(1) (3.6) is a quadratic programming problem in n+1 variables and k linear constraints. For n>k it might be advantageous to solve instead the dual problem in k variables and k+1 constraints

$$\min \frac{1}{2}\|\sum_I \lambda_i g_i\|^2 + \frac{1}{t}\sum_I \lambda_i \alpha_i \quad \textit{subject to } \lambda \in \Lambda. \tag{5.10}$$

If $\lambda^\star$ solves (5.10) then $d(t) := -t\,\Sigma\lambda_i^\star g_i$, $v(t) := -t\|\Sigma\lambda_i^\star g_i\|^2 - \Sigma\lambda_i^\star \alpha_i^k$ is the solution of (3.6); cf. the discussion following (3.7).

(2) It may be useful to add in (3) with $m_3 > m_1$: if $f(x_{k+1}) - f(x_k) \leq m_3 v(t_k)$ then start the computation of t_{k+1} with $t_{k+1}^1 > t_k$.

5.4 CONVERGENCE

We assume for the following that f is bounded below and that the stopping parameter ϵ in Figure 1 is set to 0. We put $X^\star := \{x^\star \in \mathbb{R}^n \mid f(x^\star) = \inf_{x\in\mathbb{R}^n} f(x)\}$ and will prove, that if $X^\star \neq \emptyset$, then the sequence of iterates $\{x_k\}$ converges to some $x^\star \in X^\star$. (t_k is the parameter which gives a Serious or Null Step; let $v_k = v(t_k)$).

PROPOSITION 5.3. *Suppose the algorithm* BT *generates an infinite sequence* $\{x_{k(l)}\}_{l\in\mathbb{N}}$ *resulting from Serious Steps. Then*

$$\lim_{l\to\infty} v_{k(l)-1} = 0. \tag{5.11}$$

PROOF. Suppose (5.11) does not hold. Then there is some $\nu > 0$ such that for a suitable subsequence $\{x_{k(l(i))}\}_{i\in\mathbb{N}}$

$$v_{k(l(i))-1} \leq -\nu \quad \text{for all } i.$$

Since by construction

$$f(x_{k(l(i))}) - f(x_{k(l(i))-1}) < m_1 v_{k(l(i))-1}$$

we get a contradiction to the boundedness of f. ∎

PROPOSITION 5.4. *If algorithm* BT *generates only a finite number of Serious Steps, i.e. there exists* $\bar{k} \in \mathbb{N}$ *such that* $x_k = x_{\bar{k}}$ *for all* $k \geq \bar{k}$, *then* $\lim_{k\to\infty} v_k = 0$.

PROOF. Because of lack of space we only sketch the proof: the algorithm in Section 5.2 shows that we have $t_k \geq t_{k+1} \geq \dots \geq \underline{t}$ for $k \geq \bar{k}$. Similar to KIWIEL [3] we can show that the optimal value of (5.9) tends to 0 for $k \to \infty$, provided (5.3a) holds. This implies $v_k \uparrow 0$. ∎

Proposition 5.3 and 5.4 yield:

THEOREM 5.5. *Every accumulation point of* $\{x_k\}_{k\in\mathbb{N}}$ *is optimal.*

PROOF. Let $x_{k(l)} \to \bar{x}$. Then we know from Proposition 5.3 and 5.4 $v_{k(l)-1} \to 0$, thus $\sigma_{k(l)-1} := \Sigma\lambda_i^{k(l)-1}\alpha_i^{k(l)-1} \downarrow 0$ and $z_{k(l)-1} := \Sigma\lambda_i^{k(l)-1}g_i \to 0$. Further we get from $z_{k(l)-1} \in \partial_{\sigma_{k(l)-1}} f(x_{k(l)-1})$

$f(y) \geq f(x_{k(l)-1}) + z_{k(l)-1}(y-x_{k(l)-1}) - \sigma_{k(l)-1} \geq f(x_{k(l)}) + z_{k(l)-1}(y-x_{k(l)-1}) - \sigma_{k(l)-1}$ *for all* y $\in \mathbb{R}^n$.

For $l \to \infty$ we can show $f(y) \geq f(\bar{x})$ for all $y \in \mathbb{R}^n$. ∎

To prove convergence of the sequence $\{x_k\}_{k\in\mathbb{N}}$ we need:

PROPOSITION 5.6. *Let* $\bar{x} \in \mathbb{R}^n$ *hold* $f(\bar{x}) \leq f(x_k)$ *for all* $k \in \mathbb{N}$. *Then for any* $\delta > 0$ *there exists* $n \in \mathbb{N}$ *such that*

$$||\bar{x}-x_k||^2 \leq ||\bar{x}-x_n||^2 + \delta \quad \textit{for all} \quad k \geq n.$$

PROOF. A detailed proof can be found in [8]. ∎

From Proposition 5.6 and Theorem 5.5 we get

THEOREM 5.7. *Let* $X^\star \neq \emptyset$, *then the sequence* $\{x_k\}_{k\in\mathbb{N}}$ *converges to* $x^\star \in X^\star$.

REMARK. We give some further properties of the algorithm BT, which can be found in detail in [8].

(1) If $X^\star \neq \emptyset$, then $\{f(x_k)\}_{k\in\mathbb{N}}$ converges to $\inf_{x\in\mathbb{R}^n} f(x)$.

(2) If we choose a stopping parameter $\epsilon > 0$, then the algorithm stops within finitely many steps.

(3) If f is a piecewise linear function, then the algorithm ends up with $x^\star \in X^\star$ within finitely many steps, even for $\epsilon = 0$.

(4) To make the algorithm less expensive, we can "reset" in iteration k in the following way: we choose $G_{k+1} := \{\Sigma\lambda_i^k g_i, g_{k+1}\}$ instead of $G_{k+1} := G_k \cup \{g_{k+1}\}$.

(5) We can handle linear constraints by including them into (3.6), i.e. we consider

$$\begin{aligned} &\min\ v + \frac{1}{2t}||d||^2 \\ &\text{subject to } v \geq g_i d - \alpha_i, \quad i = 1,\dots,k, \\ &A(x_k+d) \leq b. \end{aligned}$$

6 NUMERICAL RESULTS

The algorithm BT was tested as a double precision FORTRAN code on a VAX 8600. We summarize the results in Table 1. Column 1 contains the name of the problem and the reference; column 2 contains the optimal value and column 3 and 5, respectively, count the computational amount needed by BT and M1FC1, respectively, to produce the function value in columns 4 and 6.

Problem	f*	BT		M1FC1	
		#f/g	f(x)	#f/g	f(x)
MAXQUAD [5]	–0.841408	41	–0.841404	68	–0.841405
TR 48 [5]	–638565	157	–638565	435	–630340
POWELL [3]	0	3	0	8	0
SHOR [4]	22.60016	27	22.60016	87	22.60017
GOFFIN [4]	0	58	$0.7\cdot10^{-12}$	144	$0.3\cdot10^{-3}$
ROSEN-SUZUKI [4]	–44	18	–43.99998	53	–43.99998
TSP [2]					
n = 14	–3322	24	–3322	35	–3322
n = 29	–1608	33	–1608.	69	–1608
n = 442	≤ –51	200	–50.68	200	–50.44
HILBERT [8]	0	18	10^{-7}	24	10^{-4}

Obviously our algorithm performs rather well, even compared to the highly sophisticated code M1FC1. This motivates further research, e.g. for the nonconvex and the constrained case.

REFERENCES

[1] Dulleck, K.: Private communication, Bayreuth 1987.

[2] Grötschel, M.: Operations Research I. Skript Universität Augsburg (1985).

[3] Kiwiel, K. C.: Methods of Descent for Nondifferentiable Optimization. Springer-Verlag, Berlin-Heidelberg-New York-Tokyo 1985.

[4] Kiwiel, K. C.: Proximity control in bundle methods for conves nondifferentiable minimization. Preprint Systems Research Institute, Polish Academy of Sciences, Warsaw (1987).

[5] Lemaréchal, C.: Extensions diverses des méthodes de gradient et applications. Thèse d'Etat Paris 1980.

[6] Lemaréchal, C.: Construction Bundle Methods for Convex Optimization. Fermat Days 85, edited by Hiriart-Urruty, J.B., North Holland, Amsterdam 1986.

[7] Mifflin, R.: A modification and an extension of Lemaréchal's algorithm for nonsmooth optimization. Nondifferential and Variational Techniques in Optimization, Sorensen, D. C. and Wets, R. J.-B., eds., Math. Prog. Study 17 (1982) 77-90.

[8] Schramm, H., Zowe, J.: A combination of Bundle and Trust Region Methods. Preprint Universität Bayreuth 1987.

[9] Zowe, J.: Nondifferentiable Optimization – A Motivation and a Short Introduction into the Subgradient- and the Bundle-Concept. NATO ASI Series, Vol. F15, Computational Mathematical Programming, edited by Schittkowski, K., Springer-Verlag, Berlin-Heidelberg 1985.

OPTIMIZATION PROBLEMS WITH MAX-SEPARABLE CONSTRAINTS AND ADDITIVELY SEPARABLE OBJECTIVE FUNCTION

Karel Zimmermann [1]

We shall consider optimization problems of the following form:

$$\text{Minimize} \quad f(x_1,\dots,x_n) = \sum_{j=1}^{n} f_j(x_j)$$

subject to

$$\max_{1 \leqq j \leqq n} z_{ij}(x_j) \ \varrho_i \ b_i \,, \quad i=1,\dots,m \qquad \text{(P0)}$$

$$h_j \leqq x_j \leqq K_j \,, \quad j=1,\dots,n,$$

where $f_j \colon R \longrightarrow R$ are continuous, $h_j, H_j, b_i \in R$, ϱ_i is one of the relations $\leqq$, $=$, $\geqq$, and $z_{ij} \colon R \longrightarrow R$ are continuous and such that the set

$$\{x_j \mid z_{ij}(x_j) = b_i\}$$

is either a closed interval or empty. Since the equalities in (P0) can be replaced by two inequalities and the $\leqq$-inequalities can be replaced by new bounds on the variables x_j, the problem (P0) can be reduced to the following form:

$$\text{Minimize} \quad \sum_{j=1}^{n} f_j(x_j)$$

subject to

$$\max_{1 \leqq j \leqq n} r_{ij}(x_j) \geqq 0, \quad i=1,\dots,m \qquad \text{(P1)}$$

$$k_j \leqq x_j \leqq K_j \,, \quad j=1,\dots,n \,,$$

where $r_{ij}(x_j) = z_{ij}(x_j) - b_i$ for the corresponding i, j and k_j, K_j are the new bounds resulting from the old ones and the $\leqq$-inequalities (we assume $k_j \leqq K_j$ for all j, otherwise the set of feasible solutions of (P0) is empty).

---

[1] Faculty of Mathematics and Physics, Charles University
Malostranské náměstí 25, CS-118 00 Praha 1

We shall show in the sequel that (P1) can be solved via solving certain set covering problem ([2]). The following notation will be used in the sequel:

$N = \{1,\dots,n\}, \quad S = \{1,\dots,m\},$

$X_j^{(i)} = \{x_j \mid k_j \leqq x_j \leqq K_j,\ r_{ij}(x_j) = 0\}$ for all i,j

$S_j(x_j) = \{i \mid i \in S,\ r_{ij}(x_j) \geqq 0\}$ for all $j \in N$ and $x_j \in R$.

The set of feasible solutions of (P1) will be denoted by M. Note that in accordance with our assumptions the set $X_j^{(i)}$ is either empty or a closed interval; we shall set in this latter case $X_j^{(i)} = [x_j^{(i)}, y_j^{(i)}]$ (remark that it can happen that $x_j^{(i)}$, $y_j^{(i)}$ coincide, e. g. if r_{ij} is a strictly monotone function).

<u>Theorem 1.</u>

$$x = (x_1,\dots,x_n) \in M \iff x_j \in [k_j,K_j]\ \forall j \in N \ \&\ \bigcup_{j \in N} S_j(x_j) = S.$$

<u>Proof:</u>

Suppose $x \in M$. Then $x_j \in [k_j,K_j]$ for all $j \in N$ and there exists for each $i \in S$ an index $j(i) \in N$ such that $r_{ij(i)}(x_{j(i)}) \geqq 0$, so that $i \in S_{j(i)}(x_{j(i)})$. Therefore $\bigcup_{j \in N} S_j(x_j) = S$.

Suppose now the right-hand-side of the relation $\iff$ is fulfilled, and let i be an arbitrary element of S. Then there exists $j(i) \in N$ such that $i \in S_{j(i)}(x_{j(i)})$, so that $r_{ij(i)}(x_{j(i)}) \geqq 0$. We obtain therefore:

$$\max_{j \in N} r_{ij}(x_j) \geqq r_{ij(i)}(x_{j(i)}) \geqq 0 .$$

Since $i \in S$ was arbitrary, it follows from here that $x \in M$.

Q.E.D.

We shall redenote the points k_j, K_j, $x_j^{(i)}$, $y_j^{(i)}$ in the interval $[k_j,K_j]$ in such a way that we obtain an increasing sequence of $\bar{s}_j$ points $(1 \leqq \bar{s}_j \leqq 2m+2)$:

$$\bar{d}_j^{(1)} < \dots < \bar{d}_j^{(\bar{s}_j)}$$

(it is therefore $\bar{d}_j^{(1)} = \min_{i \in S} \{\min \{k_j, K_j, x_j^{(i)}, y_j^{(i)}\}\}$, $\bar{d}_j^{(\bar{s}_j)} = \max_{i \in S} \{\max \{k_j, K_j, x_j^{(i)}, y_j^{(i)}\}\}$ and the coinciding points of k_j, K_j, $x_j^{(i)}$, $y_j^{(i)}$ are denoted by the same point $\bar{d}_j^{(k)}$, $1 \leqq k \leqq \bar{s}_j$). Let $z_j^{(k)}$ be defined as follows:

$$f_j(z_j^{(k)}) = \min\{f_j(x_j) \mid x_j \in [\bar{d}_j^{(k)}, \bar{d}_j^{(k+1)}]\}$$

Let us remark that the minimum can be attained in more points and $z_j^{(k)}$ is one of them arbitrarily chosen (if we have a choice, we prefer the point coinciding with one of the two boundary points of the corresponding interval, if in such point the minimum is attained). We shall redenote now the points $\bar{d}_j^{(k)}$, $z_j^{(k)}$ for $1 \leqq k \leqq \bar{s}_j$ in a similar way as above and obtain a new sequence

$$d_j^{(1)} < \ldots < d_j^{(s_j)} ,$$

where $1 \leqq s_j \leqq (2m+2) + (2m+2-1) = 4m + 3$.

Lemma 1.

Let $d_j^{(k)}$, $k=1,\ldots,s_j$, be defined as described above. It holds then for all $x_j \in (d_j^{(k)}, d_j^{(k+1)})$:

$$S_j(x_j) \subset S_j(d_j^{(k)}) , \quad S_j(x_j) \subset S_j(d_j^{(k+1)}) .$$

Proof:

Let us introduce the following notation:

$$T_j^+ = \{i \mid i \in S,\ X_j^{(i)} \neq \emptyset,\ r_{ij}(x_j) \leqq 0 \text{ for } x_j \leqq y_j^{(i)},\ r_{ij}(x_j) > 0 \text{ for } x_j > y_j^{(i)}\}$$

$$T_j^- = \{i \mid i \in S,\ X_j^{(i)} \neq \emptyset,\ r_{ij}(x_j) \geqq 0 \text{ for } x_j \leqq y_j^{(i)},\ r_{ij}(x_j) < 0 \text{ for } x_j > y_j^{(i)}\}$$

$$\bar{T}_j^+ = \{i \mid i \in T_j^+,\ x_j^{(i)} = y_j^{(i)}\}, \quad \bar{T}_j^- = \{i \mid i \quad T_j^-,\ x_j^{(i)} = y_j^{(i)}\}.$$

Let x_j be an arbitrary point of $(d_j^{(k)}, d_j^{(k+1)})$. It is then:

$$S_j(x_j) = S_j(d_j^{(k)}) \setminus \bar{T}_j^- = S_j(d_j^{(k+1)}) \setminus \bar{T}_j^+ ,$$

which implies the assertion of our lemma.

Q.E.D.

Lemma 2.

Suppose $d_j^{(k)}$ are defined as above, $x_i \in (d_j^{(k)}, d_j^{(k+1)})$. It holds:

$$\text{either } f_j(d_j^{(k)}) \leqq f_j(x_j) \quad \text{or} \quad f_j(d_j^{(k+1)}) \leqq f_j(x_j) .$$

Proof:

It follows immediately from the construction of elements $d_j^{(k)}$, $d_j^{(k+1)}$ that the minimum of f_j over the interval $[d_j^{(k)}, d_j^{(k+1)}]$ is attained either in the point $d_j^{(k)}$ or in the point $d_j^{(k+1)}$; the assertion of the lemma follows immediately from this fact.

Q.E.D.

It follows from Lemma 1 and Lemma 2 that for an arbitrary point $x_j \in [k_j, K_j]$ there exists always an index k, $1 \leqq k \leqq s_j$, such that

$$S_j(x_j) \subset S_j(d_j^{(k)}) \quad \text{and} \quad f_j(d_j^{(k)}) \leqq f_j(x_j).$$

Therefore if $M \neq \emptyset$, the optimal solution of the problem (P1) can be found already among those feasible solutions of (P1), the coordinates of which are equal for each $j \in N$ to one of the points $d_j^{(k)}$, $1 \leqq k \leqq s_j$. In the other words, the problem (P1) reduces now to the following:

Find a point $\bar{x} = (\bar{x}_1, \dots, \bar{x}_n)$ such that $\bar{x}_j = d_j^{(p_j)}$, $1 \leqq p_j \leqq s_j$, for all $j \in N$ with the following properties:

(i) $\bigcup_{j \in N} S_j(d_j^{(p_j)}) = S$

(ii) Suppose $x = (x_1, \dots, x_n)$, $x_j = d_j^{(v_j)}$ for all $j \in N$ and $\bigcup_{j \in N} S_j(d_j^{(v_j)}) = S$. Then $\sum_{j \in N} f_j(\bar{x}_j) \leqq \sum_{j \in N} f_j(x_j)$.

We shall show now that this latter problem can be reduced to the "classical" set covering problem, which is investigated e. g. in [2]. We define the following numbers for this purpose:

$$a_{ij}^{(k)} = \begin{cases} 1, & \text{if } i \in S_j(d_j^{(k)}) \\ 0 & \text{otherwise} \end{cases} \qquad \text{for all } i \in S,\ j \in N \text{ and } 1 \leqq k \leqq s_j .$$

We shall set further $c_j^{(k)} = f_j(d_j^{(k)})$ for all $j \in N$, $1 \leqq k \leqq s_j$.

Let p be such a number ("penalty coefficient") that

$$p > 2 \sum_{j \in N} \sum_{k=1}^{s_j} |c_j^{(k)}| .$$

Our problem can now be reformulated as follows. We are given for each $j \in N$ a system of s_j $(1 \leqq s_j \leqq 4m+3)$ subsets of the set S, namely

$$S_j(d_j^{(k)}) \quad \text{for} \quad 1 \leqq k \leqq s_j ;$$

we have to choose for each $j \in N$ exactly one subset $S_j(d_j^{(p_j)})$ with $1 \leqq p_j \leqq s_j$, such that $\bigcup_{j \in N} S_j(d_j^{(p_j)}) = S$ and the element

$$\bar{x} = (\bar{x}_1, \dots, \bar{x}_n) \quad \text{with} \quad \bar{x}_j = d_j^{(p_j)} \quad \text{for all} \quad j \in N$$

with the minimal value of the sum .

$$\sum_{j \in N} c_j^{(p_j)}$$

over all such choices. It follows then immediately from the definition of $c_j^{(k)}$ that this sum is equal to $\sum_{j \in N} f_j(d_j^{(p_j)})$.

This problem can be reformulated to the following set covering problem:

Minimize

$$g(v) = \sum_{j \in N} \sum_{k=1}^{s_j} c_j^{(k)} v_j^{(k)} + p \sum_{j \in N} \left(\sum_{k=1}^{s_j} (v_j^{(k)}) - 1 \right) \qquad (1)$$

subject to

$$\sum_{j \in N} \sum_{k=1}^{s_j} a_{ij}^{(k)} v_j^{(k)} \geqq 1 \quad , \quad i \in S \qquad (2)$$

$$\sum_{k=1}^{s_j} v_j^{(k)} \geqq 1 \quad , \quad j \in N \qquad (3)$$

$$v_j^{(k)} = 0 \text{ or } 1 \qquad j \in N, \ 1 \leqq k \leqq s_j. \qquad (4)$$

We shall refer to the problem (1) - (4) as to the problem (P2).
The sense of the variables $v_j^{(k)}$ in the problem (P2) is the following: $v_j^{(k)} = 1$ if the subset $S_j(d_j^{(k)})$ is chosen in the covering and $v_j^{(k)} = 0$ otherwise. The constraints (2) ensure that each element $i \in S$ is covered, the inequalities (3) ensure that for each $j \in N$ at least one of the subsets $S_j(d_j^{(k)})$, $1 \leqq k \leqq s_j$, is chosen. We shall show further that the penalty coefficient is chosen in such a way that the optimal solution must lie among the feasible solutions with the property

$$\sum_{k=1}^{s_j} v_j^{(k)} = 1 \quad \text{for all} \quad j \in N,$$

if such feasible solutions exist.
Remark that the problem (P2) has always the nonempty set of feasible solutions if only $\bigcup_{j \in N} \bigcup_{k=1}^{s_j} S_j(d_j^{(k)}) = S$ (if this condition is not satisfied, the set M is empty according to Theorem 1). We shall assume that this condition is satisfied to ensure that the problem (P2) has always feasible solutions (and thus optimal solution too).

Theorem 2.

Let $\bar{v}$ be the optimal solution of (P2). It holds:

(a) $\sum_{j\in N}\left(\sum_{k=1}^{s_j}(\bar{v}_j^{(k)})-1\right)\geqq 1 \Longleftrightarrow M=\emptyset$.

(b) Suppose $M\neq\emptyset$ and $\bar{x}_j=d_j^{(p_j)}$ whenever $\bar{v}_j^{(p_j)}=1$ for all $j\in N$. Then $\bar{x}=(\bar{x}_1,\dots,\bar{x}_n)$ is one of the optimal solutions of the problem (P1).

Proof:

(a) Suppose $\sum_{j\in N}\left(\sum_{k=1}^{s_j}(v_j^{(k)})-1\right)\geqq 1$ and at the same time $M\neq\emptyset$.

Let $x^0=(x_1^0,\dots,x_n^0)\in M$, so that according to Theorem 1 $x_j^0\in[k_j,K_j]$ for all $j\in N$ and $\bigcup_{j\in N}S_j(x_j^0)=S$. There exists obviously an index t_j, $1\leqq t_j\leqq s_j$, such that $x_j^0\in[d_j^{(t_j)},d_j^{(t_j+1)}]$ (for each $j\in N$). We shall define the element $x^1=(x_1^1,\dots,x_n^1)$ as follows:

$x_j^1=x_j^0$, if either $x_j^0=d_j^{(t_j)}$ or $x_j^0=d_j^{(t_j+1)}$

$x_j^1=d_j^{(t_j)}$, if $f_j(d_j^{(t_j)})\leqq f_j(x_j^0)$ and $x_j^0\in(d_j^{(t_j)},d_j^{(t_j+1)})$

$x_j^1=d_j^{(t_j+1)}$, if $f_j(d_j^{(t_j+1)})\leqq f_j(x_j^0)$ and $x_j^0\in(d_j^{(t_j)},d_j^{(t_j+1)})$.

We have then

$$x_j^1\in[k_j,K_j] \text{ for all } j\in N \text{ and further } \sum_{j\in N}f_j(x_j^1)\leqq\sum_{j\in N}f_j(x_j^0).$$

According to Lemma 1 we have $S_j(x_j^0)\subset S_j(x_j^1)$ for all $j\in N$ so that $\bigcup_{j\in N}S_j(x_j^1)=S$ and therefore according to Theorem 1 $x^1\in M$.

Let us define now the coordinates of v as follows:

$$v_j^{(k)}=\begin{cases}1, & \text{if } x_j^1=d_j^{(k)}\\ 0 & \text{otherwise}\end{cases} \qquad \text{for all } j\in N,\ 1\leqq k\leqq s_j.$$

It follows immediately from the construction of x^1 that

$$\sum_{j\in N}\left(\sum_{k=1}^{s_j}(v_j^{(k)})-1\right)=0.$$

It holds then:

$$g(\bar{v})\geqq\sum_{j\in N}\sum_{k=1}^{s_j}c_j^{(k)}v_j^{(k)}+p>\sum_{j\in N}\sum_{k=1}^{s_j}|c_j^{(k)}|\geqq\sum_{j\in N}\sum_{k=1}^{s_j}c_j^{(k)}v_j^{(k)}.$$

Since $\sum_{j\in N}\sum_{k=1}^{s_j}c_j^{(k)}v_j^{(k)}=g(v)$, we obtained that $g(\bar{v})>g(v)$, which is in contradiction with the optimality of $\bar{v}$. Therefore we proved the implication $\Longrightarrow$ of the assertion (a) of our lemma. To prove the implication $\Longleftarrow$, we shall assume that $M = \emptyset$.

Suppose $\sum_{j\in N}\left(\sum_{k=1}^{s_j}(\bar{v}_j^{(k)}) - 1\right) = 0.$

It is then $\sum_{k=1}^{s_j} \bar{v}_j^{(k)} = 1$ for all $j \in N$ so that there exists for each exactly one t_j, $1 \leq t_j \leq s_j$ such that $\bar{v}_j^{(t_j)} = 1$ and the other $\bar{v}_j^{(k)}$ (for $k \neq t_j$, $1 \leq k \leq s_j$) are equal to 0. Let us consider now the system of subsets $S_1(d_1^{(t_1)}),\dots,S_n(d_n^{(t_n)})$. Since $\bar{v}$ satisfies (2), this system covers the set S, i. e.

$$\bigcup_{j\in N} S_j(d_j^{(t_j)}) = S .$$

Let us define the vector $\tilde{x} = (\tilde{x}_1,\dots,\tilde{x}_n)$ as follows: $\tilde{x}_j = d_j^{(t_j)}$ $\forall j \in N$, so that

$$\tilde{x}_j \in [k_j, K_j] \ \forall j \in N \ \& \ \bigcup_{j\in N} S_j(\tilde{x}_j) = S.$$

According to Theorem 1 we have then $\tilde{x} \in M$, which is a contradiction with the assumption that $M = \emptyset$. We proved the assertion (a).

(b) Let $x = (x_1,\dots,x_n)$ be an arbitrary element of M. We shall prove that $\sum_{j\in N} f_j(\bar{x}_j) \leq \sum_{j\in N} f_j(x_j)$. Since $x \in M$, we have according to Theorem 1

$$x_j \in [k_j, K_j] \ \forall j \in N \ \& \ \bigcup_{j\in N} S_j(x_j) = S.$$

Further there exists obviously for each $j \in N$ an index t_j, such that $x_j \in [d_j^{(t_j)}, d_j^{(t_j+1)}]$. Similarly as in the proof of (a) we shall construct the element $x^1 = (x_1^1,\dots,x_n^1)$ as follows:

$$x_j^1 = x_j \ , \quad \text{if either} \quad x_j = d_j^{(t_j)} \quad \text{or} \quad x_j = d_j^{(t_j+1)}$$

$$x_j^1 = d_j^{(t_j)}, \quad \text{if} \quad f_j(d_j^{(t_j)}) \leq f_j(x_j) \quad \text{and} \quad x_j \in (d_j^{(t_j)}, d_j^{(t_j+1)})$$

$$x_j^1 = d_j^{(t_j+1)}, \quad \text{if} \quad f_j(d_j^{(t_j+1)}) \leq f_j(x_j) \quad \text{and} \quad x_j \in (d_j^{(t_j)}, d_j^{(t_j+1)}).$$

It holds obviously that $x_j^1 \in [k_j, K_j]$ $\forall j \in N$ and $\bigcup_{j\in N} S_j(x_j^1) = S$ (in accordance with Lemma 1) so that we obtain (according to Theorem 1) that $x^1 \in M$. Let us define variables $\hat{v}_j^{(k)}$ for $1 \leq k \leq s_j$, $j \in N$ as follows:

$$\hat{v}_j^{(k)} = \begin{cases} 1 \ , & \text{if } x_j^1 = d_j^{(k)} \\ 0 & \text{otherwise} \end{cases} \qquad \text{for all } j \in N.$$

It is then obviously $\sum_{k=1}^{s_j} \hat{v}_j^{(k)} = 1$, so that $\hat{v}$ satisfies (3) from (P2). Since the system of subsets $S_j(x_j^1)$, $j \in N$ is a covering of S, the vector $\hat{v}$ satisfies also the inequalities (2) from the problem (P2). Therefore $\hat{v}$ is a feasible solution of (P2). Since $\bar{v}$ is the optimal solution of (P2), it must hold: $g(\bar{v}) \leqq g(\hat{v})$. It follows from the definition of $\hat{v}$ that $g(\hat{v}) = \sum_{j \in N} f_j(x_j^1)$. Vector x^1 was constructed from x in such a way that it is obviously $\sum_{j \in N} f_j(x_j^1) \leqq \sum_{j \in N} f_j(x_j)$.

Summarizing these considerations we obtain the following inequalities:

$$g(\bar{v}) \leqq g(\hat{v}) = \sum_{j \in N} f_j(x_j^1) \leqq \sum_{j \in N} f_j(x_j) ,$$

which gives us the optimality of $\bar{x}$, because it is

$$g(\bar{v}) = \sum_{j \in N} \sum_{k=1}^{s_j} c_j^{(k)} \bar{v}_j^{(k)} = \sum_{j \in N} f_j(\bar{x}_j) ,$$

and x was an arbitrary element of M.

Q.E.D.

Remark 1.

If we set $f_j(x_j) \equiv 0$ for all $j \in N$ and $p \equiv 1$, we can use the procedure described above to find out whether the set M is empty or not. This fact follows immediately from the assertion (a) of Theorem 2.

Remark 2.

For solving the set covering problem (P2) one of the procedures suggested for instance in [2] can be made use of.

Remark 3.

The procedure described above generalizes the results obtained in [1], [3].

References

[1] Jajou, A.F. and K. Zimmermann: Max-separable equations and the set covering. AUC, to appear in 1988.

[2] Garfinkel, R. S. and G. L. Nemhauser: Integer Programming. John Wiley and Sons, New York etc., 1972.

[3] Zimmermann, K.: Systems of max-separable inequalities and the set covering. Seminarberichte der HU Berlin, to appear in 1988.

Structural Optima In Nonlinear Programming

S. ZLOBEC[1]

"*Programming sticks upon the shoals*
Of incommensurate multiple goals,
And where the tops are no-one knows
When all our peaks become plateaus
The top is anything we think
When measuring makes the mountain shrink.
The upshot is, we cannot tailor
Policy by a single scalar,
Unless we know the priceless price
Of Honor, Justice, Pride, and Vice.
This means a crisis is arising
For simple-minded maximizing."

K. Boulding in [7]

ABSTRACT

We introduce a new kind of optimality in nonlinear programming based on the notion of stability, i.e., continuity of the feasible-set point-to-set mapping. The new optima are termed "structural optima" and they include all usual local optima as a special case. The main result is a complete characterization of structural optima for a large class of nonconvex programs. We also study continuity of the corresponding Lagrange multipliers and show how the results can be extended to multi-objective programming.

1. INTRODUCTION

Many people believe, or want to believe, that the underlying behavioural structure of management and economic systems generates basically "continuous" processes (see [1,35]). Still, surprisingly, continuity is not represented in the standard definitions of local and global optima in nonlinear programming. As a result, it appears that many optima in nonlinear programming are only of theoretical interest.

For example, consider the problem of minimizing the function $f = z_2$ over the feasible set F in the (z_1, z_2)-plane, depicted in the figure below. Suppose that the corresponding economic system is running with a production plan described in the figure by the feasible point P. The global optimum G appears useless here, because every feasible perturbation of P via SO, that improves the value of the objective function, results in discontinuity of the feasible set and in its contraction locally to a set of measure zero. Such perturbations are economically unacceptable. In this situation the point SO appears to be at least as important as the points L of the local minima. In what follows, the points, such as SO in Figure 1, will be included in the definition of "optimality".

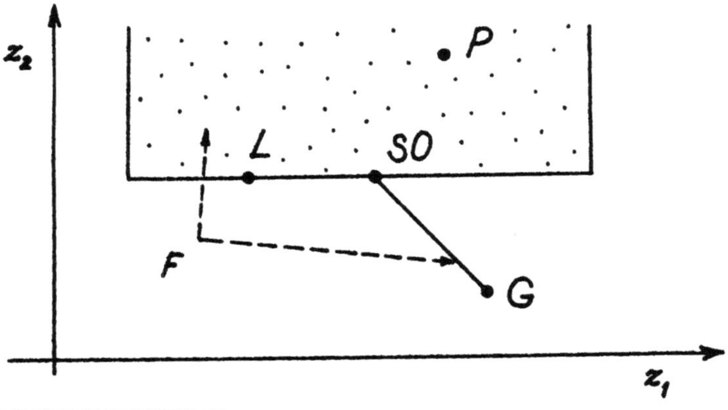

Figure 1

[1]McGill University, Department of Mathematics and Statistics, Burnside Hall, Montreal, Canada, H3A 2K6.

The main objective of this paper is to define and characterise conceptually a new kind of "optimality" that uses a notion of continuity. This will be achieved by combining some elementary ideas and results from parametric programming [2,3,7,10,11,13,14, 21,22,23,24,27], BBZ theory of convex programming [4], input optimization [25,26,28,29, 32,33,35], and point-to-set mappings [5,15,18,30].

We study the general nonlinear programs of the form

$$(P) \qquad \begin{array}{l} \min \; f^0(z) \\ s.t. \\ \qquad f^i(z) \leq 0, \qquad i \epsilon P = \{1, \ldots, m\} \end{array}$$

where $f^j : R^n \rightarrow R, j \epsilon \{0\} \cup P$ are continuous functions. After a splitting of the variable $z \epsilon R^n$ into $x \epsilon R^{n-p}$ and $\theta \epsilon R^p$, the program (P) can be reformulated as a mathematical "model"

$$(P, \theta) \qquad \begin{array}{l} \min\limits_{(x)} \; f^0(x, \theta) \\ s.t. \\ \qquad f^i(x, \theta) \leq 0, \qquad i \epsilon P \end{array}$$

where θ is thought of as a parameter. The model (P, θ), rather than program (P), seems to be a natural description of many real life situations. For example, in linear programming

$$\begin{array}{l} \min \; c^T x \\ s.t. \\ \qquad Ax \leq b \\ \qquad x \geq 0 \end{array}$$

one can specify $\theta = (A, b, c)$ or any part of these. At this stage the two formulations are equivalent, as far as optimal solutions are concerned. (See [27].)

Let us concentrate on the model (P, θ). For each θ we denote by

$$F(\theta) = \{x \epsilon R^{n-p} : f^i(x, \theta) \leq 0, \quad i \epsilon P\}$$

its feasible set,

$$\tilde{F}(\theta) = \{\tilde{x}(\theta)\}$$

the set of (globally) optimal solutions, and

$$\tilde{f}(\theta) = f^0(\tilde{x}(\theta), \theta)$$

the optimal value. As in input optimization, we consider θ to be an "input", while the triple $\{F(\theta), \tilde{F}(\theta), \tilde{f}(\theta)\}$ is the "output". Continuity of the output has been studied locally in a "region of stability" (see [16,32,35]). We say that a set $S \subset R^p$ is a region of stability at $\theta^* \epsilon S$ if the point- to-set mapping $\Gamma : \theta \rightarrow F(\theta)$ is lower semicontinuous at θ^* over S. This will now be formalized for "convex models" (P, θ), i.e., models where $f^j(., \theta) : R^{n-p} \rightarrow R, j \epsilon \{0\} \cup P$ are convex functions. (We will study mainly such models.)

1.1 DEFINITION *Consider the convex model (P, θ) at some θ^*. A set $S \subset R^p$ is a region of stability at $\theta^* \epsilon S$ if, for each open set $A \subset R^{n-p}$, satisfying $A \cap F(\theta^*) \neq \emptyset$, there is a neighbourhood $N(\theta^*)$ of θ^* such that $A \cap F(\theta) \neq \emptyset$ for each $\theta \epsilon N(\theta^*) \cap S$.*

Since the point-to-set mapping $\Gamma : \theta \rightarrow F(\theta)$ is closed (because of the continuity assumptions on $f^i, i \epsilon P$), it follows that Γ is "continuous" at θ^* with respect to every region of stability (in the sense of, say, Hogan[15]). Continuity of $\Gamma : \theta \rightarrow F(\theta)$ guarantees closedness of $\tilde{F}$ and continuity of $\tilde{f}$, for convex models.

A breakthrough in parametric optimization came with the realization that, at least for convex models, some of the regions of stability could actually be calculated. Within these regions (typically polytopes) a path is determined, connecting two inputs, that preserves continuity of the mapping Γ and improves the optimal value function $\bar{f}(\theta)$. By repeating the process, from the new input, one globally optimizes the mathematical programming model (P, θ) by stable (i.e., output–continuous) perturbations of the parameters. This is the subject of input optimization.

At the moment, about 20 different regions of stability are in use, not only in input optimization, but also in other areas, such as a study of functional equations occurring in parametric random decision systems with complete connections [31], data envelopment analysis [8], and linear programming [9]. In the construction of these regions, as well as in the formulation of optimality conditions, we use the computable objects of convex analysis such as (defined for each θ)

$$P^{=}(\theta) = \{i \epsilon P : x \epsilon F(\theta) \Rightarrow f^i(x, \theta) = 0\}$$

and

$$F^{=}(\theta) = \{x \epsilon R^{n-p} : f^i(x, \theta) = 0, \quad i \epsilon P^{=}(\theta)\}.$$

Three sample regions of stability at θ^*, for convex models and realistic objective functions (i.e., $\bar{F}(\theta^*) \neq \emptyset$ and bounded), are

$$M(\theta^*) = \{\theta : F(\theta^*) \subset F(\theta)\}$$

$$W(\theta^*) = \{\theta : F^{=}(\theta^*) \subset F^{=}(\theta), \quad P^{=}(\theta^*) = P^{=}(\theta)\}$$

and the larger of the two (but more difficult to construct)

$$H(\theta^*) = \{\theta : F(\theta^*) \subset F_*^{=}(\theta)\}$$

where $F_*^{=} : R^p \rightarrow R^{n-p}$ is a point-to-set mapping defined by

$$F_*^{=}(\theta) = \{x : f^i(x, \theta) \leq 0, \quad i \epsilon P^{=}(\theta^*)\}.$$

(This mapping was introduced with a different symbol in [2], also see [28,29].) The intersection of all regions of stability at θ^* is termed *the* region of stability at θ^*.

An optimal input is defined with respect to a region of stability as follows (see [32,33,35]):

1.2 DEFINITION *Consider the model (P, θ) at some θ^* with a realistic objective function. Let S be a region of stability at θ^*. If $\bar{f}(\theta^*) \leq \bar{f}(\theta)$ for every $\theta \epsilon N(\theta^*) \cap S$, where $N(\theta^*)$ is a neighbourhood of θ^*, then θ^* is a locally optimal input with respect to S and (P, θ^*) is the corresponding optimal realization of the model.*

It is important to note that, in input optimization, an optimal input θ^* depends on the initial input θ^0, i.e., not all optimal inputs are generally reachable from an intial (current) input θ^0 by stable perturbations of the parameters. This is illustrated by the following example.

1.3 EXAMPLE Consider the linear model

$$\begin{aligned} &\min_{(x)} \theta \\ &s.t. \\ &\qquad \theta x = 0 \\ &-1 \leq x \leq 1, \quad -1 \leq \theta \leq 1. \end{aligned}$$

Note that

$$F(\theta)=\begin{cases}[-1,1], & \text{if } \theta=0\\ 0\ , & \text{if } \theta\neq 0.\end{cases}$$

Hence the optimal input θ^* depends on the initial θ^0:

$$\theta^*=\begin{cases}0, \ \text{if } \theta^0=0\\ -1, \quad -1\le\theta^0<0\\ \text{does not exist(!)}, \ 0<\theta^0\le 1.\end{cases}$$

Note that $\theta^0>0$ and $\theta^*=0$ cannot be connected by a stable path.

The above model can also be considered as a nonlinear program, but then the subtleties of continuity are lost:

1.4 EXAMPLE The nonlinear program

$$\begin{aligned}&\min\ z_1\\ &s.t.\\ &z_1 z_2=0\\ &-1\le z_1\le 1, \qquad -1\le z_2\le 1\end{aligned}$$

has the global optimal solution: $z_1^*=-1$, $z_2^*=0$ and local optimal solutions $z_1^*=0$, $0<z_2^*\le 1$, $-1\le z_2^*<0$.

The above examples show that all optimal inputs for (P,θ) cannot be generally obtained by considering x and θ as a single variable $z=(x,\theta)$ and then by applying nonlinear programming to the corresponding (P). The point $z_1^*=0$, $z_2^*=0$ is not "detected" by nonlinear programming. Note that nonlinear programming, in its search for optimal solutions, is ignoring continuity of the feasible–set point–to–set mapping $\Gamma:\theta\to F(\theta)$.

2. STRUCTURAL OPTIMA

Since continuity (stability) is a natural and essential ingredient in model building (as many people believe), it appears that continuity of some sort should be included in the definition of optimality for the program (P). In order to avoid technicalities and concentrate on the essentials, we will study below only those programs (P) with the property that for every partitioning $z=(x,\theta)$ the sets of local and global optimal solutions $\tilde{F}(\theta)$ coincide for every fixed θ. (The ordering in x,θ, and z is irrelevant.) In view of the preceding discussion a more "natural" definition of optimality is suggested:

2.1 DEFINITION *A feasible point z^* of the program (P) is called a structural optimum if, for at least one splitting $z^*=(x^*,\theta^*)$, θ^* is a locally optimal input with respect to the region of stability at θ^* and $x^*=\tilde{x}(\theta^*)$ is a corresponding optimal solution.*

Note that by specifying $z=x$ (i.e., "empty splitting"), every "usual" optimum is necessarily a structural optimum, but as the above examples have shown, there are structural optima that are not usual optima.

In our definition of structural optima we look at the program (P) as a two–level optimization problem. A point z^* is a structural optimum if every improvement of the optimal value function $\tilde{f}(\theta)$, at the θ level, results in discontinuity (possibly emptiness) of the corresponding feasible set at the x level. In order to characterize structural optima, we have to characterize optimality of both θ^* and $\tilde{x}(\theta^*)$. We will now show how this can be done for convex models. In this case, it is enough to characterize only the optimal input θ^*. (Characterizations of $\tilde{x}(\theta^*)$, for the *convex program* (P,θ^*), are well known from the BBZ theory [4].) We use the notation $P^<(\theta^*) = P \backslash P^=(\theta^*)$ and the "restricted" Lagrangian

$$L_*^<(x,u;\theta) = f^0(x,\theta) + \sum_{i \epsilon P^<(\theta^*)} u_i f^i(x,\theta).$$

Also $R_+^{q(\theta^*)}$ denotes the non-negative orthant of the $q(\theta^*) = card P^<(\theta^*)$ dimensional Euclidean space.

2.2 THEOREM *Consider the convex model (P,θ) with a realistic objective function at some θ^*. Let $\tilde{x}(\theta^*)$ be a corresponding optimal solution, and let S be an arbitrary region of stability at θ^*. Then θ^* is a locally optimal input with respect to S if, and only if, there exists a neighbourhood $N(\theta^*)$ of θ^* and a non-negative vector function*

$$U : N(\theta^*) \cap S \to R_+^{q(\theta^*)}$$

such that, whenever $\theta \epsilon N(\theta^*) \cap S$,

$$L_*^<(\tilde{x}(\theta^*),u;\theta^*) \le L_*^<(\tilde{x}(\theta^*),U(\theta^*);\theta^*) \le L_*^<(x,U(\theta);\theta)$$

for every $u \epsilon R_+^{q(\theta^*)}$ and every $x \epsilon F_*^=(\theta)$.

PROOF: It uses a necessary condition for stability ($P^=(\theta) \subset P^=(\theta^*)$ for every $\theta \epsilon N(\theta^*) \cap S$, where $N(\theta^*)$ is some neighbourhood of θ^*, see [26,35]) and the hyperplane separation theorem. For details see [28].

The above result can be simplified by introducing various "input constraint qualifications". (In one such simplification the statement "$x \epsilon F_*^=(\theta)$" is replaced by "$x \epsilon F^=(\theta^*)$", see [29,32,35]). Simplifications also result from choosing S to be a particular region of stability. For a differentiable version of the above result see [28].

Since every optimum is also a structural optimum, the necessary conditions for a structural optimum are also necessary conditions for a usual optimum. In particular, Theorem 2.2 gives a necessary condition for an optimal solution for a large class of nonlinear programs (those that can be "convexified" by some splitting of $z = (x,\theta)$), without requiring any extraneous conditions (such as a constraint qualification). Nonlinear programs where the sets of structural and usual optima coincide are called "structurally regular". Such programs include all programs with a convex feasible set (in particular, all linear and convex programs). Among structurally "irregular" programs there are many quadratic programs, as the one in Example 1.1. Another one (less "pathological") program is borrowed from [36]:

2.3 EXAMPLE Consider the quadratic program

$$\begin{aligned}
&max \quad z_1 z_5 + z_2 \\
&s.t. \\
&\qquad z_1 + z_2 + z_3 \qquad = 1 \\
&\qquad z_1 + z_2 z_6 \quad - z_4 \quad = 1 \\
&\qquad z_5 z_6 \le 1, \quad z_2 z_6 \quad \le 1 \\
&z_i \ge 0, \quad i = 1,2,3,4; \quad z_5 \ge 0.5; \quad 0.1 \le z_6 \le 2.
\end{aligned}$$

After convexification, where $\theta = (z_5, z_6)$, we find that $\theta^* = (1,1)$ is a locally optimal input over the region of stability at θ^* (for the initial $\theta^0 = (0.5, \quad 2)$). Hence

$$z_1^* = 0, \quad z_2^* = 1, \quad z_3^* = 0, \quad z_4^* = 0, \quad z_5^* = 1, \quad z_6^* = 1$$

is a structural optimum. It is easy to show that this point is not a local optimum in the usual sense. Every improvement of the objective function from this point requires $z_6 < 1$ and this results in discontinuity of the feasible set (F contracts to a point in the (z_1, z_2) - plane).

The above theorem may be useful in situations when the standard approach fails to yield results. This will be now illustrated on a well-known program of Kuhn and Tucker [19].

2.4 EXAMPLE The nonconvex (structurally regular) program

$$\begin{aligned} & \min \quad -z_1 \\ & \text{s.t.} \\ & \qquad (z_1 - 1)^3 + z_2 \leq 0 \\ & \qquad z_1 \geq 0, \quad z_2 \geq 0 \end{aligned}$$

has a unique optimal solution $z_1^* = 1, \quad z_2^* = 0$, but no saddle point. But convexification $z_1 = \theta, \quad z_2 = x$ yields the linear model

$$\begin{aligned} & \min_{(x)} \quad f^0 = -\theta \\ & \text{s.t.} \\ & \qquad f^1 = (\theta - 1)^3 + x \quad \leq 0 \\ & \qquad f^2 = -\theta \quad \leq 0 \\ & \qquad f^3 = \quad -x \quad \leq 0. \end{aligned}$$

At $\theta^* = 1$ we find $S(\theta^*) = [0,1]$ and Theorem 2.2 produces a saddle point for

$$L_*^< = -\theta[1 + u_2(\theta)]$$

with the Lagrangian functions of the form

$$0 \leq U_2(\theta) \leq \frac{1-\theta}{\theta}$$

for every $\theta \leq 1$ and close to θ^*.

If θ is kept fixed, then Theorem 2.2 recovers a BBZ characterization of optimality. If, in addition to it, an extraneous condition (such as Slater's condition) is assumed, then the result recovers the KKT classical result of Karush [17], Kuhn and Tucker [19].

3. CONTINUITY OF LAGRANGE MULTIPLIERS

In this section we are concerned with the problem of continuity of the vector function U appearing in Theorem 2.2 at an optimal input θ^* (i.e., a structural optimum $z^* = (\bar{x}(\theta^*), \theta^*)$). Uniform boundedness of $U = U(\theta)$ can be easily proven using the arguments as in [25].

3.1 LEMMA *Consider the convex model (P, θ) with a realistic objective function at a locally optimal input θ^*, with respect to a region of stability $S(\theta^*)$. Then, for an arbitrary sequence $\theta^k \epsilon S(\theta^*), \theta^k \to \theta^*$, the sequence $U(\theta^k)$ is uniformly bounded for all sufficiently large k's.*

Uniform boundedness implies the existence of limit points of $U(\theta^k)$, as $\theta^k \epsilon S(\theta^*), \theta^k \to \theta^*$. In order to show that the limit points are in $U(\theta^*)$, we need an extra condition such as lower semicontinuity of the point-to-set mapping $F_*^=(\theta)$.

3.2 THEOREM *Consider the convex model (P, θ) with a realistic objective function at a locally optimal input θ^*, with respect to a region of stability $S(\theta^*)$. Assume that the point-to-set mapping $F^=_*(\theta)$ is lower semicontinuous at θ^* with respect to $S(\theta^*)$. Then, for an arbitrary sequence $\theta^k \epsilon S(\theta^*), \theta^k \rightarrow \theta^*$, the set of limit points of the sequence $U(\theta^k)$ is non-empty and it is contained in $U(\theta^*)$.*

PROOF: See [28].

If Slater's condition holds for the convexification (P, θ) at θ^*, then U becomes the usual Lagrangian multiplier function and we recover the well-known results on its continuity [2,10,25].

4. CONCLUSION

The above ideas can be easily extended to multi-objective programs

$$(MP) \qquad \begin{array}{ll} & \min \{\Phi^k(z) : k \epsilon Q\} \\ s.t. & \\ & f^i(z) \leq 0, \quad i \epsilon P \end{array}$$

where both P and Q ar finite index sets. Using an observation made by Charnes and Cooper [6] on Pareto solutions we can extend our definition from single to multi-objective programs as follows:

4.1 DEFINITION *Consider the multi-objective program (MP). Its feasible point z^* is a structural Pareto optimum if, for at least one splitting $z^* = (x^*, \theta^*)$, θ^* is a locally optimal input with respect to the region of stability at θ^*, and $x^* = \bar{x}(\theta^*)$ is a corresponding optimal solution for the single-objective model*

$$\begin{array}{ll} & \min\limits_{(x)} \sum\limits_{k \epsilon Q} \Phi^k(x, \theta) \\ s.t. & \\ & \Phi^k(x, \theta) \leq \Phi^k(z^*), \quad k \epsilon Q \\ & f^i(x, \theta) \leq 0, \quad i \epsilon P. \end{array}$$

Similarly one can define structural weak Pareto optima and structural strong (properly efficient) optima, using non- negative and positive weights, respectively. Then such points can be characterized using the results from, e.g., [34]. Results on continuity of the Lagrange multipliers from Section 3 can be easily adapted to the new situations. Note that the above definition can also be used to define structural optima for unconstrained problems. (Specify $Q = \{1\}$ and $P = \emptyset$).

At a structural optimum one should re-examine the current strategy for improving the value of the objective function and decide whether it is feasible and worthwhile to continue the optimization process. Of course, not all structural optima have such a dramatic effect on the optimization problem as the point SO in Figure 1. One can "slice" the feasible set of (P) in many ways and introduce "generalized" structural optima. However, the "componentwise slicings" $z = (x, \theta)$ (where θ is the parameter vector and $x = x(\theta)$ is the decision vector) appear to be the most important ones for applications.

Theory of structural optima is presently in its infancy, in spite of a systematic development of input optimization. Numerical methods for calculating structural optima barely exist, in spite of several case studies solved by input optimization. These methods require not only constructive optimality conditions but also efficient subroutines for calculating sets such as $P^=$ and $F^=$, regions of stability, and improvable stable paths within these regions. Finally, note that a different definition of continuity of point-to-set mappings may produce different structural optima.

REFERENCES

[1] Ball, R.J.: Econometric model building, in: *Mathematical Modelling in Economics and Industry*, Hafner, New York 1968, 15-30.

[2] Bank, B., J. Guddat, D. Klatte, B. Kummer, and K. Tammer: *Non-Linear Parametric Optimization*, Akademie- Verlag, Berlin 1982.

[3] Bank, B., R. Mandel, and K. Tammer: Parametrische Optimierung und Aufteilungsverfahren, in: Lommatzsch, K. (ed.): *Anwendungen der linearen parametrischen Optimierung*, Akademie-Verlag, Berlin 1979, 107-123.

[4] Ben-Israel, A., A. Ben-Tal, and S. Zlobec: *Optimality in Nonlinear Programming*, Wiley, New York, 1981.

[5] Berge, C. : *Espace Topologiques, Fonctions Multivogues*, Dunod, Paris, 1959.

[6] Charnes, A. and W.W. Cooper: *Management Models and Industrial Applications of Linear Programming*, Vol.1, Wiley, New York, 1961.

[7] Charnes, A. and W.W. Cooper: Constrained extremization models and their use in developing system measures, in: Mesarović, M.D. (ed.): *Views on General Systems Theory*, Wiley, New York 1964, 61-88.

[8] Charnes A. and S. Zlobec: Stability of efficiency evaluations in data envelopment analysis, Report CCS 560, The University of Texas, Austin, March 1987.

[9] Cojocaru, I.: Regions de stabilité dans la programmation linéaire, *An. Univ. Bucuresti Mat 34* (1985) 12-21.

[10] Fiacco, A.V.: *Introduction to Sensitivity and Stability Analysis in Nonlinear Programming*, Academic Press, New York, 1983.

[11] Gal, T.: Linear parametric programming - A brief survey, *Mathematical Programming Study 21* (1984) 43-68.

[12] Guddat, J., F. Guerra Vasquez, K.Tammer, and K.Wendler: *Multiobjective and Stochastic Optimization Based on Parametric Optimization*, Akademie-Verlag, Berlin, 1985.

[13] Guddat, J., H.Th. Jongen, and J.Rueckmann: On stability and stationary points in nonlinear optimization, *J. Austral. Math. Soc. Ser. B 28* (1986) 36-56.

[14] Guddat, J., H.Th.Jongen, B.Kummer, and F.Nožička (editors): *Parametric Optimization and Related Topics*, Akademie-Verlag, Mathematical Research, Band 35, Berlin 1987.

[15] Hogan, W.W.: Point-to-set maps in mathematical programming, *SIAM Review 15* (1973) 591-603.

[16] Huang, S. and S. Zlobec: New regions of stability in input optimization, Research report, McGill University, April 1987.

[17] Karush, W.: Minima of functions of several variables with inequalities as side conditions, M.Sc. thesis, University of Chicago, December 1939.

[18] Klatte, D.: A sufficient condition for lower semicontinuity of solution sets of systems of convex inequalities, *Mathematical Programming Study 21* (1984) 139-149.

[19] Kuhn, H.W. and A.W.Tucker: Nonlinear programming, in: Neymann, J. (ed.): *Proceedings of the Second Berkeley Symposium on Mathematical Statistics and Probability*, University of California Press, Berkeley 1951, 481-492.

[20] Nožička, F.: Ein geometrischer Beweis des Dualitätssatzes der linearen Optimierung, *Math. Operationsforsch. Statistik 2* (1971) 4, 3-17.

[21] Nožička,F.: Lineare parametrische Optimierung - ein Problem der Stabilität der optimalen Lösung, in : *Konference o matematickych metodach v ekonomii*, Praha 1972, 45-82.

[22] Nožička, F., J. Guddat, H.Hollatz, and B.Bank: *Theorie der linearen parametrischen Optimierung*, Akademie– Verlag, Berlin 1974.

[23] Robinson, S.: Stability theory for systems of inequalities. Part 1: Linear systems, *SIAM J. Numerical Analysis 12* (1975) 754-769.

[24] Robinson, S.: A characterization of stability in linear programming, *Oper.Res. 25* (1977) 435-447.

[25] Semple, J. and S.Zlobec: Continuity of a Lagrangian multiplier function in input optimization, *Mathematical Programming 34* (1986) 362-369.

[26] Semple, J. and S.Zlobec: On a necessary condition for stability in perturbed linear and convex programming, *Zeitschrift für Operations Research. Series A: Theorie.* (1987, forthcoming.)

[27] Tammer,K.: The application of parametric optimization and imbedding to the foundation and realization of a generalized primal decomposition approach, Paper in [14], 376-386.

[28] Van Rooyen, M. and S.Zlobec: A complete characterization of optimal economic systems with respect to stable perturbations, *Glasnik Matematički*. (To appear.)

[29] Van Rooyen, M., M.Sears, and S. Zlobec: Constraint qualifications in input optimization, Research Report, McGill University, July 1987.

[30] Wei, Q.-L.: Stability in mathematical programming in the sense of lower semicontinuity and continuity, *Journal of the Qufu Teachers College 1* (1981). (In Chinese.)

[31] Zidaroiu, C.: Regions of stability for random decision systems with complete connections, *An. Univ. Bucuresti Mat. 34* (1985) 87-97.

[32] Zlobec, S.: Characterizing an optimal input in perturbed convex programming, *Mathematical Programming 25* (1983) 109-121; Corrigendum, ibid *35* (1986) 368-371.

[33] Zlobec, S.: Input optimizations: I. Optimal realizations of mathematical models, *Mathematical Programming 31* (1985) 245-268.

[34] Zlobec, S.: Input optimization: III. Optimal realizations of multi–objective models, *Optimization 17* (1986) 429-445.

[35] Zlobec.S.: Survey of input optimization, *Optimization 18* (1987) 309-348.

[36] Zlobec.S.: Topics in input optimization, Paper presented at the International Symposium on the Occasion of Professor A. Charnes 70-th Birthday, University of Texas at Austin, October 1987.

ON AGGREGATION OF VARIABLES IN LINEAR PROGRAMMING

Milan Vlach

1. Introduction.

In the literature concerning optimization, the term aggregation has at least two meanings. On the one hand, some authors use the word aggregation for bringing several optimization problems into one problem. On the other hand, some authors use it for replacing a single large possibly intractable problem by a smaller more tractable problem.

We shall be concerned with the latter. Roughly speaking, the approach under consideration involves

a) constructing an aggregated problem from a given original problem by decreasing the number of variables,

b) solving the resulted aggregate problem,

c) deriving an approximate solution of the original problem by disaggregating an optimal solution of the aggregate problem into a feasible solution of the original problem.

As a rule, this process results in a nonoptimal feasible solution of the original problem. Therefore the whole approach is questionable unless some useful information on the resulting error can be obtained without solving the original problem.

Two types of bounds on the error induced by aggregation are usually studied: a priori bounds and a posteriori bounds. A priori bounds may be calculated after an aggregate problem has been formed but not solved, whereas calculation of a posteriori bounds requires knowledge of a solution of the aggregate problem. Here we are concerned with a posteriori bounds on the difference between the optimal value of the original problem and the value at the resulting approximate solution. To calculate this difference exactly requires solution of the original problem. To obtain bounds on this difference without solving actually the original problem requires only information about approximate location of at least one optimal solution. Such information is directly available in problems with special structure, or can sometimes be obtained from simple examination of the constraints of the original problem in an ad hoc manner. It turns out that recent results of Mangasarian concerning quantification of some duality relations that underlie linear and monotone complementarity problems are promising in establishing a general approach for obtaining useful information on location of optimal solutions.

2. The Zipkin bound.

Let the original problem be

$$\begin{aligned} &\text{maximize} \quad x \to cx \\ &\text{subject to} \quad Ax \leqq b \\ &\qquad\qquad\quad\;\; x \geqq 0 \end{aligned} \tag{1}$$

where $A = (a_{ij})$ is an $m \times n$ matrix, $b = (b_i)$ is an m-vector, $c = (c_j)$ is an n-vector, and $x = (x_j)$ is an n-vector of variables.

The aggregate problem of (1) is determined by an $n \times p$ matrix $Q = (q_{ij})$ as follows :

$$\begin{aligned} &\text{maximize} \quad y \to (cQ)y \\ &\text{subject to} \quad (AQ)y \leqq b \\ &\qquad\qquad\qquad\quad y \geqq 0 \end{aligned} \tag{2}$$

where y is a p-vector of variables.

It is obvious that for each feasible solution y of (2), the vector $x := Qy$ is a feasible solution of (1) whenever Qy is nonnegative.

If x is a feasible solution of the original problem and if $v = (v_i)$ is an nonnegative m-vector, then $v(b-Ax) \geqq 0$ and therefore

$$cx \leqq cx + v(b-Ax) = vb + (c-vA)x \quad .$$

It follows that

$$cx^* - c\bar{x} \leqq (c-vA)x^* \tag{3}$$

whenever x^* is an optimal solution of the original problem, v is nonnegative, and $\bar{x}$ is such that

$$vb \leqq c\bar{x} \quad . \tag{4}$$

Let $\bar{y}$ be an optimal solution of the aggregate problem and let $\bar{v}$ be an optimal solution of the dual to the aggregate problem. Obviously, $\bar{v}$ is nonnegative and (4) is satisfied for $\bar{v}$ and $\bar{x} := Q\bar{y}$, because

$$\bar{v}b = (cQ)\bar{y} = c(Q\bar{y}) \ .$$

Consequently

$$cx^* - c\bar{x} \qquad (c - \bar{v}A)x^* \tag{5}$$

whenever x^* is an optimal solution of the original problem, $\bar{v}$ is an optimal solution of the dual to the aggregate problem and $\bar{x}$ is equal to $Q\bar{y}$ where $\bar{y}$ is an optimal solution of the aggregate problem.

Zipkin (1980) has extensively studied fixed-weight aggregation, that is the aggregation given by a matrix Q of the following form.

Let $\{T_1,T_2,\ldots,T_p\}$ be a partition of $\{1,2,\ldots,n\}$. For each $k = 1,2,\ldots,p$ let q_k be a nonnegative $|T_k|$-vector whose components sum to unity. Then the corresponding matrix Q is given by

$$Q = \begin{bmatrix} q_1 & 0 & \ldots & 0 \\ 0 & q_2 & \ldots & 0 \\ & \ldots\ldots & & \\ 0 & 0 & \ldots & q_p \end{bmatrix}$$

Here we assume only that Q is such that there is an optimal solution $\bar{y}$ of (2) for which $Q\bar{y}$ is nonnegative. This is true, for example, for each nonnegative matrix, provided the aggregate problem has an optimal solution.

In order to get an upper bound on the difference $cx^* - c\bar{x}$ with the help of (5) without solving actually the original problem, we assume (following Zipkin) that the following information about location of at least one optimal solution of the original problem is available: A partition $\{S_1,S_2,\ldots,S_q\}$ of $\{1,2,\ldots,n\}$, nonnegative numbers $p_1,p_2,\ldots,p_q$, and positive numbers $d_1,d_2,\ldots,d_n$ are given such that there is an optimal solution x^* of (1) with the property

$$\sum_{j\in S_k} d_j x_j^* \leqq p_k \,, \quad k = 1,2,\ldots,q \ .$$

Using this information and denoting the j-th column of A by A_j, we obtain

$$\begin{aligned} \sum_{j\in S_k} (c_j - \bar{v}A_j)x_j^* &= \sum_{j\in S_k} \frac{1}{d_j}(c_j - \bar{v}A_j)d_j x_j^* \\ &\leqq \sum_{j\in S_k} \max\left\{0, \frac{1}{d_j}(c_j - \bar{v}A_j)\right\} d_j x_j^* \\ &\leqq \max_{j\in S_k} \max\left\{0, \frac{1}{d_j}(c_j - \bar{v}A_j)\right\} \sum_{j\in S_k} d_j x_j^* \\ &\leqq \max_{j\in S_k} \max\left\{0, \frac{1}{d_j}(c_j - \bar{v}A_j)\right\} p_k \ . \end{aligned}$$

Since

$$(c - \bar{v}A)x^* = \sum_{k=1}^{q} \sum_{j\in S_k} (c_j - \bar{v}A_j)x_j^* \ ,$$

it follows from (5) that

$$cx^* - c\bar{x} \qquad \sum_{k=1}^{q} \bar{w}_k p_k$$

where

$$\bar{w}_k := \max_{j\in S_k} \max\left\{0, \frac{1}{d_j}(c_j - \bar{v}A_j)\right\}$$

may be calculated without solving (1).

3. The Mendelssohn bound

Under the same assumptions this bound can be improved by the Mendelssohn (1980) approach. This approach results in the following bound: Define $\bar{w}_k(t)$ by

$$\bar{w}_k(t) := \max_{j \in S_k} \max \left\{ 0, \frac{1}{d_j} (c_j - t\bar{v}A_j) \right\} .$$

Then

$$cx^* - c\bar{x} \leqq (t-1)c\bar{x} + \sum_{k=1}^{q} \bar{w}_k(t) p_k$$

for each real t such that $t\bar{v}$ is nonnegative. The best result is thus obtained by choosing t as the solution of the problem

$$\text{minimize} \quad t \to (t-1)c\bar{x} + \sum_{k=1}^{q} w_k(t) p_k$$

$$\text{subject to} \quad t\bar{v} \geqq 0 .$$

This problem has been studied extensively by Frolík (1982) and Knolmayer (1983).

4. Bounds for components of optimal solutions

Until now we haven't asked the question how to obtain information on where solutions of the original problem lie without actually solving the problem. Sometimes such an information can be obtained by using special structure of the original problem. As an example, let us consider the transportation problem

$$\text{minimize} \quad \sum_{i=1}^{m} \sum_{j=1}^{n} c_{ij} x_{ij}$$

$$\text{subject to} \quad \sum_{j=1}^{n} x_{ij} = a_i \quad , \quad 1 \leqq i \leqq m$$

$$\sum_{i=1}^{m} x_{ij} = b_j \quad , \quad 1 \leqq j \leqq n$$

$$x_{ij} \geqq 0 \quad , \quad 1 \leqq i \leqq m \, , \, 1 \leqq j \leqq n$$

under the assumption that all a_i, b_j and c_{ij} are nonnegative and such that $\sum a_i = \sum b_j$.

Define $z_{jk}(i)$ by

$$z_{jk}(i) := \max \{ 0, \min \{ a_i, b_k - \sum a_k \} \}$$

where the sum $\sum a_r$ is taken over all r such that

$$c_{rj} - c_{rk} > c_{ij} - c_{ik} .$$

It turns out - see Vlach (1971) - that

$$x^*_{ij} \leq a_i - \max_{k \neq j} z_{jk}(i)$$

for each optimal solution $[x^*_{ij}]$.

For problems with $A \geqq 0$, $b > 0$ and $c > 0$, we can use the result from Klein-Holm (1983) saying that

$$x^*_j \leqq \min_{a_{ij} > 0} \frac{b_i}{a_{ij}}$$

for each optimal solution x^* of (1).

Recently, Mangasarian (1985) has shown that every feasible point of a semidefinite linear complementarity problem contains numerical information on the size of some or all components of all solutions. Specializing this result to the dual pair of linear programming problems, we obtain the following simple bounds for optimal solutions x^* of the original problem.

If L is a lower bound on the optimal value of the original problem, if u is a feasible solution of the dual to the original problem, and if $J \subset \{1,2,\ldots,n\}$ is such that $uA_j > c_j$ for all $j \in J$, then

$$\sum_{j \in J} x^*_j \leqq \frac{ub - L}{\min_{j \in J} (uA_j - c_j)}$$

Obviously, as a lower bound L we can take the value of the objective function at an arbitrary feasible solution, for example at the solution $\bar{x} := Q\bar{y}$ obtained by disaggregating an optimal solution $\bar{y}$ of the aggregate problem.

5. Extensions

For the sake of simplicity, we have considered linear programming only. However, using generalization of Zipkin's bound developed by Huberman (1983) for the differentiable convex case, and results of Mangasarian and McLinden (1984) on bounds for solutions of monotone complementarity problems, we can obtain analogous results for convex programming.

6. References

1 Zipkin P.H. (1980) : "Bounds on the effect of aggregating variables in linear programs", Operations Research 28 (1980), 403-418.

2 Mendelssohn R. (1980): "Improved bounds for aggregated linear programs", Operations Research 28 (1980), 1450-1453.

3 Frolík J. (1982): "Meze chyb agregace v lineárním programování" (in Czech), Thesis, Technical University of Prague, 1982.

4 Knolmayer G. (1984): "Effects of transformations in computing bounds for simplified linear programs", Methods of Operations Research 51 (1984), 35-46.

5 Mangasarian O.L. - McLinden L. (1984): "Simple bounds for solutions of monotone complementarity problems and convex programs", University of Wisconsin - Madison, Computer Science Technical Report # 534, February 1984.

6 Mangasarian O.L. (1985): "Simple computable bounds for solutions of linear complementarity problems and linear programs", Mathematical Programming Study 25, October 1985, 1-12.

7 Vlach M. (1971): "Bounds to optimal solutions in transportation problems", Ekonomicko matematický obzor 7 (1971), 445 - 451.

8 Klein D. - Holm S.J. (1983): "Some reduction of linear programs using bounds on problem variables" ; in: Redundancy in Mathematical Programming; (eds.: Mark H.Karwan, Vahid Lotfi, Jan Telgen, Stanley Zionts), Springer-Verlag Berlin (1983), 80-86.

9 Huberman G. (1983): "Error bounds for the aggregated convex programming problem", Mathematical Programming 26 (1983), 100-108.

Author's address : Charles University
Dept. of Computer Science
Malostranské nám. 25
118 00 Praha 1

Ursula Schulze

Mehrphasenregression

Stabilitätsprüfung
Schätzung
Hypothesenprüfung

176 Seiten – 40 Abbildungen – 14,5 cm x 21,5 cm
DDR 24,– M, Bestell-Nr. 763 622 5, ISBN 3-05-500224-5

Viele zwischen einer Einflußgröße und einer Zielgröße bestehende Beziehungen sind nicht stabil, sondern besitzen mehrere Phasen. Zur Beschreibung solcher Beziehungen werden häufig Regressionsmodelle benutzt, deren Regressionsfunktionen stückweise aus verschiedenen Funktionen, den Phasenfunktionen, zusammengesetzt sind. Diese Regressionsmodelle heißen Mehrphasenregressionsmodelle.

Modelle dieses Typs treten in vielen Anwendungsbereichen auf. Obwohl die Mehrphasenregression in der internationalen statistischen Fachliteratur einen breiten Raum einnimmt, fehlt bisher eine geschlossene Darstellung der Verfahren zur Behandlung solcher Modelle, die insbesondere auf Aspekte der Anwendung eingeht.

Das vorliegende Buch will helfen, diese Lücken zu schließen. Es wendet sich an mathematisch interessierte Naturwissenschaftler, Gesellschaftswissenschaftler und Ingenieure, an alle diejenigen, die sich mit der Auswertung von Daten beschäftigen und Mehrphasenregression anwenden wollen.

Behandelt werden Verfahren zur Stabilitätsprüfung eines Regressionsmodells (kumulative Summentests), zur Schätzung der Umschlagspunkte und Phasenparameter und zur Prüfung von Hypothesen über die Phasenparameter und Umschlagspunkte in Modellen mit gegebener Phasenanzahl. Dabei wird vorausgesetzt, daß jede Phasenfunktion unmittelbar an die vorhergehende Phasenfunktion anschließt. Die Verfahren werden mit einer Vielzahl von Beispielen illustriert.

THEORIA CUM PRAXI

AKADEMIE-VERLAG BERLIN
1987

Band 1

Inverse and Improperly Posed Problems in Differential Equations

Proceedings of the Conference on Mathematical and Numerical Methods, held in Halle/Saale (GDR) from May 29 to June 2, 1979

Edited by Gottfried Anger
296 Seiten, 1979 (vergriffen)

Band 2

Fundamentals of Computation Theory FCT '79

Proceedings of the Conference on Algebraic, Arithmetic and Categorial Methods in Computation Theory, held in Berlin/Wendisch-Rietz (GDR) September 17–21, 1979

Edited by Lothar Budach
576 Seiten, 1979

Band 3

Decidability and Generalized Quantifiers

By Andreas Baudisch, Detlef Seese, Hans-Peter Tuschik, and Martin Weese
XII, 235 Seiten, 1980

Band 4

Zur Elastizitäts- und Thermoelastizitätstheorie

Potentialtheoretische Methoden in Anwendung auf verschiedene Modelle

Von Lothar Jentsch und Johannes Maul
157 Seiten, 1980

Band 5

Systems Analysis and Simulation 1980

Proceedings of the International Symposium held at Berlin (GDR) September 1–5, 1980

Edited by Achim Sydow
496 Seiten, 1980 (vergriffen)

Band 6

Nichtlineare diskrete Optimierung

von Eberhard Girlich und Michail M. Kowaljow
218 Seiten, 1981

Band 7

Completeness Problems in the Structural Theory of Automata

By Jürgen Dassow
148 Seiten, 1981

Band 8

Darstellung von Lösungen linearer elliptischer Differentialgleichungen

Von Günther Wildenhain
92 Seiten, 1981

Band 9

Introduction to the Spectral Theory of Operators in Spaces with an Indefinite Metric

By I. S. Iohvidov, M. G. Krein, and H. Langer
120 Seiten, 1982 (vergriffen)

Band 10

Nonlinear Time-discrete Systems

A General Approach by Nonlinear Superposition

By Michael Gössel
115 Seiten, 1982 (vergriffen)

Band 11

Preprimal Algebras

By Klaus Denecke
162 Seiten, 1982

Band 12

Ergodic Theory and Related Topics

Proceedings of the Conference held in Vitte/Hiddensee (GDR) October 19–23, 1981

Edited by Horst Michel
232 Seiten, 1982

Band 13

Singularly Perturbed Differential Equations

By H. Goering, A. Felgenhauer, G. Lube, H.-G. Roos, and L. Tobiska
176 Seiten, 1983

Band 14

Leben wir in einer Volterra-Welt?

Ein ökologischer Zugang zur angewandten Systemanalyse

Von Manfred Peschel und Werner Mende
233 Seiten, 1983

Band 15

Markov Decision Problems with Countable State Spaces

Optimality Criteria – Algorithms – Clustering

By Hans Michael Dietz and Volker Nollau
174 Seiten, 1983

Band 16

Stochastische Verfahren zur Suche nach dem Minimum einer Funktion

Von Ryszard Zieliński und Peter Neumann
133 Seiten, 1983

Band 17

Invertibility and Asymptotics of Toeplitz Matrices

By Albrecht Böttcher and Bernd Silbermann
200 Seiten, 1983

Band 18

Structural Induction on Partial Algebras

Introduction to Theory and Application of Partial Algebras – Part II

By Horst Reichel
205 pages, 1984

Band 19

Algebraic Methods for Toeplitz-like Matrices and Operators

By Georg Heinig and Karla Rost
212 pages, 1984

Band 20

Frege Conference 1984

Proceedings of the International Conference held at Schwerin (GDR), September 10–14, 1984
Edited by Gerd Wechsung
408 pages, 1984

Band 21

Numerical Analysis of Selected Semilinear Differential Equations

Edited by Christian Großmann, Thomas Riedrich, and Hartmut Schönheinz
203 pages, 1984

Band 22

Singular Ordinary Differential Operators and Pseudodifferential Equations

By Johannes Elschner
200 pages, 1985

Band 23

Lotka-Volterra-Approach to Cooperation and Competition in Dynamic Systems

Proceedings of the 5th Meeting of UNESCO's Working Group on Systems Theory held on the Wartburg, Eisenach (GDR) at March 5–9, 1984
Edited by Werner Ebeling and Manfred Peschel
271 pages, 1985

Band 24

Convergence Structures 1984

Proceedings of the Conference on Convergence held in Bechyně (Czechoslovakia), September 24–28, 1984
Edited by Josef Novák, Werner Gähler, Horst Herrlich, and Jan Mikusiński
254 pages, 1985

Band 25

Parcella '84

Proceedings of the II. International Workshop on Parallel Processing by Cellular Automata and Arrays held in Berlin (GDR), September 25–27, 1984
Edited by Wolfgang Händler, Tamás Legendi, and Gottfried Wolf
239 pages, 1985

Band 26

Multiobjective and Stochastic Optimization Based on Parametric Optimization

By Jürgen Guddat, Francisco Guerra Vasquez, Klaus Tammer, and Klaus Wendler
175 pages, 1985

Band 27/28

Systems Analysis and Simulation 1985
I: Theory and Foundations
II: Applications

Proceedings of the International Symposium held in Berlin (GDR), August 26–31, 1985
Edited by Achim Sydow, Manfred Thoma, and Robert Vichnevetsky
per volume 495 pages, 1985

Band 29

Parcella '86

Proceedings of the III. International Workshop on Parallel Processing by Cellular Automata and Arrays held in Berlin (GDR), September 9–11, 1986
Edited by Tamás Legendi, Dennis Parkinson, Roland Vollmar, and Gottfried Wolf
272 pages, 1986

Band 30

Fuzzy Sets Applications, Methodological Approaches, and Results

Proceedings of the International Workshop on Fuzzy Sets Applications cosponsored by IIASA and convened in cooperation with IAEA held on the Wartburg, Eisenach (GDR), at March 3–8, 1985
Edited by Steffen Bocklisch, Sergei Orlovski, Manfred Peschel, and Yasushi Nishiwaki
233 pages, 1986

Band 31

Mathematical Methods of Specification and Synthesis of Software Systems '85

Proceedings of the International Spring School, held in Wendisch-Rietz (GDR), April 22–26, 1985
Edited by Wolfgang Bibel and Klaus P. Jantke
245 pages, 1986

Band 32

A Model Theoretic Oriented Approach to Partial Algebras

Introduction to Theory and Application of Partial Algebras – Part I
By Peter Burmeister
319 pages, 1986

Band 33

Finite-difference Methods on Irregular Networks

A Generalized Approach to Elliptic Problems of Second Order
By Bernd Heinrich
206 pages, 1986

Band 34

Modern Algorithms for Large Sparse Eigenvalue Problems

By Arnd Meyer
125 pages, 1987

Band 35

Parametric Optimization and Related Topics

Edited by Jürgen Guddat, Hubertus Th. Jongen, Bernd Kummer, and František Nožička
407 pages, 1987

Band 36

Discretization in Differential Equations and Enclosures

Edited by Ernst Adams, Rainer Ansorge, Christian Großmann, and Hans-Görg Roos
250 pages, 1987

Band 37

Factorization of Measurable Matrix Functions

By Georgi Semenovich Litvinchuk and
Ilya Matveyevich Spitkovskiĭ

372 pages, 1987

Band 38

Parallel Algorithms and Architectures

Proceedings of the International Workshop
on Parallel Algorithms and Architectures
held in Suhl (GDR), May 25–30, 1987

Edited by Andreas Albrecht, Hermann Jung,
and Kurt Mehlhorn

205 pages, 1987

Band 39

Parametric Integer Optimization

By Bernd Bank and Reinhard Mandel

138 pages, 1988

Band 40

Computer Analysis of Images and Patterns 1987

Proceedings of the II. International Conference
CAIP '87 on Automatic Image Processing,
held in Wismar (GDR), September 2–4, 1987

Edited by Leonid P. Jaroslavskiĭ,
Azriel Rosenfeld, and Wolfgang Wilhelmi

282 pages, 1987

This volume contains original and survey papers related to contemporary problems in mathematical optimization and related topics (theory and applications). The topics treated in this proceedings are widely spread. They deal with integer programming, stochastic programming, parametric optimization, duality, nonlinear programming, and solution procedure for special classes of mathematical optimization problems. This collection of articles will be of interest for those readers dealing with problems in theory and applications of mathematical optimization.

In diesem Band werden Original- und Übersichtsartikel zu aktuellen Problemen der mathematischen Optimierung (Theorie und Anwendungen) vorgestellt. Das umfangreiche Spektrum reicht vom Studium diskreter Optimierungsaufgaben über stochastische und parametrische Optimierung, Dualitätstheorie bis zu Verfahren für spezielle Aufgabenklassen. Die Sammlung von Arbeiten wendet sich an Leser, die auf dem Gebiet der Theorie oder der Anwendungen der mathematischen Optimierung arbeiten.

ISBN 3-05-500543-0
ISSN 0138-3019

Bei Fragen zur Produktsicherheit wenden Sie sich bitte an:
If you have any questions regarding product safety,
please contact:

Walter de Gruyter GmbH
Genthiner Straße 13
10785 Berlin
productsafety@degruyterbrill.com